THE MATTER OF EMPIRE

ILLUMINATIONS:
CULTURAL FORMATIONS OF THE AMERICAS SERIES

John Beverley and Sara Castro-Klarén, Editors

THE MATTER OF EMPIRE

METAPHYSICS AND MINING IN COLONIAL PERU

ORLANDO BENTANCOR

UNIVERSITY OF PITTSBURGH PRESS

Published by the University of Pittsburgh Press, Pittsburgh, Pa., 15260
Copyright © 2017, University of Pittsburgh Press
All rights reserved

Manufactured in the United States of America
Printed on acid-free paper
10 9 8 7 6 5 4 3 2 1

Cataloging-in-Publication data is
available from the Library of Congress

ISBN 13: 978-08229-6747-7
ISBN 10: 0-8229-6747-2

Cover design by Joel W. Coggins

To Verónica

CONTENTS

Acknowledgments ix

Introduction:
Imperium, Metaphysical Instrumentalism, and Potosí Mining 1

1. Grounding the Empire: Francisco de Vitoria's Political Physics 40

2. The Impasses of Instrumentalism:
Revisiting the Polemics between Sepúlveda and Las Casas 95

3. Mastering Nature:
José de Acosta's Pragmatic Instrumentalism 151

4. From Imperial Reason to Instrumental Reason:
The Ideology of the Circle of Toledo 217

5. The Exhaustion of Natural Subordination: Solórzano Pereira
and the Demise of Metaphysical Instrumentalism 284

Notes 353
Bibliography 377
Index 389

ACKNOWLEDGMENTS

This project would have been impossible without the generous guidance and tutelage of Cristina Moreiras, Gustavo Verdesio, and Gareth Williams, my professors at the University of Michigan. Seminars taught by Jossiana Arroyo, Catherine Brown, Santiago Colás, and Javier Sanjinés informed my thinking and research. This study benefited from conversations with my former classmates Luis Martin Cabrera, Manuel Chinchilla, Andrea Fanta, Patty Keller, Andrea Marinescu, Jon Snyder, Fernando Velasquez, and Marcelino Viera.

I also want to thank Susana Draper for her profound and honest observations. Anna More helped to shape my thinking through innumerable hours of intellectual exchange and generous discussions. My thanks go also to Ivonne del Valle for her feedback and constant support.

My gratitude goes to my former colleagues at the University of Southern California: Daniela Bleichmar, Roberto Ignacio Díaz, Gabriel Giorgi, Peggy Kamuf, Karen Pinkus, and Sherry Velasco. It is not possible to do justice to the many different forms in which my colleagues at Barnard College have supported this book. Thanks go to Ronald Briggs, Maja Horn, Alfred MacAdam, and Wadda Rios-Font. My colleagues at Columbia—Carlos Alonso, Patricia Grieve, Seth Kimmel, Alberto Medina, Graciela Montaldo, Alessandra Russo, Jesús R. Velasco—fostered an environment both diverse and challenging. I want to thank Seth Kimmel for the title of this book.

The project was also shaped by dialogues with Santa Arias, Ralph Bauer, John Beverley, Sara Castro-Klarén, Sibylle Fischer, Ross Hamilton, Stella Nair, Ricardo Padrón, Mary Louise Pratt, Jose Rabasa, Benita Sampedro, Freya Schiwy, Patricia Seed, Elvira Vilches, Lisa Voigt, and Nicolás Wey-Gómez. I want to thank Bram Acosta, Jon Beasley-Murray, Oscar Cabezas, Patrick Dove, Alessandro Fornazzari, Erin Graft, John Kraniauskas, Brett Levinson, Alberto Moreiras, Samuel Steinberg, and Sergio Villalobos for conference panels and conversations that have enriched my ideas. I am grateful to the class discussions I had with Santiago Acosta, Anayvelyse Allen-Mossman, Miguel Ibañez Aristondo, Jae Young Chang, Omar Durán-García, Analía Lavin, Alexandra Mendez, Cybele Pena, and Roberto Valdovinos. My special thanks go to Noel Blanco Mourelle for his patience and friendship.

Andrew Ascherl, Brian Green, and Alfred MacAdam helped me with the translation of the primary sources quoted in this book. Elizabeth Castelli encouraged me to apply for a grant from Barnard College that permitted me to revise the manuscript. I am indebted to Andrew Ascherl's help in the preparation of the manuscript and to his sharp theoretical insights. At Pittsburgh University Press, I want to thank John Beverley and Sara Castro-Klarén, the series editors, as well as Josh Shanholtzer and the two anonymous readers.

I would like to thank my parents for encouraging me to pursue my intellectual goals. And above all, I thank my wife, Verónica Laguna, for her infinite and unconditional love.

THE MATTER OF EMPIRE

INTRODUCTION

IMPERIUM, METAPHYSICAL INSTRUMENTALISM, AND POTOSÍ MINING

THE METAPHYSICAL FOUNDATIONS OF IMPERIAL INSTRUMENTAL REASON

Discovered in 1545, the Cerro Rico (Rich Hill) of Potosí immediately became the main source of silver for the Spanish Empire, fueling both its political project of a Christian monarchy and the first global economy.[1] Even as they transformed Peruvian metals into the money that kept the empire together, sixteenth-century Spaniards also understood mining through long-standing metaphysical beliefs concerning the essence of matter. This metaphysical framework assumed that the natural world was composed of a raw and defective material that had to be dominated from above and directed to a higher end. Surprisingly, this metaphysics also framed the writings on natural law that were central to Spain's justification of its empire. An examination of the interactions between early political writings and writings on mining will show that the particular confluence of Iberian imperial practices and philosophical ideas in the Americas frames technological and capitalist modernity as both an imperial and a metaphysical project.

I make this argument through a contrapuntal reading of the sixteenth-century debates on Spanish sovereignty in the Americas and treatises on natural history and mining written between 1520 and 1640. Whether political or natural-historical, these texts all invoke an ontological frame derived from a "natural order" to justify (and occasionally to question) material practices such as compulsory labor in the colonial Andes (*mita*) and refining techniques for the amalgamation of metals (*beneficio*). We trace the development of this ontological frame over the course of a century and a half, beginning with the early attempts to justify the conquest and compulsory labor in the mines and ending with texts on mining written as the Spanish Empire entered its terminal decline.

The texts along this trajectory often fall into the inherent paradox of metaphysical instrumentalism: conceiving nature as open to technical manipulation resulted in the entanglement of ends and means. For instance, Spaniards consistently justified the extraction of silver and the production of money by conceiving artificial mastery (or means) as determined by a natural teleology (or end). The metaphysical problem encountered in this collapse of ends and means was that the crass and profane material means were continually threatened with the danger of becoming an autonomous end in itself, undermining the superior ends they were supposed to obey. Thus, refining techniques and compulsory labor cost the Crown the lives of the Indian vassals, while the production and circulation of silver enriched a vast credit network that benefited competing European powers, in each case avoiding the ideal imperial end. While Spanish ideology sought to create a closed metaphysical circle that dedicated all practices to a united end, however, writers were well aware of the open-ended nature of both mining production and the global economy. As the Spanish Empire entered into decline in the seventeenth century, this dependence on material means proved ultimately incompatible with perfect ends and produced clear and endemic ideological inconsistencies.

Spanish imperial science and mining are traditionally studied separately from Spanish political theory, but here these two discourses are seen as isomorphic, interpenetrating one another at every level. By foregrounding the common Scholastic basis and the interaction between these two bodies of literature, moreover, this discussion contributes to a general reevaluation of the Scholastic roots of modernity in the fields of philosophy and the history of science. A systematic examination of metaphysical language employed in distinct disciplines allows us to narrate how the view of both nature and humans as malleable material is the result of the instrumentalist presuppositions inherent in imperial ideology. Against the assumption that scientific modernity began with the Protestant empiricists, I argue that this Western metaphysical instrumentalism is the origin of the contemporary reduction of nature to technologically disposable material.[2] This metaphysical ideology developed in the context of colonial Andean mining, and there was a specifically colonial indigenous attribution of life to the mineral world that was not exterior to but, rather, dialectically engaged with imperial metaphysics. This engagement still provides modern scholars with the basis for a critique of imperial metaphysical instrumentalism.

SCHOLASTICISM AND IBERIAN IMPERIAL IDEOLOGY

In order to examine the common Scholastic basis of imperial politics and mining, we must begin with attempts to ground the Spanish Empire in Aquinas's metaphysics. After the discovery and conquest the Spaniards tried to justify the appropriation of riches and the practice of mining in the New World. Scholasticism provided the theological and philosophical foundations for justifying the whole colonial enterprise.[3] The name of the movement that engaged in thinking contemporary politics through the work of Aquinas is the School of Salamanca.[4]

The founder of the School of Salamanca was the Dominican Francisco de Vitoria (1492–1546). Domingo de Soto (1494–1560), Melchor Cano (1509–1560), and Francisco Suárez (1548–1617) were also part of this movement.[5] The fundamental sources for Spanish Scholastics were Aristotle, Thomas Aquinas, Roman jurisprudence, civil and ecclesiastical law, and the *Decretales*, a collection assembled in the eighth century under the auspices of Pope Gregorio IX. Spanish Scholastics continued the tradition initiated by Cayetano (also known as Tomás Vio) of commenting on entire sections of Thomas Aquinas's *Summa Theologica*. The Aristotelian-Thomist tradition provided ways of confronting the threats presented by the *via moderna* and crystallized in Lutheranism, Machiavellianism, Erasmus's pacifism, and Ockham's nominalism.[6] Aquinas's "rationalism" was a perfect antidote to both Luther's theological voluntarism and Machiavelli's reason of state. The School of Salamanca opposed the pacifism of Juan Luis Vives, who saw in Charles V the triumphant unification dreamed of by Dante and Erasmus but condemned both Scholasticism and Spain's militarism by appealing to Augustine's *City of God*.

Francisco de Vitoria followed the model of the University of Paris that replaced nominalism with Thomism. Aquinas's philosophy was not only a christianization of Aristotle but also a synthesis of Aristotelianism and

Platonism. Aquinas metaphysics subsumed the theology of Augustine, Roman Law, and Cicero's natural right under a new paradigm. This paradigm accommodated empirical and factual knowledge of the world with the ontological realism of universal forms. Since Thomism defends the capacity to understand reality through the grasping of its essence, it proved useful for assigning sense to empirical facts. Aquinas's metaphysics and politics was a synthesis of Platonic doctrine of participation with Aristotelian causation.[7] The ultimate principles of Thomist ontology and theology were employed to assign sense and finality to a union of the factual (temporal) and the transcendent (eternal) realms in order to justify the evangelization and conquest of the New World. It provided a strong accountability to existing laws by grounding them in "rational" and "natural" finality. Therefore, Aquinas's providentialism provided a strong sense of legitimacy to the prince's authority by appealing to self-evident and ultimate principles capable of grounding the *imperium* as capacity to command. The political and epistemological power of Scholasticism depended on what can be summarized in the principle of subordination of the part to the whole, imperfect matter to perfect form, and material means to an immaterial end.

The task undertaken by the Spaniards was to justify their sovereignty over the newly discovered peoples by invoking their imperfect nature. Their imperfect nature, crystallized in their lack of civilization, had to be directed to their proper end, which was the common good, civilization, and salvation. The same procedure was applied to nature, which was understood as temporal means that could be used by directing it to humans' ends. Such a providentialist view of metals presupposed that available resources were a raw matter that could be employed to further Catholic expansion. This principle makes it possible to read both political writings and texts on mining through their common presuppositions, which is metaphysical instrumentalism—the ultimate ideology of the Spanish Empire.[8] In order to explain the instrumentalist presuppositions behind Aquinas's metaphysics, let us move now to the principle of the natural subordination of matter to form and means to an end.

PRINCIPLE AS ORIGIN OF DOMINATION

Let us start by explaining what a principle is. For Aristotle, and thus for Aquinas, a principle is a beginning or starting point that initiates the existence or motion of something else (*Metaphysics* 5). For Aquinas, everything existing or moving owes its existence or movement to something else. For this reason, principles surpass moving or existing things in power. In Chapter 1, Book 5, *Metaphysics* 1012b34–1013a23, Aristotle explains the notion of principle as origin or inception by using different examples. In the first example, beginning means a part of a thing "from which one would start first, e.g. a line or a road has a beginning in either of the contrary distinctions" (Aristotle, *Basic Works*, 752). According to the second example, in "learning we must sometimes begin not from the first point and the beginning of the subject but from the point from which we should learn most easily" (752). In the third example, Aristotle refers to things that have their origin inside their nature, such as the heart of an animal, or "as the keel of a ship and the foundation of a house" (752). The fourth example is that of things that have their origin outside their nature, "as a child comes from its father and its mother" (752). The fifth example refers to the origin as the will that moves something else; it locates the best examples in "the magistracies in cities, and oligarchies and monarchies and tyrannies, [which] are called *archai* and so are the arts, and of these especially the architectonic arts" (752). In the sphere of knowledge, the origin is "that from which a thing can first be known—this is also called the beginning of the thing, e.g. the hypotheses are beginning of demonstrations" (752). What all these examples have in common is "to be the first point from which a thing either is or comes to be or is known" (752). In other words, a principle as origin is something that comes first and has certain preeminence because it is more important. Since a principle involves commanding and subordinating, it is useful to examine Aquinas's commentary on the fifth example.

Before analyzing this example, however, it is instructive to say that Aquinas classifies these above-mentioned examples in two categories. According to the first sense, "a principle means that part of a thing which is first generated and from which the generation of the things begins" (Aquinas, *Commentary on Aristotle's* Metaphysics, 278). The cases of the line, the road, and the foundation of the house belong to this first sense of principle. Yet there is a second sense in which "a principle means that from which a thing's process of generation begins but which is outside the thing" (279). One example of the first would be that of the father as the origin of the child. Indeed, within this category of things that have their principle outside themselves, he finds "natural beings, in which the principle of generation is said to be the first thing from which motion naturally begins in those things, which come about through motion" (279). The second case of things that have their origin outside themselves, which is also Aristotle's fifth example, is that of "human acts, whether ethical or political, in which that by whose will or intention others are moved or changed is called a principle" (278–79). For Aquinas, both the example of the magistracies and civil power and the example of natural generation and corruptions, such as the father and the child, belong to the categories of external principles that cause the movement of something. Imperial power, which for Aquinas means sovereignty, implies this capacity to move its subjects, since "those who hold civil, imperial, or even tyrannical power in states are said to have the principal places" (279). By the will of the prince "all things came to pass or are put into motion in the states" (279). Those who have civil power "are put in command of particular offices in states as judges and persons of this kind" (279). For Aquinas, clearly, both the cases of natural movement and political subjection fall within the parameters of being moved by an external principle that precedes and exceeds the moved thing or subject. Civil power, the power of the state, is clearly an example of a principle that moves its subjects by subordinating them.

Finally, there is another example that falls under the fifth sense of principle in Aristotle and the category of external causation in Aquinas, which is the subordination of inferior arts to superior arts:

> For the arts too in a similar way are called principles of artificial things, because the motion necessary for producing an artifact begins from art. And of these arts the architectonic, which "derive their name" from the word principle, i.e., those called principal arts, are said to be principles in the highest degree. For by architectonic arts we mean those which govern subordinate arts, as the art of navigator governs the art of ship-building, and the military art governs the art of horsemanship. (279)

The example of this kind of subordination is also an example of subordination based on an external principle. This example is so important that it also appears in Chapter 1, Book 1, *Metaphysics* 981a29–981b2, where Aristotle writes, "For men of experience know that the thing is so, but do not know why, while the others know the 'why' and the cause. Hence we think also that the master-workers in each craft are more honorable and know in a truer sense and are wiser than the manual workers, because they know the causes of the things that are done" (*Basic Works*, 690). Aquinas comments on this passage, saying that "In order to understand this we must note that architect means chief artist, from *techne*, meaning chief, and *archos*, meaning art" (Aquinas, *Commentary on Aristotle's Metaphysics*, 9). The superior art is the one that "performs a more important operation" (9). Moreover, Aquinas classifies the artist's operations between disposing the material of the artifacts and directing them to an end:

> Carpenters, for example, by cutting and planing the wood, dispose matter for the form of a ship. Another operation is directed to introducing this form into matter, for example when someone builds a ship out of wood which has been disposed and prepared. A third operation is directed to use of the finished product, and this is the highest operation. But the first operation is the lowest because it is directed to the second and second to the third. Hence the shipbuilder is a superior artist compared with the one who prepares the wood; and the navigator, who uses the completed ship, is a superior artist compared with the shipbuilder. (9–10)

Therefore, just like the natural hierarchies of the physical world and the human hierarchies of the political world, the hierarchy of the arts is an example of external subordination. The subordination of the material to the artist and then the subordination of this artist to a superior artist are based on the fact that the superior artist has a clear vision of the end of the final product. What all examples of principles share is being the first thing out of which things arise and are ruled. The principle *precedes* that of which it is a principle. It precedes everything else. It is presupposed. In the case of external causation, there is always an agent that is preeminent and superior and that *commands* what is subordinated and inferior. The very notion of principle and its ultimate metaphysical character is based on presupposing that the principle is both inception and source of domination. The principle *commands*, which means it subordinates and moves. A guiding hypothesis of the present book is that the commanding character of the principle is the result of the transposition of human technical manipulation to the realm of metaphysics. Another way of framing this problem is, as will become evident in the following sections, that the intrinsic presupposition of this kind of movement is that of technical manipulation.

As explained above, natural, political, and technical subordinations are grounded on external principles. In Article 1, *Summa Theologica* IaII-ae, Aquinas joins the notion of natural order and that of political subordination by grounding both in higher principle:

> In natural order, it happens of necessity that higher things move lower things by excellence of the natural power divinely given to them. Hence in human affairs also superior must move inferior by their will, by virtue of a divinely established authority. But to move by reason and by will is to command. And so just as in the divinely instituted natural order lower things are necessarily subject to higher things and are moved by them, so too in human affairs inferiors are bound to obey their superiors by virtue of the order of natural and Divine law. (Aquinas, *Political Writings*, 58)

The hierarchical division between higher (that is, moving and ruling) things and lower (or moved and inferior) things is part of a natural order. Natural subordination includes human affairs, which include politics, where rulers govern the ruled by *commanding*, or moving by reason and will. Both natural order and political subjection share in being part of providence, the divinely instituted natural order. In order to clarify the meaning of natural order or natural subordination, let us examine some key moments of Aquinas's principles of nature, also known as the doctrine of hylomorphism. In Aquinas, there are three principles in nature: matter, form, and privation. While matter and form are principles in themselves (*per se*) because they are also positive causes, privation is an accidental principle (*per accidends*) because it cannot cause anything by itself. First, I will explain the notion of matter, since this principle also involves the principle of privation.

PRIME MATTER PRESUPPOSES INSTRUMENTAL MANIPULATION

The metaphysical status of the "prime matter" is that of a pure abstraction that separates all the sensual, empirical, and singular qualities of things by focusing on what they have in common. In his commentary to

Aristotle's *Physics*, Aquinas defines "prime matter" as a lump of amorphous, plastic, raw material that has no consistency of its own since it exists only in a composite of matter. In *Physics* 191a7–15, Aristotle writes: "This underlying nature is an object of scientific knowledge, by analogy. For as the bronze is to the statue, the wood to the bed, or the matter and the formless before receiving form to any thing which has form, so is the underlying nature to substance, i.e., the 'this' or existent" (*Basic Works*, 232). Aquinas comments on this passage, saying that the abovementioned underlying nature, "which is first subject to mutation, i.e., primary matter, cannot be known in itself, since everything which is known is known only through form" (*Commentary on* Physics, 61). This means that if matter is the imperfect, passive potency that underlies all individual material entities, form is the idea, pattern, or blueprint that gives determination and consistency to these material entities.[9] Matter is unknown and unintelligible, and only form is intelligible. This raw stuff present under every composite is pure passive potential to receive an exemplary pattern or "form" from above.[10] Aquinas continues, explaining that "prime matter is, moreover, considered to be the subject of every form. But it is known by analogy, that is, according to proportion" (61). Since this amorphous material cannot be known, it can only be understood through the mediation of analogy. The analogy of proportion can be illustrated by saying that A stands in relation to B, as C stands in relation to D. Aquinas adds, "For we know that wood is other than the form of a bench and a bed, for sometimes it underlies to one form, at other times the other" (61). We know that matter is different from form because wood is different from the bed. Aquinas thinks that experience tells us that the same wood sometimes underlies one bed and sometimes another. From there, the intellect abstracts an underlying notion of matter common to the different forms. Aquinas continues, explaining that, "when, therefore, we see that air at times becomes water, it is necessary to say that there is something, which sometimes exists under the form of air, and that other times under the form of water" (61). For something to become something else there must be an underlying substrate to both entities. Moreover, "this something is other than the form of water and other than the form of air, as wood is something other than the form of a bench and other than the form of bed" (61). The basic reasoning is an analogy according to which prime matter is other than the form, just as wood is other than the bench. Aquinas ends the paragraph saying

"This 'something,' then, is related to these natural substances as bronze is related to the statue, and wood to the bed, and anything material and unformed to form. And this is called primary matter" (61).

This is the crucial moment that explains how metaphysical thinking knows that there is an amorphous passive potential matter common to all things. It is the result of an abstraction that separates matter from all its concrete qualifications by postulating it as something that underlies already formed things. But the question that arises is, How does the intellect arrive at this idea of prime matter as an imperfect and amorphous passive potency deprived of any concreteness? By analogy with human manipulation: the prime matter stands in relation to form in the same way that bronze stands in relation to the statue. Aristotelian-Thomist metaphysics arrives at the idea of an amorphous passive raw matter by way of an analogy with human manipulation. The principle of the natural subordination of matter to form results from the transposition of technical manipulation to the natural order. Matter is subordinated to form in the same way that the bronze is subordinated to form, which is the final product, the statue:

> What is in potency cannot bring itself into a state of actuality. Bronze, for example, which is a statue in potency, does not make itself be a statue. It needs something actively working, which brings out the form of the statue from potency into act; I am speaking of the form of the generated thing, the form which we have said is the end-point of generation.... It is necessary, therefore that there be in addition to the matter and the form some principle which does something; and this is said to be what makes, or moves, or acts, or that from which the motion begins. (Bobik, 34–35)

Matter cannot produce anything because matter cannot bring itself into an actual object. It remains potential in the same way that bronze remains potential until the agent actualizes it by imposing a form on it. But form is the end-point of generation, or as Aquinas also says: "for form is the end of matter; therefore for matter to seek form is nothing other than matter being ordered to form as potency to act" (*Commentary on* Physics, 72). This means that the material means is subordinated to the form which is also the final cause. Therefore, Aquinas ties everything by saying that to matter, form, and privation there must be added an agent, which is the principle or "that from which the motion begins."

INTRODUCTION | 11

In Chapter 4e, Book 8, *Physics* 256a21–256b3, Aristotle sustains that, since every movement requires a mover, there must be a prime mover that moves itself in order to avoid an infinite regress. He exemplifies this argument by appealing to technical motion or instrumental manipulation:

> Every movement moves something and moves it with something, either with himself or with something else: e.g., a man moves a thing either himself or with a stick and a thing is knocked down either by the wind itself or by a stone propelled by the wind. But it is impossible for that with which a thing is moved to move it without being moved by that which imparts motion by its own agency: on the other hand, if a thing imparts its motion by its own agency, it is not necessary that there should be anything else with which it imparts motion, whereas if there is a different thing with which it imparts motion, there must be something that imparts motion not with something else, but with itself, or else there will be an infinite series. If, then, anything is a movement while being itself moved, the series must stop somewhere and not be infinite. Thus, if the stick moves something in virtue of being moved by hand, the hand moves the stick: and if something else moves with the hand, the hand also is moved by something different from itself. So when motion by means of an instrument is at each state caused by something different from the instrument, this must always be preceded by something else which imparts motion with itself. Therefore, if this last movement is in motion and there is nothing else that moves it, it must move itself. So this reasoning also shows that, when a thing is moved, if it is not moved by something that moves itself, the series brings us at some time or other to a movement of this kind. (*Basic Works*, 367–68)

Since everything that moves must be moved either by itself or by another agent, there must also be an agent that moves itself. This will become an important argument for proving the existence of God, the Prime Mover, a principle of all movements. Aquinas comments this passage by saying that "every mover moves something and moves by something, either by itself or by another lower mover" (*Commentary on* Physics, 551). For example, "a man moves a stone either by himself or by a stick, and the wind hurls something to the ground either by its own power or by a stone which it moves" (551). Aquinas continues, explaining that "it is impossible for that which moves as an instrument to move something without a principal mover" (551). In other words, instruments do not move themselves. Aquinas goes on: "For example, a stick cannot move without a hand." Moreover, "no one would doubt that the second mover is the instrument of the first" (551). The consequence of the incapacity of an instrument to move itself is none other than the existence of a thing that moves itself: "Just as he said above that if something is moved by another there must be something which is not moved, but not vice versa, so here he says by descending that if there is an instrument by which a mover moves there must be something which moves, not by an instrument, but by itself, or else there is an infinite series of instruments. This is the same as an infinite series of movers, which is impossible, as was shown above" (551). If there are instruments, things that are moved by human hands, then there must be a first mover since it is impossible to have a series of infinite instruments. The machine of the world requires a first mover, an external, transcendent cause that moves everything else. There is a gradation of power and capacity to move that goes from God, which is absolutely perfect (self-subsistent, self-moving) to nonliving things, which are imperfect (dependent and moved by another). In the middle there are corporeal things that are composites of matter and forms. The world is a hierarchical, natural order where matter is subjected to different forms that are intellectual, sensitive, and vegetative. While God is the ultimate agent who moves instruments, matter is the lowest imperfect principle, which is itself an instrument of the form, whether it is an intellectual, vegetative, or sensitive soul: "the whole of corporeal nature is an underlying subject to the soul, and it is related to it as matter and instrument" (Bobik, 141). In sum, the principle of the natural subordination of imperfect matter to perfect form is inseparable from an instrumental understanding of nature. To recapitulate, the three examples of principles, as such, are natural, political, and technical subordination. The three aspects are joined into one principle, which is exemplified with the example of the bronze statue.

The example of bronze provided by Aquinas and Aristotle illustrating both natural subordination and political subjection is an example borrowed from the arts. In this example, an artist (efficient cause) imposes a preexisting idea (perfect form, universal pattern, blueprint, or soul) over a prime matter (pure passive instrumental potency) in order to produce a statue (the final product). Aquinas employs the metaphor of the craftsman or the architect who imposes a preexisting rational order (forms, ideas, universal patterns, exempla) over an amorphous, chaotic, imperfect, and incomplete matter in order to achieve a perfect, complete, and self-sufficient product. Let us return to the domain of political subjection, or empire as subjection, in order to see how Aquinas joins the natural order with the compulsory character of the law. In Article 1, *Summa Theologica* IaIIae.93, titled "Whether the eternal law is supreme reason existing in God," we read:

> Just as in every craftsman there preexists a rational pattern of the things, which are to be made by his art, so too in every governor there must preexist a rational pattern of the order of the things, which are to be done by those subject to his government. And just as the rational pattern of the thing to be made by art is called art, or the exemplar of the products of that art, so too the rational pattern existing in him who governs the acts of his subjects bears the character of the law, provided that the other conditions which we have mentioned above are also present. Now God is the Creator of all things by His Wisdom, and *He stands in the same relation to them as a craftsman does to the products of his art*, as noted in the First Part. But he is also the governor of all the acts and motions that are to be found in each single creature, as was also noted in the First Part. Hence just as the rational pattern of the Divine wisdom has the character of law in relation to all the things which are moved by it to their proper end. (*Political Writings*, 102; my emphasis)

The relation of proportion here is the same as the one explained above between the bronze and the statue, where prime matter stands in relation to the final form in the same way that bronze stands in relation to a statue. Now both God and the monarch stand in relation to the subject in the same way that the artist stands in relation to the amorphous matter. These efficient causes have a preexisting end in mind that functions as a rational pattern, blueprint, or prototype. This preexisting idea lives in the mind of the Divine Artifice who then proceeds to tame the amorphous material in order to obtain a final product. In the example provided by Aquinas, the formal cause of the product of art is the preexisting idea. The efficient cause is the Divine Artifice itself. The material cause is the amorphous and plastic material, a passive potency that receives the form in the mind of the Divine Artifice.

Finally, the final cause is the product of art itself. The principle of natural subordination implies that there is a superior power that moves things to their proper end. There is a tautological performativity at work in the capacity of the principle to command amorphous matter. This tautological character resides in the fact that ends are mandatory because they have been imposed by the principle. In order to go beyond the mere tautological relation between origin and end, and to prove that the command is not just arbitrary but both rational and natural, Aquinas, following Aristotle and Plato, has to appeal to the metaphor of the artisan. Examples similar to that of the statue appear in *De regimine principum* where Aquinas states:

> That it is necessary for men who live together to be subject to a diligent rule by someone. To fulfill this intention, we must begin by explaining how the title king is to be understood. Now in all cases where things are directed towards some end but it is possible to proceed in more than one way, it is necessary for there to be some guiding principle, so that the due end may be properly achieved. For example, a ship is driven in different directions according to the force of different winds, and it will not reach its final destination except by the industry of the steersman who guides it into port. . . . Man therefore needs something to guide him towards his end. (*Political Writings*, 5)

Here, in order to explain how guiding principles must direct things to an end and how the prince must direct men to their proper end, Aquinas employs the example of how the steersman guides the ship to its proper destination. In this example, just as in the example of the artist who makes a statue, there is a clear transposition of technique to the natural order and to politics.

As a result from the use of this example borrowed from technical mastery, there is an instrumentalist presupposition in Aquinas's principle of the natural subordination of imperfect matter to the perfect end. Natural causation is preconceived as artificial causation.[11] Natural mastery, the capacity of the principle to command, is like artificial mastery, the capacity of the artist to impose form over matter, directing it to the end.[12] The reason behind this transposition of technique to nature is that principles cannot be demonstrated, because they are the origin of the demonstration. Although principles are absolutely necessary and, therefore, presupposed, they are impossible to know or demonstrate, since they are themselves the origin of demonstration. Therefore, they can only be illustrated by using an imperfect analogy. The principle's power to command—imperium itself—is illustrated by appealing to an example that backs up the principle itself. Within the frame of Scholastic metaphysics, the rational power to command is paradoxically understood as the capacity of human beings to mold an available raw material with human hands.

Therefore, Aquinas transfers the characteristics of instrumental manipulation to the natural world and political world. Despite appealing to instrumental manipulation in order to ground the capacity to command, Aquinas's Aristotelian philosophy relegates instrumental manipulation to the status of a mere passive, inert human extension. Scholasticism disavows its own transposition of artificial mastery to natural causality by reducing technique to a mere *medium*—a neutral, instrumental device that requires an efficient cause to be set in motion and directed to a preexisting end—since artifacts "have no inner impulse to change" (Aristotle, *Basic Writings*, 236). The principle of the natural subordination of imperfect matter to perfect form presupposes metaphysical instrumentalism. Instrumentalism is metaphysical because it supposes the preexistence of a supersensory idea independent of the material world already inscribed in the commanding origin. Metaphysics is instrumentalist because it borrows its apparently self-evident character from examples borrowed from instrumental manipulation, such as statue making, ship navigation, or bridle making. Since metaphysics wants to preserve its necessary and, above all, natural character, it subordinates technique to a preexisting master by relegating the instrument to the status of a passive medium. Metaphysical instrumentalism conceives nature and politics as a means to an end because it masters technique by presupposing a master that controls technique itself. Matter is a manipulatable stock, an available material instrument ready to be directed to a higher end.

MODERNITY AS TECHNOLOGICAL DOMINATION IN HEIDEGGER

The instrumentalist presuppositions of metaphysics were the object of Martin Heidegger's deconstruction of the history of Western philosophy. He is without a doubt the most influential philosopher of technology of the twentieth century.[13] Such an uncontested influence is partially based on Heidegger's insight into the mutual co-constitution between metaphysical totalizations and global technological expansion. Moreover, for Heidegger, technological domination is part of self-revelation of being itself. As Arthur Bradley explains, for Heidegger "the history of the philosophy of technology from Aristotle to the epoch of contemporary techno-science effectively becomes the history of Being's own self-disclosure—a disclosure that changes radically over time—to that being who is most equipped to receive it: Dasein" (68). The history of Western metaphysics is the attempt to legitimize technological domination of nature and human beings by endowing ultimate representations of being with a commanding power.[14] These measuring principles have a history—a rise, a productive life, and a fall. Principles are not scientific since their role is to ground science. As Reiner Schürmann explains, for Heidegger the history of Western metaphysics is the history of the rise and fall of these principles, which are representations of an ontological origin that precedes and empowers being itself. In Heidegger's words, "Metaphysics is history's open space wherein it becomes a destining that the supersensory world, the ideas, God, the moral law, the authority of reason, progress, the happiness of the greatest number, culture, civilization, suffer the loss of their constructive force and become void" ("The World of Nietzsche," in *The Question Concerning Technology*, 65). Each of these epochal principles is a failed attempt to provide a ground with normative force that would legitimate technological will to power. As Schürmann maintains, Heidegger's history of Western metaphysics is structured around a central insight, which is that philosophy has been hypnotized by Aristotle's teleology from beginning to end. As a matter of fact, Heidegger decreed that "This book [Aristotle's *Physics*] determines the warp and woof of the whole of Western thinking, even at that place where it, as modern thinking, appears to think at odds with ancient thinking. But opposition is invariably comprised of a decisive, and often even perilous, dependence" (Heidegger, *The Principle of Reason*, 63). In Schürmann's interpretation of Heidegger, Western metaphysics has been held captive by a *teleocratic* design, invented by Aristotle, that reaches its point of exhaustion with Nietzsche's doctrine of the will to power: "Both Metaphysics and logic derive from the astonishment before what our hands can make out of some material" (*On Being and Acting*, 99). Schürmann contends that, for Heidegger, the Aristotelian concepts of origin and end do not result from speculation or syllogistic logic, "from the analysis of becoming that affects material things" (99). Causality is an attempt to make intelligible becoming, or material motion. If Aristotle's *Physics* is the grounding book of Western metaphysics it is because "Causal explanation is one mode of understanding among others, although this mode has maintained its hegemony over Western philosophy" (*On Being and Acting*, 100). As Heidegger states, "the concepts of matter and material have their origin in an understanding of being that is oriented to production" (*Basic Problems of Phenomenology*, 116). Far from being an object of empirical observation, the instrumentalist gaze conceives nature in terms of making or manufacturing something out of raw materials. Matter is a basic metaphysical concept that necessarily arises when the ultimate nature of things is "interpreted within the horizon of such productive comportment" (116). From a Heideggerian perspective there is an indissoluble alliance between causal explanation and instrumental manipulation, where technical motion is initiated either by the Divine Artisan or a human subject.

Schürmann postulates that there is a third presupposition behind the Aristotelian division between self-moving things (natural, self-sufficient entities) and things moved by another (man-made, inert artifacts). The distinction presupposes causal movement and change initiated by humans that experience themselves as craftsmen, "as initiator of fabrication, that nature can in turn appear to him as moved by mechanisms of cause and effect" (*On Being and Acting*, 100). First, the philosopher finds the origin of production in himself, and then he finds it in nature and God. If the distinctive characteristic of Western metaphysics is attributing some intrinsic end to a certain origin, the "experience that guides the comprehension of origin as it is operative in the philosophy of nature is paradoxically the experience of fabricating tools and works of art, the experience of handiwork" (100). The division between things that move themselves and things that are moved by another, the division between the principle that precedes and empowers and the secondary effects that are subordinated to it presuppose the agent that moves its own hands (101). This will determine the outcome of Western metaphysics, because the foreseen end conceives the world in terms of a manipulatable stock. As Schürmann remarks, "anything, to be sure may turn into such manipulatable stock, and it may well be that, because of the exclusive emphasis on fabrication since the beginning of Western metaphysics, everything has in fact become just that" (102). The gist of Western philosophy is a "metaphysics of handiwork," of manufacture that ends up becoming artificial manipulation (104). The metaphysics of handiwork is not only teleological but also hylomorphic, since is supposes an efficient cause (a craftsman) that imprints forms (patterns or ideas) over a matter (bronze) in order to produce a final product (a statue).

Although the "metaphysics of handiwork" is at the inception of technological domination because it preconceives the world in terms of an available manipulatable stock, technological domination proper does not take place until the triumph of technological will to power.[15] Will to power triumphs when it establishes its own conditions in "values" that come to replace the former "ends." Unlike goals or ends, a value is a value if it enhances power. Also, there is no preservation of power without the enhancement of power. This means that the will to power always wills more power. In order for a value to be a value it has to produce surplus power. Schürmann explains that for Heidegger, "the teleocracy introduced into philosophy with Aristotle's *Physics* reaches the very being of all entities. But in its fulfillment, finality cancels itself" (*Being and Acting*, 188). With this triumph of subjectivism—which is also a triumph of objectivism, being both sides of the same coin—"the will to power posits itself as its own condition in positing all things as values, that is, as its own objects, in striving after mastery over the earth, in willing that everything becomes its object, what it wills is thus itself: it wills the totality of possible objects as its immanent goal" (188–89). The outcome of Aristotelian teleology is that it cancels itself. Limitless appropriation, subjection, and technological ordering dismantle teleocracy, and all that remains is a goalless will to power that wills itself by willing only more power (189). Ends cease to be given transcendent, supersensory ideas, becoming conditions or obstacles for an ever-expanding cycle of self-overpowering through technological ordering. Heidegger identifies this process with global Western expansion that implants its technological regime everywhere, indifferent to all its consequences (189).

In this goalless process of ever-expanding technological power, humans become tools of their own tools. Heidegger calls this process "enframing," which consists in revealing reality as "standing-reserve," that is to say, in the mode of ordering that challenges not only nature but also human beings, reducing them to a manipulatable stock. In Heidegger's words, "That challenging happens in that the energy concealed in nature is unlocked, what is unlocked is transformed, what is transformed is stored up, what is stored up is, in turn, distributed, and what is distributed is switched about ever anew. Unlocking, transforming, storing, distributing, and switching about are ways of revealing" (*Question*, 16). Machine technology, which makes everything available to itself, consists in arranging reality according to its orderable capacity:

> Enframing means the gathering together of that setting-upon which sets upon man, i.e., challenges him forth, to reveal the real, in the mode of ordering, as standing-reserve. Enframing means that way of revealing which holds sway in the essence of modern technology and which is itself nothing technological. On the other hand, all those things that are so familiar to us and are standard parts of an assembly, such as rods, pistons, and chassis, belong to the technological. The assembly itself, however, together with the aforementioned stockparts, falls within the sphere of technological activity; and this activity always merely responds to the challenge of Enframing, but it never comprises Enframing itself or brings it about. (20–21)

Enframing is, then, the last chapter of a history of Western metaphysics that ends up revealing everything in a one-dimensional way, reducing it to an available raw plasticity ready to be transformed, stored, and distributed. This is the defining feature of modernity, which started with the premodern metaphysics of handiwork already present in Aristotle and perfected by Scholasticism.

Heidegger provides us with a historical genealogy of the technological present in the history of metaphysics. In other words, enframing provides us with a link between technological modernity and metaphysical instrumentalism. Metaphysical instrumentalism, according to which an agent imposes a pattern over matter in order to produce a form, contains the key to understanding global technological expansion as the aimless transformation of everything into useful material. The self-expansion of the technological means becomes an end in itself. But the condition for understanding the link between metaphysical instrumentalism and modern technological reduction to standing reserve is in examining the inconsistencies that arise from metaphysical instrumentalism in its imperial dimension and colonial context. Before going into the specifics, however, let us examine another explanation that competes with Heidegger's logic of domination and that is still connected to metaphysical instrumentalism. If, according to Heidegger, Western metaphysics is a handicraft metaphysics, for Karl Marx ideological mystification, especially visible in commodity fetishism, consists of attributing an inner life to the products of human brains and hands.

MODERNITY AS CAPITALIST EXPANSION IN MARX

It is well known that for Marx, being under the spell of ideology equals not knowing what one is doing. Ideology is characterized by a split between knowledge and practice: "They do this without being aware of it" (Marx, *Capital*, 1:166–67).[16] Ideological mystification implies a misrecognition that does not simply fall under the category of false consciousness because it shapes social reality itself. Another feature of ideological mystification is the fetishistic inversion according to which the result of a network of differential relations is confused with the property of one of the elements of this network. As Marx put it in a footnote, "For instance, one man is king only because other men stand in relation of subjects to him. They, on the other hand, imagine that they are subjects because he is king" (1:49). In order understand this ideological or fetishistic misrecognition that shapes reality itself it is useful to visit Chapter 1 of *Capital*, volume 1, "The Commodity," specifically the section titled "The Fetishism of the Commodity and Its Secret." There, Marx writes: "a commodity appears at first sight an extremely obvious, trivial thing" and yet "its analysis brings out that it is a very strange thing, abounding in metaphysical subtleties and theological niceties" (163). The metaphysical subtleties of the commodity derive not from the instrumental character of the thing in question but from a larger process of valorization of the thing. Before explaining the origin of the enigmatic character of the commodity, Marx depicts that which is not enigmatic, namely, the instrumental manipulation of nature itself in order to produce a final product:

> It is absolutely clear that, by his activity, man changes the forms of the materials of nature in such a way as to make them useful to him. The form of wood, for instance, is altered if a table is made out of it. Nevertheless the table continues to be wood, an ordinary, sensuous thing. But as soon as it emerges as a commodity, it changes into a thing which transcends sensuousness. It not only stands with its feet on the ground, but, in relation to

all other commodities, it stands on its head, and evolves out of its wooden brain grotesque ideas, far more wonderful than if it were to begin dancing of its own free will. (163)

When an artisan imposes a form over a material, the product is still an "ordinary, sensuous thing." In the language of metaphysical instrumentalism, Marx seems to be saying that the telos of instrumentalism is not something transcendent, enigmatic, or "mystical." This means that the "mystical character" of the commodity is not inherent in its use value or instrumental character of the product. Only once it is inverted, when it stands on its head, does the mad dance of commodities commence. When Marx asks for the origin of the "enigmatic character" of the commodity, he unequivocally answers, "Clearly, it arises from this form itself" (164). The "form" is the form of equality that emerges through measuring commodities against other commodities and ultimately against money. The value of a commodity is expressed through the value of another commodity, and through money as the value of value:

> The coat, therefore, seems to be endowed with its equivalent form, its property of direct exchangeability, by nature, just as much as its property of being heavy or its ability to keep us warm. Hence the mysteriousness of the equivalent form, which only impinges on the crude bourgeois vision of the political economist when it confronts him in its fully developed shape, that of money. He then seeks to explain away the mystical character of gold and silver by substituting for them less dazzling commodities, and with ever-renewed satisfaction, reeling of a catalogue of all the inferior commodities which have played the role of the equivalent at one time or another. He does not suspect that even the simplest expression of value, such as 10 yards of linen = 1 coat, already presents the riddle of the equivalent form for us to solve. (149–50)

The exchangeability of the commodity is not an intrinsic property of the thing but an effect of a differential relation, which produces the perspective illusion of a thing possessing value in itself. It is useless to retranslate money into how many commodities it can buy because it still presupposes the equivalence it has to explain. In other words, the secret is the mystifying power of the form of equivalence itself.

Marx argues that "the mysterious character of the commodity form consists therefore simply in the fact that the commodity reflects the social characteristics of men's own labor as objective characteristics of the products of labor themselves, as the socio-natural properties of these things" (*Capital*, 1:165). The sensuous products of labor become "supersensible" (165). As a result, only by analyzing the "metaphysical and theological" subtleties of money as *medium* of exchange does Marx arrive at his well-known definition of commodity fetishism: "It is nothing but the definite social relation between men themselves which assumes here, for them, the fantastic form of a relation between things. In order, therefore, to find an analogy we must take flight into the misty realm of religion. There the products of the human being appear as autonomous figures endowed with a life of their own, which enter into relations both with each other and with the human race. So it is in the world of commodities with the products of men's hands" (165). In a fetishistic transposition, the products of human brains and hands acquire a transcendent, supersensible life of their own. A commodity is both the embodiment of social relations and a magical object endowed with an autonomous life of its own. Marx's theory of fetishism is basically an explanation of how, in a bottom-up fashion, a market-based network of relations generates an economic arrangement that in the end takes on a life of its own. These economic patterns are virtual formations or metaphysical ideals that transcend the individual members of the social network, achieving a certain autonomy that subordinates these individual members to the same virtual formations. The metaphysical subtleties of the commodity arise out of the ground of material labor, the products of hands and brains. The alienating effects of capital as identified by Marx are due to the fact that social dynamics are no longer regulated by the interests of those who generated that capital in a bottom-up way but, rather, by the drive to generate money out of money. The problem is that once the fetish acquires a life of its own, it is no longer possible to reduce its "misty" theological character to the material production or the antagonisms of the social life. The point of the critique of political economy is not to reduce the fetishistic dimension to the "real," material, empirical world. The point is that we cannot understand the real, material, empirical world without grasping how it is shaped by the metaphysical and theological character of the fetish. It is necessary to go through the mystification of the form of equivalence in order to demystify commodity or money fetishism. The critique of political economy means that it is necessary to go through the metaphysical subtleties and theological niceties that are taken for granted by an ideology that intervenes directly in reality. No empirical description, no matter how exhaustive it appears to be, can perform this critique: "Moreover, in the analysis of economic forms neither microscopes nor chemical reagents are of assistance. The power of abstraction must replace both"(1:90). Capitalism is a system of beliefs, a metaphysics that also imposes its own form over reality.

Imposing its form over reality is a process that is essential to modernity, although in a form different from that articulated by Heidegger. I am referring to the emergence of the capitalist world system in the sixteenth century. Marx explained the origin of modernity as capitalism both in terms of processes of production and in terms of processes of circulation. On the one hand, when prioritizing the process of production, he looks for the origin of capitalism in the inner contradictions of feudal society and the rise of manufactures launched by independent farmers (*Capital*, 3:455). On the other hand, Marx also prioritizes the process of circulation when he writes in Chapter 4 of *Capital*, titled "The General Formula of Capital": "The circulation of commodities is the starting-point of capital. The production of commodities and their circulation in its developed form, namely trade, form the historic presuppositions under which capital arises. World trade and the world market date from the sixteenth century, and from then on the modern history of capital starts to unfold" (*Capital*, 1:247). This world market emerged with the connection of the international economies of the Baltic and Mediterranean in the fifteenth century, and with the connection of Europe, America, and Asia in the sixteenth century (Karatani, *Structure of World History*, 159). As Paul Sweezy and Immanuel Wallerstein stress, European capitalism would have been impossible without the emergence of the world market.[17]

In order to explain the historical presuppositions of capitalism with the emergence of the world market in the sixteenth century, Marx moves on to explain what he considers to be the matrix of capitalism—the movement of money generating more money out of itself. Before going into specifics he explains that, when we disregard the material content of the circulation of commodities and consider only the economic forms, we find that the ultimate product of circulation is money (*Capital*, 1:247). Moreover, "this ultimate product of commodity circulation is the first form of appearance of capital" (247). Marx continues, analyzing the consequences of the transformation of simple exchange (where commodities are exchanged for money in order to buy commodities, C-M-C, into the circulation of money as capital, where the goal of the exchange is to produce more money, M-C-M′). On the one hand, in C-M-C, "the simple circulation of commodities—selling in order to buy—is a means to a final goal which lies outside circulation, namely the appropriation of use values, the satisfaction of needs" (253). On the other hand, Marx writes that in M-C-M′, "the circulation of money as capital is an end in itself, for the valorization of value takes place only within a constantly renewed movement. The movement of capital is therefore limitless" (253). With the inversion of C-M-C into M-C-M′ we witness the practical emergence of the fetishism of money. Money starts to occupy the place of power in the hierarchical structure. Money is not money because it possesses the quality of being intrinsically money but because we treat it as money when its accumulation becomes an end in itself. Money and commodities do not have the same power because money can buy commodities, while commodities are not necessarily exchanged for money. In other words, money has the power of exchangeability.[18] Exploitation of workers in industrial societies is structurally the same as the activity of merchants who buy and sell at different locations, since in both cases the labor commodity and the commodities in general are subordinated to the process of self-valorization of capital. As Marx would say, "we do not need to look back at the history of capital's origins in order to recognize that money is its first of appearance. Every day the same story is played out before our eyes" (*Capital*, 1:247).

The above-mentioned asymmetry between the limitless power of money and the instrumental value of commodities gives rise to the perverse drive to accumulate money for the sake of the accumulation of money, a "boundless drive for enrichment, this passionate chase after value" (Marx, *Capital*, 1:254). This drive to accumulate money produces an "increment or excess over the original value I call surplus value" (251). It is important to stress that this excess cannot be reduced to the instrumental manipulation and domination of nature or people. In other words, it is not reducible to Heidegger's logic of domination or to a logic of nihilistic equality. Moreover, the production and appropriation of surplus value escapes domination. Its excess is closer to what Deleuze and Guattari call "deterritorialization," a constant self-revolution where everything that is solid then melts into the air. Money always escapes. And it escapes by subordinating everything to itself. The logic of money-producing-money is not reducible to technological enframing, since it involves an excessive element—surplus value, the value against which all values disintegrate—that appears to engender itself in a continuous process of self-valorization: "Value is here a subject of a process in which, while constantly assuming the form in turn of money and commodities, it changes its own magnitude, throws off surplus value from itself considered as original value, and thus valorizes itself independently. For the movement, its valorization is therefore self-valorization. By virtue of being value, it has acquired the occult ability to add value to itself. It brings forth living offspring, or at least lays golden eggs" (255). Value, as it arises out of the form itself, is *intrinsically metaphysical and theological* because it is a pure leap of faith based on the promise that the cycle of production and circulation *will produce profit*.[19] With this leap of faith, the *salto mortale* of a postponed retroactive valuation, capitalism prospers on future credit, with no guarantee in reality except the ungrounded belief that the cycle of circulation will produce more value. In other words, the leap of faith—the split by means of which something with instrumental value jumps into a commodity as a transcendent and supersensible thing—is strictly theological or metaphysical.

Until now I have been examining how Aquinas's Scholasticism, which was the metaphysical basis of imperial ideology, had instrumentalist presuppositions. I have also shown how Heidegger understands these metaphysical presuppositions as being part of a metaphysics of handiwork whose outcome is the modern drive to transform humans into tools of their tools. In addition, we have seen how this drive to dominate by means of the imposition of rational scientific patterns over the world cannot explain the emergence of an excess that escapes control such as the drive to produce surplus value. The question then arises of how to understand these two logics of modernity, with their different aims and procedures. Both logics have a common origin—transforming reality with human hands—but completely different outcomes. The logic of domination and the logic of self-revolutionizing excess are two aspects of modernity, two irreconcilable narratives that exist only by each one's ignoring the other. Capitalism is not the by-product of the technological domination of nature, because the latter is also inseparable from market circulation and the generation of surplus value. There is no superior synthesis or neutral metalanguage that would provide an all-encompassing narrative of these two contradictory aspects of modernity.[20] Heidegger provides a history of philosophy whose outcome is will to power and technological domination, which also influenced Foucault's notion of biopolitics as the regulation and administration of life, as well as Adorno's search for answers in the logic of instrumental rationality. Marx provides a history of the excessive movement of capital that dissolves all bonds and disrupts all values by measuring them against money as the value of value. The logic of surplus power, manipulating people by apparatuses of subjection, and the logic of surplus value, the logic of production and appropriation of an excess in continuous self-revolution, are not identical and cannot be translated into one another.

Nevertheless, the wager of the present book is that the dual character of modernity, the division between the logic of domination and its logic of deterritorializing excess, emerges out of the contradictions of metaphysical instrumentalism, which can be historically articulated by examining the relations between natural law and mining. An examination of the complex interrelations and tensions between natural law and mining in the sixteenth and seventeenth centuries shows how these two logics are the two sides of the same coin, the two irreducible aspects of imperial expansion. By examining the double articulation between natural law and mining it is possible to redouble the gap between the logic of domination and the logic of capitalist deterritorialization within metaphysical instrumentalism. The Heideggerian critique of Western technological domination can be historically articulated with a Marxian critique of commodity fetishism by means of a critique of the inner impasses of Aquinas's metaphysical instrumentalism in a colonial context. The inner impasses of instrumentalism generated a certain excess—a self-valorizing value—that is impossible to control, an instrument that subordinates all instruments to itself and therefore undermines its own metaphysical presuppositions.

To recapitulate, what at first sight looks like a merely formal or logical deadlock in metaphysical instrumentalism is symptomatic of a more radical impasse that is inherent in the colonial situation itself. The two logics of modernity are narratives that still see their historical development as European achievements while foreclosing their origin in Hispanic imperial instrumentalism as applied to the lands and peoples of the Americas. The only way of explaining the historical genesis of these two logics is through the inner contradictions of metaphysical instrumentalism, but this must be on the condition that we do not consider its universal ambition as a formal, empty container but, rather, as unresolved aporias and antagonisms. The impasses of imperial instrumentalism are born in the interstices between different communities, in the conflictive and asymmetrical space of a colonial context, as a result of the attempt of one community to enjoy the property and the fruits of another community's labor. The imperial dimension of metaphysical instrumentalism remained virtual until it was actualized in a colonial relation between communities, when Spaniards saw Amerindians as an imperfect matter that should be molded and guided to the only ends common to all humanity—civilization and salvation. Although the official imperial discourse appealed to civilization and conversion as ends in themselves, an examination of the arguments invoked by the apologists of the Spanish Empire show how the aim of these ends was to mobilize the indigenous labor force and accumulate riches.

The present volume is a work of intellectual history concerning the inconsistencies behind attempts to justify the Spanish mining enterprise in its economic, political, and technological dimensions. All the texts examined here belong to the imperial school of thought and represent an imperial perspective that was later disavowed by the Enlightenment, despite being central to the development of capitalist and technological modernity. These texts were hegemonic; they created reality by influencing the Crown's policies. The decision to limit the analysis of metaphysical instrumentalism to texts that fall squarely within the imperial tradition is based on the need to find a solution to the problem of the separation between the fields of political ideology and history of science in colonial studies. The decision to focus exclusively on what is hegemonic, dominant, or the rule rather than the exception is based in an attempt examine the presuppositions of imperial reason, the consequences of which are still visible today. By examining texts that were dominant and influenced the decisions made by the Crown, it is possible to fill a gap in the study of the Spanish Empire, namely, the separation between political philosophy and history of science. While Spanish imperial science and mining are traditionally studied separately from Spanish political theory, here we make the case that these two discourses are isomorphic and intertwined. It is not simply that natural law provides the theory for mining practice in a unidirectional way, since they are both opposite sides of the same metaphysical instrumentalism. Their relation is one of dynamic interpenetration and mutual constitution. Natural law justified mining by providing it with a goal, and mining changed natural law by becoming an end in itself. Far from limiting itself to being a material means, a passive instrument, mining proved to be more indispensable, and also more destructive, than the goals invoked by natural law. Mining literally denaturalized natural law by forcing it to face its own disavowed presuppositions.

Natural law and mining are isomorphic because they share the same hylomorphic and teleological structure based on the principle of the natural subordination of matter to form and means to an end. It is not possible to reduce natural law to a theater of appearances and mining to an objective, material, socioeconomic process. Both natural law and mining share the same metaphysical principles and presuppositions: the transposition of artificial manipulation to the sphere of natural order and political mastery. Both consider their objects (nature and human populations) as imperfect matter that needs to be directed to a final end by means of the imposition of rational patterns. The relation between both natural law and mining is one of mutual co-constitution and irreducible tension. But the irreducible tensions between them are internal to metaphysical instrumentalism itself and derive from the transposition of artificial manipulation to natural order. Metaphysical instrumentalism depends formally on the transposition of artificial manipulation to the natural order only in order to proceed to subordinate artificial subordination to a higher master. This disavowal has practical consequences: the means effectively dominate the whole process, mediating between the invoked ends and the community in which the end is to be realized. If, from an idealist or metaphysical perspective, the end dominates the means, from a materialist and historical perspective both the end and the object are materialization of the means that subordinates the end. As Heidegger would put it, the means, or the instrument, is not an inert neutral device but a process of domination over nature. As Marx would put it, the end of the whole process is not the satisfaction of needs but the endless drive to accumulate more riches. The ultimate inner contradiction of metaphysical instrumentalism is that the end is the means of the means.

THE STRUCTURE OF THE PRESENT BOOK

The present book is structured around a contrapuntal reading of this entanglement within the sixteenth-century debates on Spanish sovereignty in the Americas and writings on natural history and mining written between 1520 and 1640. Each chapter shows a slight shift of perspective with respect to the previous chapter. Each shift is the result of previous failed attempts to find a solution to the problem of how to ground the Spanish dominion and the practice of mining. The Spanish dominion invoked the principle of subordination of the material means to the transcendent end, which contained irreducible contradictions. Therefore, the works discussed in each chapter brought new contradictions, which end up producing new failed attempts to cope with these contradictions. The arrangement of the chapters follows the demands of the problems addressed by each of the authors analyzed here and the various impasses generated by metaphysical instrumentalism in its imperial dimension. The chapters are also arranged chronologically, beginning with an examination of the earliest attempts to justify Spanish dominion—intellectual efforts in which mining was addressed abstractly—and ending with an overview and analysis of the crisis of the imperial enterprise that emerged from the material contradictions introduced by the mining practices founded on imperial ideology.

Chapter 1, "Grounding the Empire: Francisco de Vitoria's Political Physics," begins with the history of the principle of subordination of matter to form and of means to an end by examining Francisco de Vitoria's (c. 1492–1546) *relecciones*, key lectures that were later transcribed and edited by students. Vitoria was the jurist theologian who systematized the first attempts to justify the conquest and colonization of the New World through a discussion of the problems of the sovereignty of indigenous peoples and the appropriation of their material riches by the Spanish Empire. The chapter starts with an examination of what Vitoria considered the only valid reasons for subjecting the indigenous peoples to the Spanish Crown as they appear in *On the American Indians* (1539). In this work, he states that trade, commerce, and the extraction of metals (in places that do not belong to anybody) are universal rights grounded on the laws of nations (*ius gentium*). If Indians deny this right to Spaniards, they commit injury against the latter. As a result, Spaniards can make just war on them using the natural right to repel force with force. In *On the Law of War* (1539), Vitoria explains that only a "perfect" (i.e., self-sufficient) community with dominion and civil power over its subjects can declare war and repel force with force. The problem that arises is that of the metaphysical foundations of a civil power with self-sufficient capacity to declare war and protect the interests of its subjects. Vitoria examined these ontological foundations of civil power in *On Civil Power* (1528) and in *On the Power of the Church* (1532). In *On Civil Power*, Vitoria grounds the legitimate power of an autonomous commonwealth on the four Aristotelian causes (material, formal, efficient, and final). Political mastery over lands and subjects depends on the principle of the natural subordination of matter to form, means to ends, and the imperfect to the perfect.

A close examination of Vitoria's arguments shows that, despite all attempts to political dominion on a natural order, this natural order still presupposes technical mastery. In other words, the principle of natural subordination presupposes artificial subordination, since God stands before the machine of the world in the same way that an architect or artisan stands in front of raw material before making a final product. In one and the same move, dominion borrows mastery and control from technique and labor and subordinates the transformative power of the craftsman (i.e., the art of bridle making) to a superior power (i.e., the art of war) to mention the examples used by Vitoria himself. Ultimately, natural right presupposes the complete opposite of a natural given order: political mastery and autonomy presuppose *technical* mastery and autonomy, artificial manipulation by human hands. Therefore, it is safe to conclude that the ultimate aim of imperial ideology is to justify the technical means it uses to achieve its ends. The law that dictates that war can be declared in order to protect trade, circulation, evangelization, and the exploitation of metals is grounded on a notion of mastery that presupposes technical mastery, which is ultimately the art of war itself. This irresolvable contradiction plagues the very ontological foundations of Iberian expansion. In the attempt to justify the subordination of one community to another, Vitoria employs all the resources of metaphysical instrumentalism. Vitoria's doctrine of just war contains a technical and instrumentalist conception of nature and law that lies at the center of Iberian imperial ideology. This fissure in the center of imperial ideology reappears in the debate of Valladolid, which is the problem of the second chapter.

Chapter 2, "The Impasses of Instrumentalism: Revisiting the Polemics between Sepúlveda and Las Casas" explores the way in which these authors attempted to fill the breach opened in the structure of Vitoria's theoretical edifice. Although Sepúlveda has often been cited for his use of philosophical arguments in order to declare that the Amerindians were slaves by nature, scant attention has been given to his use of a Thomist vocabulary and the principle of the "natural subordination" of matter to form and means to ends. Sepúlveda considers the basis of all dominion to be the principle of the subordination of imperfect matter to perfect form. As a matter of fact, for Sepúlveda, the mere disobedience of Amerindians (who are compared with amorphous matter) to the superior command of the Spaniards (identified with forms) is sufficient reason for declaring war and correcting the deviations of natural order. An examination of Sepúlveda's *Demócrates Segundo* (1544) makes explicit the instrumentalist and imperial presuppositions of his use of this principle by which he radicalizes the line of Vitoria's thinking. For instance, while Vitoria hesitates to infer that the Indians are natural slaves, Sepúlveda does not hesitate to say it because they stand in relation to the Spaniards in the same way that "imperfect matter" stands in relation to "perfect form."

Las Casas deconstructs this argument by appealing to the literal sense of the principle of the natural subordination of matter to form. Since every individual substance is a composite of matter and form, the form of one entity cannot be imposed onto the matter of another entity because each individual entity has its own form and its own matter. This means that imperial instrumentalism cannot be used to justify the subordination of one community by another. For this reason, I argue that the Valladolid debate represents a radical impasse in metaphysical instrumentalism that was also potentially present in Vitoria. Hylomorphic instrumentalism simultaneously can and cannot be used to subject one community to another. In his attempt to grant autonomy and the capacity of self-government to Amerindians, Las Casas ignores the identification of the people with the material cause (Vitoria) and proceeds to identify them with the efficient cause. This move allows him to strengthen the conclusion of the first part of Vitoria's lecture *On the American Indians*, which argues that Amerindians were self-sufficient masters and possessors and that they should be restituted of all their stolen lands and riches. The theoretical deadlock produced by the inner inconsistencies of metaphysical instrumentalism would find more practical solutions in José de Acosta, who is the object of analysis of the third chapter.

Chapter 3, "Mastering Nature: José de Acosta's Pragmatic Instrumentalism," examines José de Acosta's *De procuranda Indorum* (1588) and his *Natural and Moral History of the Indies* (1590). Acosta's corpus includes both writings on the debates about Spanish sovereignty over the Indians and writings on science and mining techniques. Acosta's texts are an elaborate adaptation of late Scholastic principles to the evangelical mission and ideology of Spain's early-modern "universal monarchy." *De procuranda Indorum* revisits the arguments of Vitoria, Sepúlveda, and Las Casas about the conquest and colonization of the West Indies. Although Acosta accepts that the Spanish Empire is founded on unjust conquest and violence, he maintains along with Vitoria that circulation and evangelization are sufficient reasons for subjecting the Indians to Spanish rule. Moreover, he uses Thomist teleological arguments in order to justify the mita, which was a compulsory system of indigenous work in the mines of Potosí. In his *Natural and Moral History* he offers a program of rational and causal explanation of the particularities of the natural phenomena of the New World. In this work, Acosta introduces the subject matter of mining in book 4, where he discusses the composition of metals. He uses the Scholastic principle of the subordination of matter to form and means to an end in order to explain the providential role of metals, which were placed in the bowels of the earth by God in order to fulfill the ends of the universal monarchy. Acosta does not only use metaphysical instrumentalism to justify the Spanish presence in the Americas; he also uses the principle of natural subordination to provide a causal explanation of technical and material practices such as mining.

The impasses of metaphysical instrumentalism reappear in Acosta when he discusses the instrumental role of metals. While he employs the principle of subordination of matter to form to argue that metals are the lowest and least perfect entity, whose role is to serve the superior entities such as plants, animals, and humans, he also argues that precious metals are not just one particular thing among others but, instead, something that is virtually everything since it can buy all other things. Money ceases to be merely one instrument among others and becomes the measure of everything else, subordinating everything to its own power. Acosta attributes this capacity of money to the inherent qualities of precious metals, which are incorruptible and extremely negotiable, not subject to the wear and tear of all other commodities. As a result, Acosta participates in the commodity-fetishist belief in precious metals as possessing inherent value. Moreover, he argues that mining is essential for supporting the whole imperial enterprise, including its evangelical and administrative spheres. The search for profit, for surplus value, is the only glue that keeps the empire together.

Nevertheless, when he describes the interior of the mines of Potosí, he employs a gloomy language that depicts the underground world as something inhuman. For this reason, I argue that Acosta's pragmatic instrumentalism disavows the undesirable violence of mining as an essential part of Spanish dominion: at this point, the imperial subject knows very well what he is doing, but he keeps doing it. Despite appealing to Vitoria's arguments, Acosta finds out that there is no valid origin of the empire, and thus he is forced to ground it retroactively. This means that, like Vitoria, Acosta thinks it is not expedient to abandon the Indies. Yet going beyond Vitoria, he knows that the mines are what sustains the empire. The aim of the program of causal explanation by means of the principle of natural subordination is to justify the extraction of profit. Acosta fills the gap opened by the impasse between Las Casas and Sepúlveda (and present in Vitoria's inconsistencies) with the notion of money as the Spanish Empire's new end.

Chapter 4, "From Imperial Reason to Instrumental Reason: The Ideology of the Circle of Toledo," explores the ideological framework behind the reforms introduced by Francisco de Toledo, the infamous viceroy of Peru from 1569 to 1582. Toledo's fame as "supreme organizer of Peru" is based on a series of radical measures he implemented, including the introduction of the *mita* (a regime of tributary work, simultaneously obligatory and nominally remunerated), a revolutionary method of amalgamation (the *beneficio*), and the establishment of *reducciones* (places where Andeans were relocated for the purpose of introducing them to Christianity and to a Western way of life). These measures strengthened the authority of the colonial state and established a mercantilist economy in a moment of economic and political crisis in the Andes. In this chapter we analyze the ideological background of the Toledan reforms. First, we examine the *Anónimo de Yucay* (1571), a radical defense of the Toledan policies structured around an attack on Las Casas, in which the authors argue that mining is central to the Spanish Empire within the context of the same teleological frame employed by Acosta. While in Vitoria the absolute necessity of the ends were still prospective, in the *Anónimo* the employment of ends is completely retrospective. Contingent historical facts, such as the coincidence between the formation of a Holy Alliance against the enemies of Catholicism in Europe and the discovery of the mines of Potosí, are interpreted within a providential and teleological frame whose ultimate aim is to justify the continued centrality of mining.

Next is an analysis of Juan de Matienzo's *Gobierno del Perú* (1567), which inspired the Toledan introduction of a rotational system of tributary labor. Matienzo followed Sepúlveda in appealing to the Indians' "servile nature." He claimed that, instead of freely joining the market economy, selling commodities, and offering their services for wage labor, they simply retreated into a regime of mere subsistence. Matienzo argues that this lack of economic interest was due to both the absence of individual property within the Indian community and their collective subjection to the despotic authority of the Incas. As a result, the peasantry had little incentive to seek any improvement in their condition, since the wages they earned were often appropriated for communal purposes. Matienzo provided another reason for compelling the Indians to work in the mines: the natives were inclined to vice, capable of learning only mechanical rather than liberal arts, and yet disinclined to work. Like Vitoria and Sepúlveda, Matienzo drew heavily upon Aristotle's dualism between material body and substantial form, claiming that the indigenous peoples were fit only to obey and not to command because they have strong bodies but weak understanding. Finally, we examine José Luis Capoche's *Relación general de la villa imperial de Potosí* (1585), a text often overlooked in colonial literary studies, which provides a retroactive justification of these Toledan policies. Capoche, an owner of several *ingenios* (amalgamating mills) and mines, narrates the transition from the system of the refinement of precious metals by *huayras* to the method of amalgamation in such a way that the mita appears as a necessary consequence of this technological advance. His arguments are a mixture of some technical pragmatism with a rhetoric of catastrophe that gives mining an air of necessity, thus implying that the entire colonial enterprise depends on it. Capoche argues that there is no evangelization without mining, no mining without amalgamation, and therefore no amalgamation without the mita or compulsory labor. Capoche's verdict was unequivocal: the situation gives the Spanish the choice of either forcing the Indians to work in the mines and amalgamating mills or facing a catastrophic and total breakdown of the interconnected networks of economic relations.

Since these texts continue to use a philosophical and theological vocabulary, imperial discourse regarding the possession of mineral wealth in the New World begins to crack under the pressure of the earlier contradictions in metaphysical instrumentalism. The conflation between the artificial and the natural parallels the contradictions between attempts to ground the business of empire on a transcendent final cause (the "common good" of the natives) and the material dependence on precious metals and indigenous labor that proves to be incompatible with the ends invoked. Reinscribing mining into a broader ideological picture introduces a slight displacement of perspectives: whereas the classic debates on Spanish sovereignty during the period of conquest grounded "transcendent ends" on a preexisting onto-theology, at this later stage "practical ends" are deprived of any external ground, becoming necessary only under the threat of an undesirable catastrophe. The principle of natural subordination, which once served to justify the "just war" and the dispossession of the natives, suffers a slow productionist metamorphosis when adopting the task of "administering" and "conserving" the Indies. The means cease to be subordinated to the end. The means becomes an end in itself. The chapter ends with a reflection on how these texts can be read as an expression of a logic of domination and subjection and an expression of the logic of the disruptive power of surplus value, as they are the two contradictory sides of the same metaphysical instrumentalism.

Finally, in Chapter 5, "The Exhaustion of Natural Subordination: Solórzano Pereira and the Demise of Metaphysical Instrumentalism," we analyze the work of the jurist Juan de Solórzano Pereira (1575–1655), author of the most influential treatise on Spanish imperial law, which is also the last great defense of the Spanish government, *Política indiana* (1647). This work exemplifies the exhaustion of the principle of natural subordination, in which the irreducible gap between the instrumental means and the political ends parallels the political and economic crises of the empire at the beginning of the seventeenth century. When revisiting the juridical foundations of the Spanish possession of the mineral riches and the disposition of Amerindian labor, Solórzano employs the same philosophical vocabulary used by his predecessors, with the predominant influence of Acosta, while opposing Vitoria's natural law arguments with a more Roman and retroactive justification of the Spanish Empire. In the *Política indiana*, the impasses of instrumentalism and the social deadlock they represent appear under the form of an examination of the arguments in favor of and against the mita introduced by Toledo. The arguments in favor of the mita invoke the principle of subordination of matter to form and means to an end, as well as the classical corporatist image of society as a harmonious social body so dear to Aquinas. The arguments against the mita invoke the failure of the mita to achieve its ends, since the means themselves are destructive of indigenous lives and incompatible with the goals of civilizing and evangelizing them.

Solórzano faces an irresolvable impasse: the Spanish Empire is impossible both *with* and *without* mining. Mining, understood as the articulation of mita and the process of amalgamation, is an ambivalent material support that maintains the life of the empire while undermining its legitimacy. Potosí becomes the symptom of the contradictions of imperial instrumentalism: the Spanish Empire became dependent on a surplus of silver extracted from Potosí that disrupted the very lives of the Andeans and undermined the moral basis of the kingdom. Solórzano's solution to this impasse is to argue that, although it is impossible to defend the mita without unleashing a backlash, nevertheless, the mita should be practiced until the arrival of a better "judgment" or solution. As a matter of fact, the arguments against the mita are transformed into arguments in favor of continuing to mine while also keeping in mind the true ends of a Catholic monarchy. Therefore, mining cannot become an end in itself, because it has to remain a by-product of the true transcendent ends of the Catholic empire. Solórzano represents the exhaustion of natural subordination because, despite using a metaphysical vocabulary to frame the problem, he fails to see how the problems associated with mining were introduced by imperial instrumentalism itself.

I argue that this irresolvable contradiction at the heart of the Spanish Empire was present from the very inception of imperial reason in the work of Vitoria. The principle of the subordination of material means to the final product presupposed a disavowed productionist gaze: on the one hand, it results from the transposition of technique to physics, and on the other hand, it subordinates technique to higher ends. While in natural law, the dependence on disavowed instrumentalism remains virtual and abstract, after Potosí and the Toledan reforms, the dependence on such a disavowed instrumentalism becomes actual and material. The virtual contradictions in Vitoria become actual irresolvable contradictions materialized in the mines of Potosí. While at the beginning of the present book, with the analysis of Vitoria, we can see how natural right presupposes some disavowed will to mastery and transformation of reality, by the end, with the examination of Solórzano, we can see how mastery and the production and appropriation of surplus have to remain by-products subordinated to the transcendent ends of the Spanish Empire.

The most paradoxical outcome of metaphysical instrumentalism is the idea of metals as living entities capable of reproducing themselves in the mines. This belief was shared by both Andeans and Spaniards. Solórzano, for instance, writes that, since metals are fruits that grow under the ground, they belong to the Crown, which has right of possession over them and can let particulars enjoy them by means of *mercedes*, or royal grants. José de Acosta also considered metals as possessing a certain kind of self-generating power that allowed them to serve superior entities in a natural order in which the imperfect is subordinated to the perfect. Metallurgists such as Álvaro Alonso Barba, inventor of a method of amalgamation in copper cauldrons and author of the *Arte de los metales* (1640), defended the idea of metals as having a certain power of mutation of their own due to their capacity to instrumentalize their four elements. In other words, the belief in the self-reproducing capacity of metals is not only compatible with but also necessary for instrumentalism since it produces the fantasy of a never-ending source of riches.

In the final section of Chapter 5 we analyze Andean beliefs about the self-reproductive capacity of metals as they have been considered by contemporary scholars. According to these Andean beliefs, the world is a living whole that requires care and respect. The earth, *Pachamama*, is a divine source of universal life. Everything, including metals, is a living fruit of the earth. The idea of this living whole implies a certain harmonic balance between humans and the earth as an immanent source of life.[21] Since everything is alive and a continuous source of nourishment, humans have to repay the earth for its services. For the representatives of decolonial theory such as Walter Mignolo, the view of the earth as a living whole and all things as fruits of the earth is a system of understanding that is completely different from modern and Eurocentric systems of development. Andean beliefs and practices are expressions of an indigenous episteme associated with a political autonomy and an ideal of communality that are irreducible to Western values. Understanding the earth as a living entity implies an ideal of harmony and plenitude with the goal of the regeneration of life.[22]

The first aim of this analysis is to demonstrate how certain attempts to romanticize Andean vitalism, such as some expressions of decolonial theory, remain blind to the problematic convergence between Andean and imperial beliefs on the self-reproductive capacity of metals. Second, it demonstrates how the most serious academic attempts to determine the specificity of these Andean beliefs have to rely on an implicit or explicit philosophical vocabulary that is intrinsically hylomorphic. This means that Andean vitalism becomes readable only retroactively, through the fissures and paradoxes of metaphysical instrumentalism. Some scholars explain this belief by using a philosophical vocabulary, such as the division between essence and appearance and between matter and form. Moreover, contemporary conceptualizations of Andean uses of money oscillate between considering money either as a neutral instrument useful for satisfying human needs or as a gift irreducible to the logic of exchange, a means of contributing to the circulation of and reproduction of money as cycles and reproduction of life. The idea of the capacity of metals to generate more metals inside the mines ends up being uncannily similar to the valorization process of capital, M-C-M′, in which money becomes a fetish that generates more money out of itself. As Carmen Salazar-Soler demonstrates in "Encuentro de dos mundos," these Andean beliefs, far from being simple obstacles to modernity, were conditions of possibility of the integration of indigenous workers into the modern world of the mines. Andean vitalism cannot be relegated to the status of a pure and untouched alterity that resists Eurocentrism, because *it is the other side of instrumentalism*, which becomes readable once metaphysical instrumentalism exhausts its possibilities by being deprived of a higher end. Andean vitalism—the self-reproductive power of something that does not have any higher purpose than to be life that generates more life—is the name of an excess that is internal to metaphysical instrumentalism and can only become readable through the impasses of metaphysical instrumentalism.

Decolonial thinking reduces the difference between life and what exploits and subordinates life to other ends to a distinction between local ways of knowing and Eurocentric (or global) power structures. Although it is necessary to refuse to romanticize life by identifying it as a local knowledge irreducible to the Eurocentric matrix of power, this does not mean that metaphysical instrumentalism exhausts Andean vitalism. Moreover, Andean vitalism is simply instrumentalism turned against itself once this excess is elevated to the status of an end in itself. When instrumentalism is turned against itself, we witness a movement that is contrary to the one going from the useless to the useful, as in the retroactive readings of metaphysical instrumentalism. We witness a movement that goes from something that was initially useful for the emergence of modernity to something that is useless and even harmful for utilitarian instrumentalism. From one perspective, vitalism was not an obstacle to technological capitalism but the condition of possibility of the integration of indigenous beliefs into modernity. For another perspective, vitalism was also associated with an organic view of nature that was inseparable from the return of the Inca and anti-colonialist emancipation. With the colonial structure emerges a dysfunctional kernel inherent in instrumentalist metaphysics itself, since the condition of possibility of technological modernity, an excess of life inseparable from the imperfection of matter, is also a threat. When associated with the persistence of Andean beliefs, the dysfunctional kernel of metaphysical instrumentalism can override the interests dictated by technocapitalism, on the condition that they persist even beyond their goal-oriented, survivalist utility. Such nonfunctional attachment that clings to these beliefs, even when they outlived their purpose or usefulness, gives body to a maladaptive colonial antagonism that is inseparable from the failure of metaphysical instrumentalism itself. The ultimate product of the impasses of metaphysical instrumentalism is Andean vitalism as a remainder that escapes modernity's double logic of technological domination and capitalist deterritorialization. This ambivalent and archaic remainder, which becomes readable only retroactively through the colonial situation, is the result of the inner contradictions of the imperial principle of the natural subordination of the material means to the transcendent end.

By examining the interpenetration between early political writings and texts on mining, the argument here is that the impasses of instrumentalism triggered by the confluence of colonial practices and the material force of ideas in the Americas frame technological and capitalist modernity as both an imperial and a metaphysical project. Modern concepts such as *human* and *natural resources* emerge as a result of the violence of abstraction inherent in the metaphysical instrumentalism that shaped the perception of the New World in terms of prime matter. This ambivalent product of metaphysical instrumentalism—both its condition of possibility and its inherent obstacle—contains the potential for thinking and historicizing a non-instrumentalist thinking, but that would be the object of another book. The wager of this book is that we cannot narrate the history of Andean mining without first going through the defiles of metaphysical instrumentalism.

1

GROUNDING THE EMPIRE

Francisco de Vitoria's Political Physics

Within a year of setting foot on the shores of America in 1492, Christopher Columbus was already leading the first mining expedition on the island of Hispaniola. Although mining on the island was a failure in the long term, the Caribbean became the primary source of precious metals for the Spanish Empire during the first two decades of the sixteenth century (Kamen, 88). The search for metals in the Caribbean was the first step toward the creation of a global economy. Since the discovery, "the Americas accounted for the overwhelming majority of the world's bullion production" (Cross, 397).[1] Moreover, the New World bullion "greatly accelerated the process of settlement in the Americas and largely financed the expansion of European trade with the rest of the globe" (Cross, 397). In fact, the fourteen tons of gold extracted from the islands during this period saved Charles V from the edge of bankruptcy (Kamen, 88). And yet, the precious metals extracted from the colonization of the Caribbean in these twenty years amounted only to "a third of the *annual* New World bullion production a century later" (Cross, 401). As Kamen explains, extraction of precious metals, commerce, and taxation were central to the imperial politics of Charles V, whose goal was the creation of a trans-oceanic network of mobile capital, "without which imperial power could not function" (53). Equally important, the precious metals from the New World already circulated within an international debtor-creditor web. Charles V used the promise of precious metals arriving from the Americas to set up credit with the Augsburg, Genoa, and Antwerp bankers (88). This meant that a high percentage of American gold and silver belonged to foreign banks years in advance of being mined (88). Charles also levied a tax on the extraction of precious metals from America called the "fifth" (89). The empire was a transnational capitalist enterprise made possible by the conjunction of indigenous workers, precious metals, and a financial system. Precious metals were the glue that kept everything together. The question that arises is how the Spanish justified their appropriation of Amerindian riches.

Before the reign of Charles V, the exploitation of the natural and human resources of the Indies motivated the creation of several institutions. One of them was the Casa de Contratación, established in 1503 for the purpose of monopolizing and rationalizing the exploration, exploitation, and trade of raw materials from the New World. From 1504 to 1519 Nicolás de Ovando, the governor of Hispaniola, cemented the social and economic structure of the Spanish colonies with the introduction of another vital institution, the *encomienda* system. The encomiendas were titles granted by the Crown to conquistadors through which the latter were allowed to extract tribute and labor from Amerindians, ostensibly in exchange for protection and religious instruction.

These momentous events provided the backdrop for Francisco de Vitoria's studies under the guidance of Peter Crockaert in the University of Paris. Crockaert had been a dedicated nominalist, but he converted to Thomism in 1507 and replaced the study of Peter Lombard's sentences with Aquinas's *Summa Theologica*, just as his illustrious student would later do at the University of Salamanca. Burdened by the political, ethical, and theological dilemmas regarding its possessions in America, the Spanish Crown decided to bring together a group of theologians and jurists in Burgos to discuss the legitimacy of the conquest of the New World. In his *Secundum librum sententiarum* (1510), the Scottish theologian John Mair argued that the inhabitants of the New World were the barbarians and slaves by nature that Aristotle had described in the *Politics*. Two condemnatory sermons delivered by the Dominican Antonio de Montesinos in 1511 on the island of Hispaniola provided the decisive moment in these polemics over the conquest of the New World and the exploitation of its resources. Inspired by Montesinos's condemnation of the encomenderos and conquistadors, Bartolomé de Las Casas would join the Dominicans' campaign against the encomienda and the encomenderos'mining policies. The Juntas de Burgos held in 1512 was meant to address this issue amid the emergence of some of the most well-known Renaissance critiques of politics: Erasmus of Rotterdam's *The Praise of Folly* (1509), Machiavelli's *The Prince* (1514), and Thomas More's *Utopia* (1516). As Ignacio González Casasnovas explains, the ideological debate concerning the Spanish presence in the New World was from the beginning linked to the ethical, legal, and political problem of appropriating precious metals and exploiting Amerindian labor, particularly in the mines (González Casasnovas, 4). The problem of the encomienda was accompanied by the sinister and infernal image of mining inherited from the Greco-Roman tradition and projected upon the Spanish settlers of Hispaniola and the Antilles.

The participants in the Juntas de Burgos, convened by Ferdinand the Catholic, declared that the inhabitants of the Indies were free vassals of the Crown, granting the Crown a degree of political supremacy over the newly discovered nations while denying that the Amerindians were slaves by nature. The Laws of Burgos, issued in 1512 and 1513, attempted to reconcile the encomienda system established by Nicolás de Ovando with the freedom of indigenous peoples and their Christian indoctrination. Among the participants in these early debates were the Dominican Matías de Paz and the jurist Juan de Palacios Rubios. Both Paz, in his *Dominio regum Hispaniae super indios* (1512), and Palacios Rubios, in his *De insulis oceanis* (1512), defended the political dominion of the Spanish monarchy over the Indies. This authority was based on the concession of sovereignty over the New World granted by Pope Alexander VI and derived from medieval theocratic doctrine that defended the universal authority of the church in the temporal affairs of Christian and infidel territories. However, the authority of the church over temporal politics did not seem to provide a sufficient justification for the violent conquest and subjections of the inhabitants of the New World. As a matter of fact, the creation of an imperial transoceanic network of commerce and search for profit and surplus demanded a radically different foundation.

In 1526 Charles V allowed for the discovery and exploitation of mines by any of his vassals without prior knowledge or consent from royal authorities. Subsequently, the conquest of Peru by a small mercenary army led by Francisco Pizarro made the legendary Incan mines available to Spanish exploitation, and in 1539 Illán Suarez granted Pizarro and his brother permission to use up to five hundred Amerindians in the extraction of precious metals (González Casasnovas, 8). In 1540 the discovery of mines in Zacatecas and Guanajuato in the viceroyalty of New Spain likewise allayed the Spanish Crown's fiscal crises, leading up to the 1545 discovery of Potosí, what would become the greatest source of silver in the empire.

Vitoria returned to Spain in 1523, just as the Castilian monarchy was asserting its political hegemony in the wake of the Comunero revolt. From Paris, Vitoria brought with him the Thomist philosophy and theology that would ignite a revival of Scholasticism in Spain and revitalize the universities of Salamanca, Coimbra, and Alcalá.[2] Vitoria was an essential figure in the creation of an intellectual climate that experienced a radical change during the 1520s and 1530s. As Pagden explains, this was the origin of a new direction in Spanish intellectual history, the beginning of the School of Salamanca, or Second Scholasticism (*Spanish Imperialism*, 15).[3]

The most distinguished members of this movement were Vitoria, Domingo de Soto (1494–1560), Melchor Cano (1509–1560), and later the Jesuits Francisco Suárez (1548–1617) and Luis de Molina (1535–1600). This group's influence transcended the borders of Spain and ultimately defined the thought of all Catholic Europe. The interests of these thinkers likewise surpassed the boundaries of theology, delving into economics, physics, metaphysics, and international law. Between 1537 and 1539, Vitoria, then a professor at the University of Salamanca, gave a series of *relecciones* (dissertations), which were later transcribed and edited by his students and colleagues, dealing with the New World polemics that had occupied Spanish jurists and theologians since the beginning of the century. The *relección* was an academic literary genre that consisted of a thorough examination of a concrete problem. The aim of the examination was to offer a definitive resolution of the problem, "an inquiry into its concrete points that explored its propositions in greater detail" (Adorno, 109).[4]

The university's prestige is evident in the fact that the Crown convened a group of jurists and theologians from the school (Vitoria, Soto, and Cano) in order to clarify the ethical and political conundrums concerning the legality and justice of the Spanish conquest of the Americas (Adorno, 109).[5] As Adorno explains, the ascendance of the School of Salamanca in discussions concerning the legitimacy of conquest and colonization waned in the wake of the two debates between Las Casas and Sepúlveda in 1550 and 1551, "when it was deemed the complex, academic consideration of such issues and the counsel provided presented more problems than they resolved" (109). These deliberations would have no parallel in the histories of other European empires (Pagden, *Spanish Imperialism*, 5). Pagden thinks that these debates belonged to the "tradition of ritual legitimation" employed by the Castilian Crown since the Middle Ages whenever confronted with "uncertain moral issues" (5). The Spanish Crown was able to confer upon itself "the self-appointed role as guardian of universal Christendom" because, unlike in France and England, the Catholic monarchs and their successor Charles V enjoyed a relatively unified realm undisturbed by regional autonomy and particular interests, with the exception of the Comunero revolt in 1521 (5). While Las Casas's and Vitoria's conclusions did not exactly coincide with the interests of the Crown, these inquiries nevertheless formed part of the "ideological armature of what has some claims to being the first European nation-state" (5). This propensity toward the inquiry into political legitimacy established the habit of critically examining the state by assigning limits to empire with the aid of the fundamental principles of Scholastic philosophy (6). Vitoria's attempt to derive dominion from reason as opposed to grace would avoid any attempt to ground Iberian expansion and possession of the new territories in the context of the former Christianized Roman Empire. In other words, it is safe to assert that Vitoria's corpus is a critique of imperial reason, an attempt to delimit a Christian empire within the limits of reason in order to provide a more rigorous and solid foundation.[6]

Vitoria has been the object of attention of different scholars and political thinkers. Brian Tierney discusses Vitoria, Suárez, and Grotius in the context of the medieval tradition of nature as the ultimate source of dominion.[7] Annabel S. Brett sees late Scholastics in general and Vitoria in particular as antecedents of a political order based on libertarian principles.[8] Quentin Skinner interprets Vitoria's neo-Thomist natural law as an antecedent of modern constitutionalism, the subjective theory of natural rights, and John Locke's contractual theory of political obligation.[9] Richard Tuck argues that theories of possession did not originate in liberalism but in authoritarian theories such as the medieval theories of *dominium* as total control of the world.[10] For instance, Aquinas's notion of dominium attempted to limit the Franciscan theories of poverty, according to which the Franciscans were allowed to enjoy commodities without having property rights or dominium over them (Tuck, 20). Monica Brito Vieira explains that the whole pattern of Grotius's *Mare Liberum* follows Vitoria's questions in *De Indis*.[11] Grotius argues in *Mare Liberum*, effectively, that the Dutch have the right to wage a war against the Portuguese based on Vitoria's main propositions such as the open access of all nations to the globe, the illegality of expropriating nonbelievers of their public or private rights of ownership, the incapacity of exclusively possessing the sea by any particular nation, and the universal right to trade (Vieira, 364). According to Toy-Fung Tung, when Skinner, Daniel Deckers, Brian Tierney, Annabel Brett, and Anthony Pagden see Vitoria's theories through the secular Enlightenment and post-Enlightenment "lenses of individual rights and constituted nation-states, they find inconsistencies" (Tung, 47).[12] Tung explains that these scholars interpret Vitoria's "apparently contradictory endorsement of both the *lex regia*—whereby secular sovereignty is established by the people's transfer of authority (*auctoritas*) to the ruler—and the Pauline-Augustinian principle that all power (*potestas*) is from God" (47).[13] As Tung observes, these apparent contradictions are resolved once we locate Vitoria's politics within his late medieval framework.[14]

Carl Schmitt explicitly declares that in Vitoria, "the *liberum commercium* was not the liberal principle of free trade and free economy in the sense of the 'open door' of the 20th century; it was only an expedient of the pre-technical age" (114).[15] By separating Vitoria from his theological context, liberal readers misread him and systematically ignore his intellectual milieu (Schmitt, 115). For Schmitt, liberal appropriations of Vitoria do not take into consideration the *missionary mandate*: the *liberum commercium* and the *ius peregrinandi* depend on the *potestas* of the pope who ordered the evangelization of the New World. The missionary mandate is issued by *potestas spiritualis*, a mandate that is both institutionally stable and intellectually self-evident (Schmitt, 119). Despite praising Vitoria for providing the first "rational" discussion about the land possession of the New World, Schmitt thinks that Vitoria is not a fully modern political thinker because he adheres to the notion of a *just war* grounded in a *just cause*. Despite conceiving a nondiscriminatory concept of war, one that does not discriminate between believers and nonbelievers, he did not expand his position into one of *justus hostis,* thus remaining within the frame of the medieval just war (Schmitt, 122).

My own contribution to the understanding of Vitoria consists of an exploration of the ontological presuppositions of his thought. I will argue that Vitoria's defense of freedom of commerce and the appropriation of precious metals of the New World are backed up by Aquinas's metaphysical instrumentalism. In this sense, I follow the line of thought opened by the authors mentioned above who emphasize the continuity between Scholasticism and liberalism, but I focus on the ideological role of the imperial principle of "natural" subordination of matter to form and means to an end, a principle that remains silent, implicit, and invisible in the liberal celebration of the power and control of the autonomous individual over nature and technology. By emphasizing Vitoria's metaphysical presuppositions, I highlight not only the instrumentalist dimensions of his thought but also the role of the Spanish Empire in the genesis of a capitalist and technological modernity. Instead of following a chronological succession that sees Vitoria's later dissertations as superseding his earlier political thinking, I will begin by examining Vitoria's consideration of the valid reasons for subjecting the Amerindians, in *On the American Indians* (1539), and then will continue with the attempts to ground dominion and civil power on the principle of natural subordination, in *On Civil Power* (1528), *On the Power of the Church* (1532), and his commentaries on Thomas Aquinas (1533–1534).[16] This will not be a chronological reading. Instead, I will provide a retrospective reading that begins with Vitoria's dissertations on the American Indians and on the laws of war and ends with the theoretical foundations of his relecciones on civil power. I will begin with an examination of what Vitoria considered the first valid argument in favor of the subjection of the Indians, which will lead to a discussion of the way Vitoria grounds the entire colonial enterprise in metaphysical instrumentalism.

THE VALID TITLES IN *DE INDIS*

In the first part of *De indis* (*On the Laws of War*; 1539) Vitoria inquires "by what right (*ius*) were the barbarians subjected to Spanish rule?" (*Political Writings,* 233). The second part of the dissertation is dedicated to the "powers" of the Spanish kings over their Amerindian subjects "in temporal and civil matters" (233). In the third section, Vitoria attempts to define the power that both king and church exercise over the Amerindians in "spiritual and religious matters" (233). Vitoria begins by clarifying that the Spanish Crown is in no way compelled to justify anew the "rights and titles" on the basis of which the territories of the New World are held in "pacific possession," despite the fact that these territories were already occupied by Amerindians (233). Vitoria says that only on four possible grounds would it be possible to deny that the natives were "true masters" (had dominion or ownership) before the arrival of the Spaniards. These four grounds are that the Indians were "sinners [*peccatores*], unbelievers [*infideles*], madmen [*amentes*], or insensate [*insensati*]" (240). He discusses if sinners can be masters in Question 1, Article 2, and he concludes that people in mortal sin can be true masters. In Article 3 he concludes that unbelievers can be true masters. In Article 4, he states that irrational creatures cannot have dominion, for dominion is a legal right (247). In Article 6 he discusses whether madmen can be true masters, and they can, because they can suffer injury. Vitoria states that barbarians cannot be enslaved or robbed because they had true dominion or ownership over their possessions:

> The conclusion of all that has been said is that the barbarians undoubtedly possessed as true dominion, both public and private, as any Christians. That is to say, they could not be robbed of their property, either as private citizens or as princes, on the grounds that they were not true masters (*ueri domini*). It would be harsh to deny to them, who have never done us any wrong, the rights we concede to Saracens and Jews, who have been continual enemies of the Christian religion. Yet we do not deny the right of ownership (*dominium rerum*) of the latter, unless it be in the case of Christian lands which they have conquered. (251)

Vitoria asserts that the political power of the "barbarians" was grounded in natural and divine law because they were the lords of the land *before* the arrival of the Spaniards. If the Amerindians' possession is based on the eternal, immutable primary causes (common utility), then no reasonable empire can justify its alienation unless they have committed an injustice that sufficiently warrants a just war.

LEGITIMATE TITLES

For Vitoria, the New World would pass under the rule of the Spaniards by seven unjust titles and by seven or possibly eight "just titles." Vitoria first focuses on the seven unjust titles. In Question 2, Article 1, he rebuffs the idea that the emperor can be the master of the whole world. In Question 2, Article 2, he examines the possession of the New World based on the pope's authority. Question 2, Article 3, focuses on possession by right of discovery, which is the core topic of the third unjust title. The refusal of the indigenous people to embrace the Christian faith is discussed in the fourth unjust title, which is examined in Question 2, Article 4. In Article 5 of the same question, Vitoria denies the possibility of Christian princes coercing the barbarians on the basis of their sins against the law of nature. In Question 2, Article 6, he doubts the ability of barbarians to have voluntary choice. Finally, in Article 7, he refused the idea that the New World is a gift from God.[17]

Vitoria concludes that the natives had true mastery or control of their lands and possessions, at both the public and the private level. However, he reopens the problem that was apparently closed by the end of the first question by then providing "just titles" for subjecting the indigenous peoples of the New World. The first two titles appeal to the right to travel and propagate the Christian faith. The third, fourth, fifth, and sixth titles involve the capacity of the Christians to protect innocent newborn Christians against the tyranny of infidels. The eighth "possible" title is the incapacity of the Indians to govern themselves. Again, the common core of all these titles is their being rooted in natural law rather than in faith. My claim is that these titles are backed up by the principle of the natural subordination of material means to the transcendent ends.

The basis for rejecting the unjust titles is their common ground, which is the attempt to justify imperial rule in a Christian theocratic frame. For Vitoria, natural law and reason are the only possible grounds for dominion, civil power, and a just war. In other words, the lectures on the American Indians provide a critique of imperial reason that shows how every time empire tries to ground itself on faith it falls outside the limits and jurisdiction of reason itself. Nevertheless, Vitoria's edifice has a fissure that results from the incapacity of reason to ground itself. This fissure is visible in the silent presupposition of the metaphysical principles of the conquest: natural order results from the transposition of artificial subordination to nature.

Once Vitoria has rejected the notion of a just war based on these motives, he then attempts to establish the legitimacy of the first title, which is the right to defend the "natural partnership and communication" (278). This argument starts by establishing the Spaniards' right to "travel and dwell in those countries, so long as they do no harm to the barbarians" (278). No positive law can forbid international travel, commerce, or evangelization because these are fundamental rights of natural law. Thus, if the Amerindians were to attempt to deny these rights to the Spaniards, they would commit an offense against them (282). Vitoria constructs the argument in negative terms. Since nobody can forbid what natural law does not forbid, it is legal to travel to the New World, engage in trade and commerce, appropriate precious metals that do not belong to anybody, and finally defend this right by military force after persuasion fails (278).

Vitoria continues, saying that the law of nations either is or derives from natural law. Moreover, the law of nations is what natural reason establishes among nations (278). The law of nations considers it inhuman to treat strangers and travelers badly without special cause—unless, of course, "travelers were doing something evil" (278). Moreover, from the "beginning in the world, when all things were held in common, everyone was allowed to visit and travel through any land he wished" (278). The "division of property" does not take away the right to travel because natural law does not "prevent men's free mutual intercourse with one another by this division" (278). According to natural law, "all things which are not prohibited or otherwise to the harm and detriment of others are lawful" (278). Traveling does not harm, therefore, it is lawful (278).

If natural and divine law allow traveling, then positive law cannot prohibit what natural law allows. Positive law can never contravene its natural foundation: "But if there were a human enactment (*lex*) which barred them without any foundation in divine or natural law, it would be inhumane and unreasonable, and therefore without the force of the law" (279). Here one can see how Vitoria turns a loophole in divine and natural law into a condition of possibility for traveling and engaging in commerce.

Vitoria then jumps to the conclusion that both natural and divine laws justify the Spaniards' right to use the abundance of precious metals in the

New World. This right, according to Vitoria, can also be enforced through violent means: "My second proposition is that *the Spaniards may lawfully trade among the barbarians, so long as they do no harm to their homeland*. In other words, they may import the commodities which they lack, and export the gold, silver, or other things which they have in abundance; and their princes cannot prevent their subjects from trading with the Spaniards, nor can the princes of Spain prohibit commerce with the barbarians" (279; original emphasis).

Natural law allows trade, which is a private enterprise consisting of an exchange of things that one party lacks for the things the other party possesses in abundance. In other words, the right to free exchange is inseparable from the imbalance of lack and surplus of material resources. As a result, the Indies will provide their surplus of silver and gold, and the Europeans will provide the commodities that are lacking in the Indies. Commerce presupposes the international division of roles according to the imbalance between the lack of commodities and the surplus of material wealth in the Indies. Native authorities cannot prevent private enterprise because it involves the mutual consent of private parties. The first argument is based on the right of commodity exchange between private parties that have not entered yet in relations of domination and servitude. Moreover, Vitoria's "barbarians" cannot put limits on trade and commerce. Understood this way, trade and commerce is a more powerful force than civil power.

Vitoria states that "the proof follows from the first proposition" (279). The Spaniards have the right to trade and import commodities they lack such as gold and silver *because* it is not lawful to put limits to free circulation. The law of nations (*ius gentium*) and natural law permit travelers to trade "so long as they do not harm to the citizens" (278). Divine law also allows trade, therefore positive law cannot prohibit it: "therefore any human enactment (*lex*) which prohibits such trade would indubitably be unreasonable" (279–80). As a result, it is "unreasonable" to prevent free circulation of commodities such as silver and gold. By natural law, the barbarian princes are obliged to allow the Spaniards to travel and engage in commerce "and therefore cannot prohibit them without due cause from furthering their own interests, so long as this can be done without harm to the barbarians" (280). To illustrate his point, Vitoria explains that the Spaniards have no right to impede the profits of the

French by forbidding commerce between the nations. Such a law would be unjust and contrary to Christian charity (280). The right to enjoy the search for profit is unconditional, and any obstacle to this search should be removed. Commerce is not only *allowed* by the *ius gentium*: it is a universal and necessary right.

Vitoria explicitly addresses the problem of mining, which took place from the very beginning of colonization: "My third proposition is that if there are any things among the barbarians which are held in common both by their own people and by the strangers, *it is not lawful for the barbarians to prohibit the Spaniards from sharing and enjoying them*" (280; original emphasis). For example, Vitoria poses that barbarians are not allowed to prohibit Spaniards from searching for riches "if travelers are allowed to dig for gold in common land or in rivers or to fish for pearls in the sea or in rivers" (280). Nonetheless, Spaniards are capable of searching for riches "without causing offence to the native inhabitants and citizens" (280). If the ban of trade and commerce is certainly "unreasonable," then it would be no less unlawful to forbid the search for precious metals in places that are held in common by Spaniards and natives.

The basis for appropriating things that are held in common lies in the right to travel and conduct trade. Possessing what belongs to nobody would then not be regarded as an act of dispossession:

> Secondly, in the law of nations [*ius gentium*] *a thing which does not belong to anyone* [*res nullius*] becomes the property of the first taker; therefore, if gold in the ground or pearls in the sea or anything else in the rivers has not been appropriated, they will belong by the law of nations to the first taker, just like the little fishes of the sea. And there are certainly many things which are clearly to be settled on the basis of the law of nations, whose derivation from natural law is manifestly sufficient to enable it to enforce binding rights. But even on the occasions when it is not derived from natural law, *the consent of the greater part of the world is enough to make it binding, especially when it is for the common good of all men.* (281; my emphasis)

The law of nations, which derived from natural law, allows mining, which is basically searching for "gold in the ground" that belongs to the "first taker." The Spaniards are able to claim as their own the gold they find because it is hidden underground and therefore does not belong to the natives. This act of appropriation is mandatory because it is allowed by the common good and the natural law. All laws, including eternal law, natural law, and human law, consist of directing things to the common good by rational means. In the *Summa Theologica* IaIIae.90, Aquinas states that law is always directed to the common good: "The first principle in practical matters, which are the object of practical reasoning, is the final end" (Aquinas, *Political Writings*, 79). Therefore, the basic presupposition behind the right to mine is the subordination of means to the final end of the "greater part of the world," which is "common good" (79).

In *On Civil Power* (1528), Vitoria argues that the whole world is like a republic, and its basic unit would consist of *ius gentium*: "The whole world, which is in a sense a commonwealth, has the power to enact laws which are just and convenient to all men; and these make up the law of nations" (Vitoria, *Political Writings*, 40). Vitoria also states that "those who break the law of nations, whether in peace or in war, are committing mortal crimes, at any rate in the case of the grave transgressions such as violating the immunity of ambassadors" (40). As a result, "no kingdom may choose to ignore this law of nations, because it has the sanction of the whole world" (40). The consensus of the majority of the international community and the imperative of the common good mandate that the Amerindians allow Spaniards to search for precious metals. Impeding the common utility of the totality would hamper the realization of the ultimate end of every society and the whole global community. The principle of natural subordination at the core of natural law supports the circulation of commodities and the exploitation of the natural resources of the Indies. It is crucial to remember the retroactive character of this justification: commerce and mining were already taking place, and immense quantities of precious metals were already fueling the network of commerce. Ultimately, Vitoria is making an apology for the events that took place before he delivered his lecture and his grounding of them after the fact in the principle of natural law and the laws of nations.

Vitoria states that if the natives attempt to deny the natural right of the Spaniards to usufruct the natural resources and commodities of the New World, the "Spaniards ought first to remove any cause of provocation by reasoning and persuasion" (281). The Spaniards should "demonstrate with every argument at their disposal that they have not come to do harm, but wish to dwell in peace and travel without any inconvenience to the barbarians. And they should demonstrate this not merely in words, but with proof" (282). Violence can be used only when reason and persuasion fail: "But if reasoning fails to win the acquiescence of the barbarians, and they insist on replying with violence, the Spaniards may defend themselves, and do everything needful for their own safety" (282).

Reason and persuasion must be backed up with violence: "It is lawful to meet force with force" (282). Vitoria alludes to the famous principle *vim vi repellere licet* (Watson, *The Digest of Justinian*, 98), which will, in *On the Laws of War* (1539), become the only true and solid foundation for declaring war on the natives.[18] Vitoria adds that the failure of persuasion and reason are not the only case where Spaniards have the right to repel force with force. They can also use force "if there is no other means of remaining safe they may build forts and defenses" (*Political Writings*, 282). If Spaniards suffer an offense "they may on the authority of their prince seek redress for it in war, and exercise the other rights of war" (282). The rights of war include, of course, dispossession and subjection of the natives. It is significant to note that Vitoria looks for proof of his argument in Aquinas: "The proof is that the cause of the just war is to redress and avenge an offence" (282). He adds, "But if the barbarians deny the Spaniards what is theirs by the law of nations, they commit an offence against them. Hence, if war is necessary to obtain their rights [*ius suum*], they may lawfully go to war" (282). Vitoria unequivocally concludes that the Spaniards can make war on the natives if they deny them what is "theirs by law of nations," which is the right to travel, conduct trade, and engage in commerce, including the right to appropriate precious metals that do not belong to anybody.

After stating that the "barbarians" are fearful and act cowardly, Vitoria states that, if they attack out of fear of the strong and armed Spaniards, the latter must defend themselves "within the bounds of blameless self-defense" (282). Once safety is secured, they may not exercise the other rights of war against barbarians such as killing them or looting or occupying their communities (282). The Spaniards must secure their own safety "but do so with as little harm to the barbarians as possible since this is a merely defensive war" (282). As one can notice, for Vitoria the conquest is "defensive war" that merely protects what is allowed by natural right.

Vitoria's sixth proposition is that, once all measures to secure safety have been exhausted, the Spaniards can proceed to conquer and subjugate the native's communities. He grounds this proposition in the following proof: if "the aim of war is peace and security" then "it becomes lawful for them to do everything necessary to [attain] the aim of war, namely, to secure peace and safety" (283). Once again, war is a means that has to be subordinated to the final end. However, Vitoria's argumentation seems to indicate that the aim is to justify the instrumental means.

Vitoria reaches the point where he is able to combine the conclusion of his first question (that before the Spaniards' arrival the "barbarians" possessed true dominion, in both public and private affairs) with the conclusion of the second question (although they were true masters, Christians were empowered to take possession of their territory):

> My seventh proposition goes further: once the Spaniards have demonstrated diligently both in word and deed that for their own part they have every intention of letting the barbarians carry on in peaceful and undisturbed enjoyment of their property, if the barbarians nevertheless persist in their wickedness and strive to destroy the Spaniards, they may then treat them as no longer innocent enemies, but as treacherous foes against whom all rights of war can be exercised, including *plunder, enslavement, deposition of their former masters, and the institution of new ones.* (283; my emphasis)

It is clear and unequivocal that for Vitoria there is no contradiction between the dominion of the indigenous peoples and the deposition of their mastery over their lands and possessions. The ontology of mastery and dominion as natural subordination is a double-edged sword: it grounds both the autonomy of the natives and the right to dispossess them of their dominion if they threaten the Spaniards' right to trade and search for profit and surplus.

Vitoria qualifies his statement with some limits by asserting that "all this must be done with moderation, in proportion to the actual offence" (283). He brings the whole issue to a close by a categorical assertion: "The conclusion is evident enough: if it is lawful to declare war on them then it is lawful to exercise to the full the rights of war. And it is confirmed by the fact that all things are lawful against Christians if they ever fight an unjust war; the barbarians should receive no preferential treatment because they are unbelievers, and therefore can be proceeded against in the same way. It is the general law of nations that everything captured in war belongs to the victor" (283). Once again, Vitoria's desire to distance the justification of the conquest from any kind of attempt to ground power on grace parallels his desire to give a solid ground to the conquest: mastery or dominion over precious metals that still do not belong to anybody and the free circulation of commodities. Just war is a means to an end, and here the end is the common good, which protects commerce and the extraction of mineral wealth that belongs by right to the first to seize it. The ultimate foundation of the right to make war is the autonomous capacity of the commonwealth to reject force with force (*vim vi repellere vicet*) in order to defend the preexisting right to travel, engage in commerce, and evangelize in the name of the "common good" of the whole world. Vitoria throws the violence of conquest out the door only in order to let it enter through the window as the defense of the inalienable right to circulate, extract, and exchange the natural resources and commodities found in the New World. Ultimately, Vitoria grounds the subjection of the Indians in the right of private entrepreneurs to circulate and search for profit.

Vitoria maintains that, if the Spanish have the right to travel to the Indies and to trade with their inhabitants, then they must also have the right to spread the Gospel, and the Spanish are thereby required to "correct and direct them [the barbarians]" (284). It is incumbent upon all Christians to "instruct them in the holy things of which they are ignorant" (284). If the Amerindian princes should refuse to admit the Christians' right to preach the Gospel then the latter *"may preach and work for the conversion of that people even against their will*, and may if necessary take arms and declare war on them, insofar as this provides the safety and opportunity needed to preach the Gospel" (285; original emphasis). The Spanish could declare war and bring to submission any Amerindian princes who "try to call them [the barbarians] back to their idolatry by force or fear" (286). As soon as the Amerindians are converted to Christianity, even by force, *"the pope might have reasonable grounds for removing their infidel masters and giving them a Christian prince, whether or not they asked him to do so"* (287; original emphasis).

The Indians have public and private dominion over their things and lands, and yet they can be subjected to Spanish rule. The first just title involves the unconditional possibility of engaging in commerce and the extraction of precious metals. This title can be protected by the use of force and military violence. On this basis, the conquest is grounded in natural law: it is natural to circulate, trade, and produce profit, and it is also natural to use force in order to remove any obstacles.

VITORIA CONCLUDES THAT COMMERCE MUST GO ON

After considering the rest of the valid titles, Vitoria concludes that, "if all these titles were inapplicable, that is to say, if the barbarians gave no just cause for war and did not wish to have Spaniards as princes and so on, the whole Indian expedition and trade would cease, to great loss to the royal exchequer, which would be intolerable" (291). He is saying that *even* if every title failed to provide solid grounds and sufficient reasons, trade and the extraction of tribute should not cease. This proposition is working as a meta-justification that works outside the seven bad reasons and eight good reasons. This meta-justification is unconditional and shows that the true aim, the silent presupposition, is not the benefit of the Indians but the production of profit. In other words, trade and the Indian expedition *cannot* cease ("it is intolerable") even if these are not based on just titles. Let us remark that Vitoria does not mean that there are no just titles—to the contrary, he offers seven (and possibly eight) just titles, but he is reminding his audience of the ultimate goal of the Spanish presence in the Indies, which is to conduct trade and to extract tribute by drawing the consequences of a hypothetical failure of every solid ground for making war on the natives.

His first reply is that "trade would not have to cease" (291). This is simply to remind the reader of the first valid title for subjecting the indigenous peoples, the right to repel force with force in case of the injury done to free circulation and commerce: "As I have already explained, the barbarians have a surplus of things which the Spaniards might exchange which they lack" (291). Vitoria argues that the surplus of natural resources and the lack of technology of the natives would continue to fuel commerce. He adds that "they have many possessions which they regard as uninhabited, which are open to anyone who wishes to occupy them" (291). The natives cannot contravene the law of nations that allows the Spaniards to occupy what does not belong to anybody. Vitoria continues this second reply saying: "Look at the Portuguese, who carry on a great and profitable trade with similar sorts of peoples without conquering them" (292). In sum, for Vitoria, the ideal empire is an international commonwealth transnational network of christianization, of commerce, and of the extraction of riches.

Vitoria insists on maintaining that the royal revenues will not cease:

"A tax might just as fairly be imposed on the gold and silver brought back from the barbarian lands" without violating the principles of justice (292). This tribute would consist of the fifth part—or of a greater amount depending on the value of the merchandise (292). The levying of tribute would be entirely justified by "reason," given that "the sea passage was discovered by our prince, and our merchants would be protected by his writ" (292). Vitoria's metaphysical instrumentalism grounds commerce and capitalist mercantilism in a teleological order, which is protected by the state through natural law.

Finally, in his third reply, he insists on the pertinence of the second valid title, the right to circulate and to evangelize the natives: "it is clear that once a large number of barbarians have been converted, it would be neither expedient nor lawful for our prince to abandon altogether the administration of the territories" (292). The three responses to the mere possibility of not acknowledging the validity of his "just titles" appeal to the unconditional and necessary character of the right to circulate, trade, and evangelize.

If trade and commerce are universal and necessary rights and a means for increasing royal revenue, and if the right of evangelization involves universal "communication" and circulation around the globe, in *De iure belli* (*On the Law of War*) war is the means for defending "common utility." But only a valid civil power with true imperium (sovereignty) and dominium (mastery and control over things or subjects) can declare war.[19]

ONLY A PERFECT COMMUNITY CAN DECLARE WAR

In *On the Law of War* (1539), Vitoria examines the conditions of just war by revisiting some of the arguments that he first developed in *On Civil Power*. What we need to know for the present purpose is that he begins this relección by distinguishing between the private individual who can defend himself but cannot take revenge and the Republic, which has the right to both defend itself and demand retribution for injuries it has received. Invoking Aristotelian doctrine, Vitoria argues that without the ability to wage war "the commonwealth cannot sufficiently guard the public good and its own ability unless it is able to avenge injuries and teach its enemies a lesson, since wrongdoers become bolder and readier to attack when they can do so without fear of punishment" (300). He likewise appeals to the authority of Augustine, for whom only the prince can declare war within the "natural order" (301). Vitoria argues that a republic is a "perfect community," which in turn leads us to the question of the nature of the "perfect" community.

Vitoria explicitly draws his idea of political power from the metaphysical foundations of the classical concept of the perfect community:

> Let us begin by noting that a "perfect" thing is one in which nothing is lacking, just as an "imperfect" thing is one in which something is lacking: "perfect" means, then, "complete in itself" [*quod totum est, perfectum quid*]. A perfect community or commonwealth is therefore one which is complete in itself; that is, one which is not part of another commonwealth, but has its own laws, its own independent policy, and its own magistrates. Such commonwealths are the kingdom of Castile and Aragon, and others of the same kind. It does not matter if various independent kingdoms and commonwealths are subject to a single prince; such commonwealths, or their princes, have the authority to declare war. (301)

In the tradition of Aquinas, Vitoria defines perfection as wholeness and imperfection as incompleteness. Entities that move themselves obtain a degree of perfection in proportion to their relative autonomy. Those entities that are moved by others obtain their relative perfection from the completeness of the beings that move them. Moreover, the perfect community functions by organizing and subordinating matter to an end; the imperfect must be subordinated to the perfect, and the part must be subordinated to the whole.

Just war cannot be waged because of religious differences or due to the desire for territorial expansion. War for the "personal glory or convenience of the prince" is also illegitimate because the prince receives his authority from the republic and he must therefore guide his actions with a view to the "common good of the commonwealth" (303). As a result, Vitoria's fourth proposition declares, "*the sole and only just cause for waging war is when harm has been inflicted*" (303; original emphasis). Vitoria again returns to the problem of dispossession addressed in the first proposition: "*in the just war one may do everything necessary for the defense of the public good*" (304). The ultimate end of war is the "defense and preservation of the commonwealth" (304). Vitoria reinforces his argument by again appealing to the individual's right to self-defense: "We have proved that this is lawful in the case of a private individual in his own defense, and therefore it must be all the more so in the case of commonwealth or its prince" (304).

War is the ultimate means that an autonomous commonwealth has of preserving itself and avenging injuries done from outside. In other words, declaring war and defending itself is an inherent right of the perfect and self-sufficient community. Let us remember that the first just title for declaring war on the natives is to repel the obstacles placed in the way of universal trade, communication, and the common utility of the whole world. Putting up obstacles to the extraction of metals and the production of profit not only is irrational but also represents an injury done to the natural rights of the Spaniards to circulate and evangelize. The Spanish Empire is a network of production of profit backed up by the right of the Spanish Crown to declare war on those who represent an obstacle to trade. Ultimately, imperium is nothing but the self-sufficient, autonomous community defending the right of its vassals to produce a profit for the sake of the "common utility" of the globe. In order to examine the ontological presuppositions behind Vitoria's just titles and the perfect community's right to repel force with force, it is necessary to explain the four causes of the complete and self-sufficient commonwealths. These four causes are the fundamental and ultimate principles of civil power and the binding power of the law. Vitoria grounds civil power in the principle of "natural subordination" of imperfect matter to perfect form and means to an end, but this principle results from the transposition of the sphere of artificial production to natural production. The ontology of political dominion, mastery, or control borrows its persuasive power from the form of technical mastery, dominion, and control.

Vitoria concludes that the ultimate justification for making war on the Indians is to repel injuries because the necessity of taming division and disobedience by means of the art of war is the silent presupposition behind *first principles of natural right itself*. The principle of natural subordination cannot be demonstrated, but it can be illustrated by transposing the technique of repelling force with force to the ontological structure of reality. In Vitoria, the proof that the law of nations derives from eternal law is that inferior crafts are subordinate to superior ones, "as bridle making is subordinate to the art of war."

THE PRINCIPLES OF CIVIL POWER

It is no coincidence that Vitoria begins his analysis of the foundations of civil power in *De potestate civili* with a consideration of the Aristotelian doctrine of causes: "And since, as Aristotle says, 'men do not think they know a thing till they have grasped the *why* or primary cause of it'" (Vitoria, *Political Writings*, 4).[20] I shall consider my brief fulfilled if I first investigate the causes of civil and lay power, which is to be the subject of this whole relection. Once the causes are understood, the potential and effects of power itself will become evident" (4). An analysis of the problem of civil power within the framework of Aristotelian-Thomist physics precedes the consideration of the problem of civil power in relation to the sovereignty of the *bárbaros*, the legitimacy of just war, and the universality of *ius gentium*. But why begin a discussion of civil power with the *physical* doctrine of Aristotle and Aquinas? For Vitoria, in order to explain who has the authority to command it is necessary to first explain who has the power to move those who are subject to the law. In Article 1, *Summa Theologica* IIaIIae.104, titled "Whether one man is bound to obey another," Aquinas writes that he who obeys is moved by the command (imperium) of he who orders *just as objects are moved by physical or natural causes*:

> As the actions of natural things proceed from natural powers, so do human actions proceed from the human will. In the natural order, it happens of necessity that higher things move lower things by the excellence of the natural power divinely given to them. Hence in human affairs also superior must move inferior by their will, by virtue of a divinely established authority. But to move by reason and will is to command. And so just as in the divinely instituted natural order lower natural things are necessarily subject to higher things and are moved by them, so too in human affairs inferior are bound to obey their superior by virtue of the order of natural and Divine Law. (Aquinas, *Political Writings*, 58)

In this passage Aquinas explains the principle of natural subordination in terms of commanding or *moving* just as physical things move each other based on their God-given power. Let us keep in mind that physics is the study of material and natural things, their principles and causes. Aristotelian physics will provide the framework for analyzing both political power and the generation and corruption of precious metals in the New World. Martin Heidegger—one of the most important philosophers of technology for having elevated technology to the character of "first philosophy"—sees Aristotle's physics as the foundational book of Western metaphysics and the "principle of reason." In Heidegger's words, "Aristotle's *Physics* remains the fundamental book of what later is called metaphysics. This book determines the warp and woof of the whole of Western thinking, even at that place where it, as modern thinking, appears to think at odds with ancient thinking" (Heidegger, *The Principle of Reason*, 63). Aristotle's physics, with his four causes (material, efficient, formal, and final), will remain the ultimate presupposition both of Aquinas's metaphysics and theology and of Vitoria's politics. In two of the most cited and debated works in sixteenth-century Spanish intellectual circles, *De regimine principum* and the *Secunda secundae*, Aquinas grounds dominion (mastery or control) on the Master Guarantor or origin (an all-encompassing and preexisting unity) that subordinates effects to causes, matter to form, and means to ends. In other words, political power should be founded on an essentially natural and rational law because it issues from the rational will of the divine architect. This divine craftsman possesses a preexisting idea (form or end) in his mind before imprinting it on matter, which is a pure, passive, and plastic potentiality deprived of determinacy. Just as Vitoria employs the framework of Scholasticism's latinized and Platonized Aristotelian physics in his deliberation on the problem of civil power, he likewise uses these premises in his analysis of the sovereignty and rationality of the inhabitants of the New World, in which he applies the four causes (material, formal, final, and efficient) along with the three principles of physics (matter, form, and privation).

Vitoria adopts Aristotle's doctrine from the *Physics*, according to which "not merely in the physical sciences but in all human sciences as well: that *necessary causes*, the first and most potent of all causes, must be considered as functions of purpose" (*Political Writings*, 4). Vitoria reminds us that the doctrine of finality as the cause of all causes, was established by Aristotle and Plato, and "it has proved a mighty tool in philosophy, shedding light on all subjects" (4). The concept of final cause frames Vitoria's ideas on the generation and corruption of inanimate matter as well as his reflections on human actions. According to Aristotle, the preexisting origin implies both a beginning and an act of domination, as it is a command to shape matter in conformity with a final end product.

Consequently, only an inquiry into the cause of all causes, the *finality* as principle or origin, can validate the physical and theological foundations of civil authority.

Notwithstanding the importance of finality, the idea of matter also vied for the place of ultimate foundation and first principle in traditional Scholasticism. Vitoria maintains that not only average philosophers but also those of the first rank mistakenly "saw necessary causes as inherent in matter" (4). To refute this claim, Vitoria turns to the trope of the house, which, along with the tropes of the statue and the ship, was also used by Aristotle and Aquinas to explain the generation and corruption of natural and artificial entities. Those who defended the force of matter as principle, "argued that the necessary cause in the construction of a house was not its purpose, the use for which its human occupants might design it, but the fact that heavier matter naturally tends to sink downwards; this, in their view, was why the stones and foundations are placed below the ground while the lighter wooden superstructure is placed on top" (4–5). By introducing this example, Vitoria tries to demonstrate that the necessary cause is the "purpose" and "use" and not the forces of matter. Vitoria adduces Aristotle's classic example of the construction of a house in order to refute the hypothesis of the force of matter: it would be absurd, he maintains, to attribute the necessity of the construction of a house to the weight of its foundations. The reason behind this choice is that the necessity of use or purpose seems self-evident in the construction of a house. In the construction of a house there is an agent, a master guarantor that has a certain idea or pattern (form) that imposes on an amorphous multiplicity (matter) in order to obtain a final product (purpose, use). The final cause is necessary because the house in question cannot exist without a purpose: people make houses in order to live in them. Thus, Vitoria performs a double transposition from artisanal technique to nature and then from nature to politics in order to attribute form and finality to a substance that would otherwise remain indeterminate, plastic, amorphous, and mutable. In other words, the foundational principle is not the material *from which* the house is made but, rather, *the purpose for the sake of which* it is constructed. Dominion as mastery or control derives from the final necessary cause to which the means are subordinated. Purpose reigns. Telos commands.

By confusing the concepts of necessity and matter, philosophers "could not give a proper explanation of the smallest thing, let alone comprehend with their philosophy the fabric and mechanism of more complex structures" (5). According to Vitoria, the force of matter does not sufficiently mark the boundary between the human and the nonhuman, for it claims "that man must walk erect, while brute animals creep on their bellies, not for any inherent purpose or utility, but because the physical nature and condition of animal matter is different from ours!" (5–6). As it is not a foundational principle, the force of matter cannot explain human uniqueness, intellectual activity, or free will; therefore, the force of matter cannot be the necessary cause of political authority. The natural origin of civil society—already understood in terms of technical transformation or attribution of purpose—is further determined by the distinction between men and beasts. The limit separating humans and animals is ultimately the limit between finality (utility) and purposeless matter. Vitoria attributes a preexistent and preordained utility to the necessity of matter, and he assails the doctrine of Lucretius and Epicurus for espousing the belief that the force of matter is a sufficient condition for the existence of a house or the capacity to contemplate the heavens. The atomists erroneously maintained that "everything is the chance product of aimless combinations of atoms crashing into each other in the infinite void" (6).

This, according to Vitoria, is an egregious error that has been refuted by Cicero in his *De natura deorum* and by Lactantius in *De opficio dei* (6). As atomism and the force of matter reduce necessity to the contingent effect of an arbitrary interaction of inanimate particles in an "infinite void," Vitoria refutes this doctrine as an insufficient explanation of the finality and utility of the law in his attempt to legitimate civil power. Thus, Vitoria opposes the concept of finality as cause of all causes to the concept of the force of matter. Here Vitoria repeats Aquinas's ontological and political solution, that is, his decision to transpose finality from the sphere of artisanal technique to the sphere of politics and ultimately to nature itself.

The necessity of a divine architect who governs nature is inseparable from technical finality: God relates to his creation in the same way that the artist relates to his work. Since Vitoria preunderstands the material world as a means to an end, he has no place for the random convergence of atoms (their attraction and repulsion, or the negative and differential antagonism of forces). Heraclitus's river of continuous change is repudiated in the name of the final cause. Vitoria's summary rejection of the antagonism inherent in the chaotic and contingent combination of atoms is the other side of the metaphysical instrumentalism that underlies his theory of civil power. However, Vitoria rejects the arbitrary violence implied by the force of matter and its antagonism (repulsion and attraction) of matter in order to naturalize it under the guise of the defense of autonomy.

Once the contingent arbitrary encounter of atoms has been rejected, Vitoria proceeds to affirm that heavens, earth, and man are guided by a *necessary* purpose. Moreover, "every single atom exists for some use and purpose; and that everything must be as it is because of that purpose or *final cause*, which is the true reason and *necessary cause* of all things" (6). The foundational principle of the totality of all entities, their true reason and necessary cause is *the origin as the principle of subordination of matter to form and means to ends*. Nothing on earth or the "orbis terrae" is without its reason, that is to say, its finality. Finality is the cause of all causes.

If everything has a final cause, then, according to Vitoria, "We must now inquire and investigate *what is the purpose or final cause for which the power under discussion is constituted*" (6; original emphasis). Civil power consists of the capacity to dictate human laws solidly grounded on natural law. The legitimacy of the law is not a result of it simply being mandated but, rather, a function of its preexisting finality ("use and purpose"). In order to explain the finality of mankind, Vitoria makes a series of unacknowledged quotations from De opificio dei (*The Workmanship of God*), written by Lactantius, a Christian apologist from the fourth century.[21] In one of these quotations Lactantius (and Vitoria) state that: "She [Mother Nature] provided each species with its own defense against attack, giving stronger creatures weapons to fight off the attacker, weaker ones the ability to escape danger by fleetness of foot, [and those lacking both strength and speed] the ability to protect themselves with cunning or by taking cover in a burrow; so some animals have wings to fly, or hooves to run, or horns, others have teeth or claws for fighting, and none lacks defenses for its own protection" (7). While animals are naturally equipped for conflict, "to mankind Nature gave 'only reason and virtue,' leaving him otherwise frail, weak, helpless, and vulnerable, destitute of all defense and lacking in all things, and brought him forth 'naked and unarmed like a castaway from a shipwreck'" (7). Vitoria extracts both this phrase and the comment above from Lactantius's book, which is a meditation on Lucretius and the doctrine of atomism within the context of Neoplatonic Christian theology.[22] In fact, this is a reframing of atomism by using the Thomist doctrine of *De regimine*. Other animals *naturally* possess defensive prostheses for their protection, while man is likewise equipped with only his reason and virtue. According to this view, man's reason consists of the ability to use rules to assess concrete circumstances with a view to their ends or ultimate utility. Therefore, reason is to man what natural defenses are to animals, namely, a medium for repelling the aggression of others and guaranteeing one's autonomy.

While there is an analogy between the attributes of animals and those of humans, nature nevertheless fails to provide humans with adequate means for defending themselves. Human beings are essentially deprived of natural means for their own defense, thus they are forced to employ artificial supplements devised through the use of reason and labor. Vitoria, therefore, dismisses the force-of-matter hypothesis in order to demonstrate that only final causes are necessary. Then he proceeds to establish a certain equivalence between animals' natural means of protection and human beings' specific attributes, an equivalence that naturalizes the concept of "defense." Human weakness is ameliorated through unity, association, and the formation of societies: "since it is agreed that the soul is composed of two parts, understanding and the will" (8). As Aristotle teaches in *Nichomachean Ethics* 1103a14–18 (*Basic Works*, 952), Vitoria says, "understanding can only be perfected by training and experience," and both can be gained by living in isolation from other human beings (8). The necessity of association for human subsistence is natural, thus men "seem to be at a disadvantage compared to brute animals, for whereas they are able to understand the things that are necessary for them on their own, men cannot do so" (8). Thus, association is the remedy, or supplement, for the absence of the natural means of defense that are bestowed upon the animal kingdom.

Aquinas's *De regimine principum* further develops this principle of human association, which is an implicit foundation of Vitoria's arguments:

> Now each man is imbued by nature with the light of reason, and he is directed toward his end by its actions within him. If it were proper for man to live in solitude, as many animals do, he would need no other guide towards his end; for each man would then be a king unto himself, under God, the supreme King, and would direct his own actions by the light of reason divinely given to him. But man is by nature a social and political animal, who lives in a community more so, indeed, than all other animals; and natural necessity shows why this is so. For other animals are furnished by

nature with food, with a covering of hair, and with the means of defense, such as teeth, horns or at any right speed in flight. But man is supplied with none of these things by nature. Rather, in place of all them reason was given to him, by which he might be able to provide all things for himself, by the work of his own hands. One man, however, is not able to equip himself with all these things, for one man cannot live a self-sufficient life. It is therefore natural for man to live in fellowship with many others. (Aquinas, *Political Writings*, 5–6)[23]

Animals defend themselves by repelling force with force. In the first place, Aquinas establishes an analogy between animal prostheses and defensive weaponry. In the place of corporeal means of defense (horns or teeth), nature has endowed man with reason as a substitute or supplement of his physical deficiencies. Thus, reason is both a means of directly procuring that which human beings lack as well as a means of crafting instruments to fulfill other needs.

Aquinas adds that a *solitary man* is incapable of providing for his own defense and sustenance, clearly implying that the need for self-sufficiency, as a means of perfection and fulfillment, precedes the need for defense and for the tools fashioned by human ingenuity that function as supplements to animal prostheses. Reason is patently associated with the fabrication of "works of his own hands": the lack of defensive prostheses and the correlative need to fashion tools for subsistence is fundamental to human nature. At the same time, an isolated human being cannot provide himself with these tools; life *among* conspecifics is thus inextricably linked to the innate deficiency of the lone man. The axiom that an individual cannot lead a perfect or self-sufficient life in isolation is based on this need to create instruments in order to supplement a natural inadequacy. To live perfectly and self-sufficiently is to employ one's reason in the manufacture of tools that substitute for animal prostheses.

Comparing both texts, one can observe that Vitoria elides much of Aquinas's argument by proceeding directly from the premise of man's need for defense to the use of reason and the need to live in communities. Both thinkers, however, posit autonomy, perfection, and self-sufficiency of the political community as the ultimate necessities. Vitoria neglects the human need to manufacture tools and instead focuses his attention on the need for political association. As they both maintain that finality is fundamental to the rational order of the universe, Aquinas and Vitoria sustain that law dictates the natural subordination of the part to the whole, of matter to form, and of artificial means to the natural ends. In order to prove the natural subordination of individuals to the final cause (common utility), both thinkers argue that individual being is incomplete (lacks means of self-defense), imperfect (incapable of living a self-sufficient life), and needs to supplement this incompleteness by artificial means (reason in conjunction with the work of his own hands).

The silent presupposition of both texts is that natural subordination supplements a lack in human nature that is filled by artificial supplements. There is a tension between this silent presupposition—the need to fulfill human incompleteness by artificial needs—with Aquinas and Aristotle's theory of technique. Despite using technical examples in order to illustrate physical causality, Aristotelian Thomism reduces the artifact to an essentially inert and neutral tool with no capacity to move itself. For Aristotle and Aquinas, while nature has an inner capacity to move itself, a technical artifact such as "a bed and a coat and anything else of that sort, *qua* receiving these designations—i.e. insofar as they are products of art—have no inner impulse to change " (Aristotle, *Basic Works*, 236). When commenting on this passage, Aquinas explains, "but things which are not from nature, such as a bed and clothing and like things, which are spoken of in this way because they are from art, have in themselves no principle of mutation, except *per accidens*, insofar as the matter and substance of artificial are natural things" (*Commentary on Aristotle's* Physics, 75). As a result, "insofar as artificial things happen to be iron or stone, they have a principle in motion in them, *but not in so far as they are artifacts*" (75; my emphasis). In other words, artifacts do not have a causal power of their own, being moved only by an efficient cause that deploys them for some preexisting and ruling end. They are neutral, passive, and inert devices deprived of any formative power of their own, and yet they are constitutive of what is most intrinsically human, making life in society not only possible but necessary. In other words, if Heidegger, for example, is capable of saying that Aristotle's *Physics* is a foundational book of metaphysics, it is because it is the origin of an instrumentalist conception of technology, that is, a way of thinking technique in terms of a neutral medium "that can be utilized for good or ill depending upon who or what happens to wield it" (Bradley, 5).[24] The irreducible contradiction of natural right and its axiom, natural subordination, is the need to supplement the incompleteness of nature by means of artifacts made by human hands, opening the way to the mutual co-constitution of the natural and the artificial, predetermining nature to denaturalization.

While Vitoria uses the example of an artificial entity in order to dismiss the force-of-matter hypothesis, he nevertheless maintains that the city itself is not an artificial structure. In the following passage he presents a stark dichotomy within the nucleus of political ontology:

> The clear conclusion is that the primitive origin of human cities and commonwealths was not a human invention or contrivance to be numbered among the artifacts of craft, but a device implanted by Nature in man for his own safety and survival.
>
> It follows immediately from this reasoning, *that the final and necessary cause of public powers is the same*. If assemblies and associations of men are necessary to the safety of mankind, it is equally true that such partnerships cannot exist without some overseeing power of governing force. Hence the purpose and utility of public power are identical to those of human society itself. If all members of society were equal and subject to no higher power, each man would pull in his own direction as opinion or whim directed, and the commonwealth would necessarily be torn apart. The civil community (*civitas*) would be sundered unless there were some overseeing providence to guard public property and look after the common good. (9; original emphasis)

This intricate passage warrants a thorough analysis. Vitoria invokes the classical metaphysical foundation of Scholastic political thought, which is the concept of an origin understood simultaneously as an inception and an act of domination. First, he affirms that the origin of the city is the natural (and divine) law and not merely a human artifice. This does not mean that he believes the city was literally created by God but, rather, that the origin of the city and the state is an effect of the natural order and that living in society is a final and necessary cause. Final cause is natural or necessary because it is not based on an artificial or contingent decision. Political sovereignty is derived from the principle of origin and domination, an unknowable and indemonstrable supposition that is nevertheless illustrated using the example of the artist's technique and the artisan's manipulation of matter. On one hand, Vitoria rejects Lucretius's materialism (the hypothesis of the force of matter) in the name of teleology and "natural subordination" transposing the sphere of artisanal technique to nature. On the other hand, he defends the natural origin of civil power, erasing his transposition of technique to nature and maintaining that human society is not a product of craft. The silent presupposition behind final causality is artisanal technique, which functions to illustrate how domination derives from self-evident and indemonstrable principles. In order to dispel the force-of-matter hypothesis (that is, the contingent violence of random occurrences), Vitoria's instrumentalist metaphysics of domination employs a metaphor based on technique that ultimately reveals the self-evident, tautological, and performative necessity of a final causality. The repudiation of the chaotic force of material flux is accomplished by invoking a necessary and rational order that erases its own silent presupposition, which is the transposition of technique to nature.

The rejection of the force-of-matter hypothesis is likewise a rejection of the equality that brings chaos to the hierarchical structure of the natural order: equality inevitably leads to destruction. Thomist metaphysical instrumentalism forces the reader to choose between either the destructive chaos inherent in a random confluence of individuals or a hierarchical subordination to a final cause. The necessity of unity and hierarchy precludes the threat of division and destruction that is inherent in the concept of equality. The impossibility of conceiving equality without order and the concomitant need for hierarchies reflects the double bind inherent in the principle of hylomorphic and teleological subordination: order and hierarchy must be imposed or else chaos and division will reign freely.

Using the analogy of the human body, Vitoria opposes the idea of anarchic equality with that of a hierarchical structure containing various degrees of power: "Just as the human body cannot remain healthy unless some ordering force (*ius ordinatrix*) directs the single limbs to act in concert with the others to the greatest good of the whole, so it is with a city in which each individual strives against the other citizens for his own advantage to the neglect of the common good" (10). The traditional corporatist view of society according to which individual powers (single limbs) must be subordinated to the whole (human body) is the consequence of the necessary subordination of the means to the final cause. Vitoria closes the section affirming that the final and "most potent" cause of civil power is a "utility and necessity so urgent that not even the gods can resist it" (10). In other words, the iron necessity of the origin of civil power is predicated upon the inherent "utility" and "purpose" of society as opposed to the contingent and variable force of matter. Vitoria's political theology, however, is not a kind of Benthamite utilitarianism, for the principle as origin and final cause of political life is "common" utility, or the harmonious unity of the community. The multitude must submit to the principle of utility as a transcendental end.

Thus, in his repudiation of the force-of-matter hypothesis, Vitoria not only invokes a final causality that is traced from the causation of artifacts but also draws an analogy between the defensive prostheses of animals and human beings' natural endowments. By highlighting the silent presuppositions of Vitoria's text it is possible to make visible that the transposition of technique to nature and politics is a retroactive imposition of sense or finality upon preexisting chaos or division. The aim of the ideological mystification of founding the political will upon a utility as natural end is to justify technical or instrumental means.

THE EFFICIENT CAUSE AND CIVIL POWER

After demonstrating that the final cause (common utility) is a necessary cause, Vitoria then inquires about the efficient cause of political power. Given that God is the author of natural law, "it is evident that public power is from God, and cannot be over-ridden by conditions imposed by men or by any positive law" (*Political Writings*, 10). God is the origin of wisdom and so he ordered all things graciously (10). Vitoria finds this principle affirmed in the writings of Cicero: "nothing is more acceptable to the deity who governs the universe and who created everything in the world than the assemblies and councils of men duly banded together which we call cities (*civitates*)" (10). If cities and societies are not mere artifacts or products of human craft but natural things made by God, natural order is God's product. Indeed, the divine decree marks the limit separating a rational natural order and arbitrary or tyrannical artifice.

Without the foundation of a preexisting transcendent cause, a divine artisan-master that guarantees purpose and sense, human actions would seem to originate in violence rather than rational laws, just as Aquinas himself asserts. The need of a transcendent ground arises from the impossibility of founding human law upon its own authority, thus turning to a structuring metaphor that bridges the gap between divine and human law. Just like material and formal causality, efficient causality also requires a metaphor that connects it to artisanal technique. This transposition from artisanal technique to politics is erased by some omnipresent, sovereign origin that dictates ends. This transposition, however, cannot be explicit without subverting the finality of the law, that is, its solid foundation and universal necessity. Consequently, he must hide or repudiate this transposition. Nevertheless, human law (positive or natural) cannot be founded on an artificial, accidental, or human form. God acts as the Supreme Architect who orders, frames, and directs the multiplicity of human forms guiding them toward preordained ends.

To conclude this proposition Vitoria asserts, "And if commonwealths and cities are founded on divine or natural law, so too are civil powers made by divine law, without which such commonwealths cannot survive" (10). In order for societies to "subsist" and maintain any degree of longevity, they must submit to the origin and finality ordained by the divine artificer for the common utility. Vitoria maintains that it is necessary to prove the function of efficient causality by appealing to both authority and reason. Vitoria appeals to Aristotle's *Physics*, where it is explained that "lighter and heavier bodies are set into motion by no other cause than the natural inclination to motion with which the First Mover endows them" (10). If divine efficient causality "was responsible for endowing men with the necessity and inclination which ensure that they cannot live except in partnership (*societas*) and under some ruling power, we must conclude that partnership and power are themselves God-given" (10). For both Aristotle and Aquinas, all that moves is moved by another entity and has an end beyond itself (Aquinas, *Political Writings*, 39).

Action follows essence because the generation of entities (both natural and artificial) consists of assigning forms and preexisting ends to beings. Actions and entities are predetermined by an efficient cause toward an external end, or they are products of an indiscriminate violence that contradicts the natural order. Entities receive their disposition toward movement and their habits from God. As everything is moved by something else along a chain of causes, all of these motives eventually return to the Prime Mover himself. The law can overcome its violent and arbitrary foundations only by grounding reason in God, thus stopping the vertiginous infinite regress to which it is subject. Thus, the supposition of God as the only certain foundation of being is essentially a negation of flux, time, and matter. This is an opportune moment to emphasize that these physical-causal propositions (all bodies are moved by another body) are not self-evident axioms but, rather, the product of an arbitrary transposition from the physical to the political sphere. This gesture originated in Aristotle's physics and was bolstered by the medieval Latinization of the Greek philosopher. In Vitoria's relecciones this principle assumed a geopolitical, planetary dimension. The law is theological and transcendent, as it is founded on an origin and preexisting finality that is beyond the historical, finite, and immanent world of human reason and labor.

Natural difference, whether specific or generic, is decreed by God: "structure" and "form" infuse the nature of things with being: "For things which are natural to all creatures must themselves be created by God, the author of nature, since he who gives the creatures their form and structure, as Aristotle again says, must also be responsible for the consequential things entailed by that form" (*Political Writings*, 10). Vitoria explicitly merges hylomorphic doctrine with Saint Paul's well-known condemnation of those who resist temporal power (11). God grants form, order, and organization, and whoever resists these also resists the Almighty. Moreover, Vitoria provides a hylomorphic interpretation of Saint Paul: "whoever therefore resisteth power, resists the ordinance of God" (174). Therefore, to resist civil power is consummate to resisting substantial form, transgressing the limits of the species, and contradicting the rational will of God. To bestow form is to confer a duty, assign a place, and allot a rank.

THE MATTER OF THE COMMONWEALTH

Vitoria proceeds from the analysis of final and efficient causes to a consideration of material cause: "But the material cause on which this naturally and divinely appointed power rests is the commonwealth. The commonwealth takes upon itself the task of governing and administering itself all its power to the common good" (*Political Writings*, 11). Apparently the material cause embodied in the commonwealth has the autonomous power to administer itself and govern itself. Vitoria offers the following proofs for the commonwealth taking upon itself the task of directing its powers to the necessary cause: "Divine and natural law require there to be some power to govern the commonwealth, and since in the absence of any divine law or human elective franchise (*suffragium*) there is no convincing reason why one man should have power over another, it is necessary that this power be vested in the community, which must be able to provide for itself" (11). Since individuals do not have power over each other, power is inherent in the community.

The material cause is the matter from which an entity is produced. The multitude itself, with its capacity for self-government, is the material from which the perfect community is constituted. Matter as pure passive potentiality is the capacity of the multitude for self-determination that depends on the regulating force of God. As a pure passive potentiality awaiting the imprint of form, the material cause is nevertheless the least determinate of all the causes. The multitude, then, is the matter from which God, as efficient cause, generates political authority by imposing the form of the law through the determination of a universal end (common utility) thus producing a sovereign community.

In order to understand the material cause of civil power in Vitoria we must abstract the power *over* the material cause in order to isolate the pure passive potency *previous* to the composite of matter and form. Also, it is important to remember that Vitoria denies that necessary causes are material causes. Therefore the necessary use and purpose inherent in the final cause do not belong to the material cause. In this paragraph, Vitoria simply states that, in the absence of power from above (whether it comes from divine law or *suffragium*), there is no reason for an individual to have power over another. In other words, previous to the imposition of the law from above there is only the right to repel force with force inherent in the individuals themselves. Therefore, since Vitoria denies that the force of matter is a necessary or natural cause of civil power, the material cause previous to its subordination to the form from above (the proper essence of power) is composed of autonomous individuals who do not have power over each other. The commonwealth is the material cause because individuals do not have power over each other.

Thus, there is no hierarchy before consensus, and the material of the multitude is an anarchic mass before being determined by form: If no one was superior to any other before the formation of cities (*civitates*), there is no reason why in a particular civil gathering or assembly anyone should claim power for himself over others (11). Abstracting the material cause from the form makes evident that previous to the elective franchise there are only autonomous individuals with the right to repel force with force. This is the first "proof" that the material cause of civil power is the communal power of the commonwealth.

The second proof consists of reasoning that, if individuals have autonomous disposition of their own bodies, then, being superior to the sum of individuals, the commonwealth must have the same right to coerce its own individuals:

> Further, every man has the power and right of self-defense by natural law, since nothing can be more natural than to repel force with force. Therefore the commonwealth, in which "we, being many, are one body, and every one members one of another" as the Apostle says (Rom. 12:5), ought not to lack the power and right which individual men assume or have over their bodies, to command the single limbs for the convenience and use of the whole. Individuals may even risk the loss of a limb if this is necessary for the safety of the rest of the body; and there is no reason why the commonwealth should not have the same power to compel and coerce its members as if they were its limbs for the utility and safety of the common good. (11)

The point of the argument is that, if individuals have the power to defend themselves and subordinate their members, the commonwealth also has the same power to defend itself and subordinate its individuals. Previous to the imposition of form over matter, individuals have the right to defend themselves, *especially if we abide by the principle that there is nothing more natural than repelling force with force*. Individuals can command their inferior limbs for the benefit and "use of the whole." The commonwealth cannot lack the power of the individuals to defend itself, also because nothing is more natural than repelling force with force. Therefore the commonwealth has the same power to command and coerce its own members (the same individuals who can repel force with force before the imposition of law over them) in order to direct them to the utility and safety of the common good. In other words, the right of individuals to repel force with force is subordinated to the final end of the social body. As a result, the commonwealth can subordinate individuals in the same way that individuals can subordinate their own limbs. But, according to the first proof, prior to the imposition of form there is only a dispersed multiplicity of individuals that do not have legitimate power over others yet are in need of self-defense. This fantasy of a really existing organic whole that directs individual elements to a higher end disavows the inner antagonism of society, where individuals are endowed with "hands and reason" in order to repel force with force, just as animals have natural equipment designed to defend themselves from aggressors. The further

Vitoria descends into the realm of matter, which is pure passive potency, the more we find conflict and contradiction: the formless plane of atomistic individuals endowed in need of self-defense, which is similar to the disavowed force of matter.

After explaining that the material cause on which the "naturally and divinely appointed power rests is the commonwealth," which is to say that only the commonwealth has the power to kill, Vitoria proceeds to explain how the material power of the commonwealth is invested in sovereigns. Vitoria explains how the commonwealth needs magistrates. He argues that, although the power belongs to the commonwealth, it has to delegate power to rulers and magistrates (12). He asserts that the "greatest and best of all forms of rule and magistracy is monarchy or kingship," and he tries to refute those who deny the legitimacy of the monarchy (12). Vitoria attacks his opponents, saying that "it is hardly surprising that these men, who have already apostatized from God and his church corrupted by their vicious ambition and pride, should also stir up sedition against the rulers" (12). Vitoria's attempt to ground power on reason is inseparable from this defense of constituted power from sedition. Vitoria unequivocally asserts that the sovereign power of the monarch does not come from the commonwealth. Let us remember that the material cause is not a necessary (final) cause but a contingent one, and in the absence of natural and divine law all that exists are individuals in need of self-defense. But did he not say that the material cause of the civil power was the commonwealth *because* it could "take upon itself the task of governing and administering itself" (11)?

Vitoria reconciles the power of the commonwealth over the individual with the impossibility of deriving the power of the monarch from the commonwealth by saying that the commonwealth cannot practically rule itself: "Though the commonwealth has the power by divine law over the individual members of the commonwealth, as has been proved, it is nevertheless quite impossible for this power to be administered by the commonwealth itself, that is to say by the multitude. Therefore it is necessary that the government and administrations of affairs be entrusted to certain men who take upon themselves the responsibilities of the commonwealth and look after the common good" (14). Compared to the necessity of the common utility (rule from above), the material cause is a pure passive potency that cannot cause itself. Vitoria rejects the bottom-up causality of the materialists to embrace a top-down causality of the final cause. Although the commonwealth has power over individuals (an individual does not have power *over* another individual because they all have the right to repel force with force), it cannot administer itself directly but only through a higher power. It is clear that the commonwealth as material cause has power *over* individuals, but not directly. In other words, the commonwealth could subordinate individuals in the same way that individuals could subordinate their own limbs. The role of the material cause is to subordinate individuals, but it is itself subordinated to a higher power.

This argument presupposes that prime matter lacks determination, perfection, and form. Prime matter is imperfect and needs to be subordinated to the end, which is a self-sufficient perfection. It is not surprising that the commonwealth cannot rule itself, since matter receives its perfection and self-sufficiency from an external preexisting master and guarantor of purpose who stands in front of the matter in the same way that an artist stands in front of the bronze. Matter cannot rule in the same way that bronze cannot produce a statue out of itself. There is a strict homology between the arguments about natural servitude and the material nature of the commonwealth: matter cannot rule itself, and it has to be *moved* by a master. The relevance of hylomorphism, when justifying hierarchical divisions between those who command and those who obey, cannot be sufficiently emphasized.

Moreover, for Vitoria it is irrelevant whether the commonwealth is subordinated to a "number of men, as in an oligarchy, or a single man, as in a monarchy" (14). If the power of the commonwealth is not tyrannical, the power of the monarch will be just too, "for it is none other than the commonwealth's power administered through the sovereign" (14). Vitoria categorically states that "the commonwealth as such cannot frame laws, propose policies, judge disputes, punish transgressors, or generally impose its laws on the individual, and so it must necessarily entrust all this business to a single man" (14). In order to affirm that the commonwealth cannot administer itself for lack of practical means, Vitoria has to rely on a concept of matter as pure passive potency that needs to be regulated from above. The incapacity of matter to regulate itself is the exact correlate of how final cause rules. The product (common utility) is more important than the material production (commonwealth, multitude).

The multitude cannot assert its right to self-determination without a prince, just as material multiplicity cannot become a statue without a power originating in a god. In order to clarify this point, Vitoria introduces the distinction between power (received from God) and authority (which belongs to the commonwealth and is transferred to the monarch):

> Let us conclude by leaving these matters and go back to clarify a point I mentioned before, namely that royal power is not from the commonwealth, but from God himself, as Catholic theologians believe. It is apparent that even though sovereigns are set up by the commonwealth, royal power derives immediately from God. For example, the pope is elected and crowned by the church, but nevertheless papal power does not come from the church, but from God himself. In the same way, the power of the sovereign clearly comes immediately from God himself, even though kings are created by the commonwealth. That is to say, the commonwealth does not transfer to the sovereign its power (*potestas*) but simply its own authority (*auctoritas*); there is no question of two separate powers, one belonging to the sovereign and the other to the community. (16–17)

Pagden maintains that the distinction between *auctoritas* and *potestas* is an attempt to meet the problem caused by the tension between power derived from God and power derived from the people.[25] "Auctoritas" or "jurisdiction" comes from the community, and the community transfers it to the king by electing it. Power as potestas (capability) comes from God and not from the people. The community cannot transfer the potestas because it does not have it. All it can do is elect a king, while the necessity of having a king is a divine mandate. The community does not retain *its own auctoritas, but it transfers it* to the monarch, who has both power and authority over the commonwealth. As a result, civil power cannot be abolished by popular consensus, as Vitoria will argue when examining the formal cause (Question 1, Article 7, *On Civil Power*). For Vitoria, there are not two powers, auctoritas and potestas, but two modes of one and the same power, which always comes from above: "And therefore we must say about royal power exactly what we have asserted about the power of the commonwealth, namely that it is set up by God and by natural law" (*Political Writings*, 17).

In the last instance, both the power of the commonwealth and the king's power come from a God who subordinates matter to form. Given that political power can be wielded either by a group or an individual, "those authors who concede that the power and jurisdiction of the commonwealth derives from divine law but deny that the same is true of regal power appear to be in error" (17). Vitoria distrusts the commonwealth's tendency to disruption and anarchy: "In commonwealths where many men share government, it is inevitable that rivals for office should spend their efforts in quarrels and seditions, tearing the commonwealth apart with different policies" (20). Only unity can tame and dominate the disruptive antagonism within matter: "therefore the best form of government is monarchy, just as the universe is controlled by a single Lord or Ruler" (20). The metaphysics of dominion is one of subordination of the potentially insubordinate field of forces of matter to a superior unity.

To conclude, it is possible to assert that in Vitoria's political physics, the material cause is a multiplicity of dispersed individuals. Before entering into a social contract, the multitude (as material cause) remains an indeterminate substance composed of equal and free, albeit diverse, individual members. The right of individuals to repel the force of other individuals with their own force would go on ad infinitum unless the unifying superior master (God and the monarch) introduced sense and order into the force of matter. Vitoria's political physics is built upon the disavowal of matter and the commonwealth: *the force of matter cannot be a foundation for civil society because it is not a necessary cause (only the end is a necessary cause); matter is an antagonistic chaos where unequal forces tend to struggle*. Vitoria's and Aquinas's choice of monarchy against the seditious deviations of the rivals for office is based in a metaphysics where "the universe is controlled by a single Lord and Ruler" (20).[26] Only a preexisting master guarantor of necessary sense and purpose can control and subordinate material multiplicity, directing it to its proper end, which is the common utility of the whole.

THE FORMAL CAUSE OF CIVIL POWER

Vitoria deduces the form of public secular power from the other three main causes in the "Proof of the Initial Proposition that a Legitimately Constituted Power Cannot Be Abolished by the Popular Consensus." Vitoria says that "these are the three main causes, final, efficient, and material, of public secular power, from which we may easily deduce its form. This is nothing other than the essence of power itself, which may be expressed in the following definition, as formulated by the authorities on the subject: *public power is the authority or right government over the civil commonwealth*" (*On Civil Power*, in *Political Writings*, 18). The "form" is a "pattern" or "idea" used by the preexisting external master artisan in order to mold an amorphous material multiplicity. Once the multiplicity is formed, it cannot reject the form from above, even by its own will or consensus. A commonwealth that abolishes a legitimately constituted power would be like the matter of a statue that rejects the imposition of the form by the artist. Consensus cannot abolish public power just as matter cannot avoid form: matter simply cannot determine (administer) itself in the same way that bronze does not make a statue out of itself without being molded by the form that comes from above.

Moreover, Vitoria asserts, "If a man cannot give up his right and ability of self-defense and of using his own body for his own convenience because his power belongs to him by natural and divine law, by the same token the commonwealth also cannot by any means be deprived of the right and power to guard and administer its affairs against violent attack from enemies, either from within or without" (18–19). This explanation parallels his discussion of the material cause of civil power where he considers it illogical to attribute to individuals the power to command their bodies while depriving the commonwealth of the same power. The commonwealth cannot renounce its right of self-defense from the injuries of sedition or of foreign nations without also losing its "public power" (19). Vitoria concludes, "even if all the members of the commonwealth were to agree to share this power freely among their number without restraint of law or obedience to magistrates, their agreement would be null and void as contrary to natural law, which the commonwealth of itself cannot abolish" (19). The inability of the commonwealth to disobey its magistrates is the corollary of the principle of subordination of matter to form, because matter that cedes to division and external aggression would be matter that disobeys the dictates of the form from above.

To recapitulate, a close reading of *On Civil Power* makes evident that natural mastery, the capacity to command grounded on a necessary end, presupposes artificial mastery. Vitoria employs the final causality of Aristotelian physics in order to attack materialists. His main argument is that matter cannot generate civil power out of itself, in the same way that the weight of the foundations of a house cannot explain the design of a house. Since only final causes are necessary and grounded on the preeminent efficient cause, which is the Divine Artifice, then material causes such as the commonwealth cannot rule themselves. Matter is an imperfect principle that needs to be governed by a superior principle that confers unity onto what otherwise would be a mere conflictive combination of individuals with the right to self-defense. Vitoria's metaphysical instrumentalism disavows matter for being antagonistic, amorphous, and prone to sedition and disruption. Without the imposition of the form, matter remains amorphous and antagonistic, a mere multiplicity that needs to be governed from above. His metaphysical instrumentalism is fetishistic, since, although civil power is not an artifact, he treats it "as if" it were when he is attacking the materialists.

Vitoria's relección on civil power precedes his relecciones on the American Indians not only chronologically but also logically. It provides the metaphysics that grounds sovereignty and imperium in the capacity to command and declare war. In his later relecciones he states that only injury, such as the disruption of commerce, can be a motive for declaring just war. Commerce and circulation has to be backed up by the power of the state, which is grounded in natural causality, the four causes that give legitimacy to civil power. The ultimate presupposition of the valid titles for subjecting the Indians is the principle of the natural subordination of matter to form and means to an end. Next, this chapter will examine the instrumentalist presuppositions of civil power and its capacity to declare war. Then, it will analyze the problem of metaphysical instrumentalism both in *On the Power of the Church* and in Vitoria's lectures on Aquinas's corpus to show that the inner core of natural law, the way it derives from reason itself, presupposes technical manipulation and subordination.

TECHNICAL SUBORDINATION

This section considers the relationship between the self-sufficiency of civil power and its subordination to the spiritual ends of the church. In *On the Power of the Church* Vitoria argues that, although temporal power does not depend completely on spiritual power, it is somehow subject not to the pope's temporal power but to his spiritual power. This subjection is illustrated using the example of the way inferior crafts are subordinated to superior crafts: "The temporal power does not depend completely on the spiritual, that is, in the way that a craft or faculty depends on a superior one, as the craft of making bridles depends on horsemanship, or boat-building on seamanship, or armory on soldiering (89). Vitoria raises this proposition to address the problem of the subordination of temporal to spiritual power using the analogy of the subordination of inferior crafts. According to Vitoria, temporal power is not completely subordinated to the spiritual: "civil power is not exclusively ordered for spiritual, as a craft is *exclusively ordered for its superior, and therefore the analogy is not at all exact*" (89; original emphasis). He goes on to argue that "if there were no soldiering there would be no armourers, just as there would be no bridle makers without horsemanship" (89).

He continues, saying that "All crafts of this kind are organic, or instrumental; if the purpose is lacking, the instrument becomes useless, or indeed non-existent" (89). The instrument depends on the whole and exists only for the sake of that on which it depends. There is no use without purpose. The purpose dictates the useful character of the means. Therefore, such analogy has limits because "that is not the case, however, with civil power in relation to spiritual power. Even if there existed no spiritual power, nor any supernatural felicity, there would still be some kind of order in the temporal commonwealth, some kind of power, as there is in natural things" (89). Temporal power, just like any natural power, would exist even if there were no transcendent and supernatural goals.

As a result "we are not to suppose that one type of power depends on the other, or exists exclusively because of it, as a *sort of instrument or part of it*" (89; original emphasis). Instruments exist for the sake of the superior causes while the civil power has a certain degree of self-subsistence. The unavoidable conclusion is that the power of the magistrate is part of royal power, "but is nevertheless self-sufficient and perfect in itself, existing immediately for its own purpose" (89). Vitoria's statement about the inexact character of the metaphor should not blind us to its instrumental presuppositions.

Vitoria is trying to preserve some immanent autonomy of the commonwealth. He refuses to see the state as nothing *but* something that exists only for the sake of the superior cause. While the means is useless if it is not subordinated to the higher purpose, the state has its own end, which is temporal. Making elbow room for the autonomy of civil power does not mean losing sight of the higher end. The metaphor is inexact, because civil power enjoys a certain autonomy. In other words, the ends of civil power are not mere means of superior power and cannot be entirely subsumed or subordinated to the higher ends of spiritual power. There is no doubt that Vitoria is transforming means into ends in themselves.

However, Vitoria's declarations about the inexact nature of the metaphor of the subordination of some technical skills to other technical skills do not mean that such subordination is not instrumentalism. On the contrary, it makes visible the irreducible contradictions within instrumentalism: there are means or temporal powers that are also ends in themselves and which cannot be completely subsumed in transcendent ends. Vitoria's assertion that the metaphor of technical subordination is inexact is not proof that metaphysics is not instrumentalist. On the contrary, it is a proof that there is an irresolvable tension within instrumentalism since he uses the same metaphor to explain how temporal power is subordinated to spiritual power (as will become evident shortly).

Although Vitoria declares that the metaphor is not exact, the inferior craft is still subordinated to the superior craft. As a matter of fact, he uses the same example derived from craftsmanship in order to illustrate that "temporal power is in some way conformable to spiritual power, as explained above" (89). Moreover, the political subordination of the temporal power to the spiritual one is still illustrated in terms of artificial subordination. In Question 5, Article 6, of the same work he argues that "civil power is somehow subject, not indeed to the pope's temporal power but to his spiritual power" (88). According to Vitoria, "the proof of this proposition is that if the purpose of any faculty depends on the purpose of another faculty, then that faculty depends on the other" (90). Any faculty or power depends on the purpose of another power and "the purpose of temporal power depends on spiritual power" (90). Vitoria adds that "the assumption in the premises is proved as follows: human happiness is imperfect, and ordered towards the perfection of supernatural felicity, *just as the craft of armoury is ordered towards soldiering and generalship, shipbuilding toward sailing, or the manufacture of ploughshares for agriculture, and so on*" (90; my emphasis). The principle of subordination of one power to the purpose of another depends on the subordination of the imperfect to the perfect. The proof of such subordination of inferior power to its transcendent end is that the imperfect is subordinated to the perfect *just as* the inferior crafts are subordinated to the superior crafts. Paradoxically, the subordination of the temporal ends to the transcendent ends does not contradict the unity of the state: "So it is not correct to think of civil and spiritual powers as two disparate and distinct commonwealths, like England and France" (90). Vitoria concludes *"hence civil power is subject in some way to spiritual power"* (90; original emphasis). The unity of the spiritual and temporal ends does not preclude a relative independence of the temporal ends: "In a single body, everything is connected and subordinated to one another, the less noble parts existing for the most noble. So too, in the Christian commonwealth, all offices, purposes, and powers are subordinated and interconnected; but it can in no sense be said that spiritual things exist for temporal ones" (91).

Vitoria put limits on the metaphor of craftsmanship, when giving certain autonomy to civil power, but only to immediately reintroduce the same metaphor when subordinating civil power to spiritual power. The subordination of temporal power to spiritual power cannot be exhausted by the metaphor of craftsmanship, only to the extent that the instrumental cause exists *for the sake* of the superior cause. Considering the instrumental cause as something that exists for the sake of a superior cause is the result of a disavowal of technique that attempts to master technique itself. By disavowing instrumentalism, Vitoria makes some room for civil power's autonomy: civil power, when considered as a whole and perfect community, is not the slave, a living yet separate tool, of the church. Unlike a prosthesis or a tool, its purpose is not exhausted by the master's purpose. And yet it somehow depends on the master, in the same way that an inferior craftsman depends on a superior one. Therefore, the principle of subordination remains an unconditional principle, and civil power is subordinated to spiritual power because it cannot contradict the transcendent final causes of the spiritual power.

The evident tension between the autonomy of the state and the subordination to supernatural ends witnesses the emergence of a certain *means that is an end in itself*. The irreducible contradiction between the means and the end results from the transposition of technique to the sphere of natural causality. The tautological circularity between principle and end has one target: to give the air of natural order to the artificial subordination of the imperfect sum of individuals to the perfect whole of civil power. In order to justify technical dominion (administration, management, and the exploitation of lands), imperial reason posits a preeminent first principle. This ultimate and irreducible contradiction is inherent in the ultimate ground of natural law, since natural subordination as such borrows its persuasive power from technical subordination.

In order to examine the problem of subordination as such let us examine how every law derives from the eternal law. In his commentary on the section of Aquinas's *Summa Theologica* dealing with eternal law, Vitoria explains how every human law must be founded upon the divine law:

"Does every law derive from the eternal law? Aquinas replies that they do. The proof is that inferior crafts are subordinate to superior ones, as bridle making is subordinate to the art of war. God is the supreme legislator" (168). This is a commentary on Article 3, *Summa Theologica* Ia–IIae.93, titled "Whether every law is derived from eternal law," where Aquinas responds to those who do not think that human law derives from eternal law using the following argument:

> Law denotes a kind of reason directing acts toward an end, as stated above. Now in all cases where there are movers ordered in relation to one another, the power of the second mover must derive from the power of the first mover, since the second mover does not move except in so far as it is moved by the first. *Hence we see the same thing in all who govern: that the plan of government is derived by secondary governors from the first governor, so that the plan of what is to be made is derived from the king by way of this command to subordinate administrators; and again, when things are to be made, the plan of what is to be made is derived from the designer to the lower craftsmen who work with their hands.* Since then, the eternal law is the plan of government in the Supreme Governor, all plans of government which are in lower governors must necessarily be derived from the eternal law. And these plans of lower governors are all other laws apart from the eternal law. Therefore all laws are derived from the eternal law in so far as they participate in the right reason. (Aquinas, *Political Writings*, 106)

In this passage there is an analogical relation between the sovereign's act of commanding and the way the architect commands inferior workers.[27] Following the principle of natural subordination, inferior craftsmen are subordinated to the designer in the same way that human law is subordinated to natural law. In metaphysical instrumentalism, natural subordination is like technical subordination. There is a transposition of technical subordination to the natural order of things. Therefore, there is an inconsistency at the very core of natural law, because the act of subordinating matter to an end serves as a proof of the binding force of the law. Moreover, in Article 1, *Summa Theologica* IaIIae.93, titled "Whether the eternal law is supreme reason existing in God," Aquinas argues that the force of law that subordinates subjects to government is analogous to the preexisting rational pattern of things in the mind of the artist who proceeds to dominate matter (*Political Writings*, 102). The sovereign's faculty to command and the architect's faculty to oversee the manual workers are grounded in the principle of natural subordination, which presupposes the technical mastery of the subordinated craftsmen.[28] God stands in front of the world just as the craftsman stands in front of the material because the principle of natural subordination presupposes technical mastery. Vitoria's instrumental reason is the disavowed presupposition of metaphysical ontology: the binding power of the law is inseparable from the capacity to transform matter. Natural law uses technique to give the appearance of necessity to the law, while it disavows technique by subordinating the manual labor of the craftsman to the architect or designer. The contradictory impasse arising from this tension is the result of the distance between the idea and the material example. The material example—far from being an imperfect copy of the preexisting idea—has an explanatory power that threatens to subvert the exemplar idea. This excess of artificiality that is intrinsic to natural law undermines the hierarchical order it enables. Technical mastery is both the condition of possibility and the ultimate obstacle of natural law. The difference between the universal designer and the particular craftsman is internal to the Divine Artifice because the latter is conceived in the image of the inferior artisan that molds shapeless matter into a final product.

In order to clarify this silent presupposition, let us revisit the four final causes of civil power. God (the efficient cause) has a rational pattern in mind (form as the essence of public power) and commands matter (the commonwealth, the set of individuals), directing it to the final end (autonomy and the common good). Political subordination is exemplified by appealing to technical mastery, yet in the case of Vitoria, the specific technique transposed into natural law is the art of war as *self-defense technique*, since the "proof" that law is subordinate to eternal law "is that inferior crafts are subordinate to superior ones, as bridle making is subordinate to the art of war" (168). The proof behind the way all laws derive from eternal law is the manual art of war, whose inherent function is to repel force with force. Vitoria thinks that there is nothing more natural than repelling force with force. After all, animals are equipped to repel force with force. The proper technique for repelling force with force is the ultimate support of the law, the law behind the law, the material example that undermines eternal law itself. It is the violent origin of natural law itself.

Let us recall that Vitoria uses the same example to explain how civil power is subordinated to ecclesiastical power. The power of the universal empire is subordinated to the ecclesiastical power. In his search for a metaphysical origin of civil power, Vitoria presupposes an imperial reason grounded in instrumental calculation and military organization for securing the control over a territory.[29] Ultimately, the right to declare just war is the right of the Spaniards to repel *force with force* and defend the right to engage in commerce and extract precious metals.[30] The inner contradiction of natural law—the incapacity to ground the law in something other than itself, some preexisting external standard—is the source of the contradiction dwelling at the core of imperial reason itself: the true final cause of the conquest is the deployment of the means itself. Commodity exchange and extraction of precious metals are backed up by war, which is the prerogative of the state. Vitoria's reasoning puts in motion an ideology of expansion, a machine of war and appropriation of the riches of the New World with the aim of producing a royal fifth for the Crown.

SLAVES BY NATURE

Before discussing the first valid title, and immediately after concluding that the natives were true masters of their lands and things, Vitoria answers the possible counter-argument that these barbarians are insufficiently rational to govern themselves:

1. Aristotle certainly did not mean to say that such men thereby belong by nature to others and have no rights of ownership over their own bodies and possessions (*dominium sui et rerum*). Such slavery is a civil and legal condition, to which no man can belong by nature.
2. Nor did Aristotle mean that it is lawful to seize the goods and lands, and enslave and sell the persons, of those who are by nature less intelligent. What he meant to say was that such men have a natural deficiency, because of which they need others to govern and direct them. It is good that such men should be subordinate to others, like children to their parents until they reach adulthood, and like a wife to her husband. (Vitoria, *Political Writings*, 251)

Vitoria explains that what separates servants and masters, or those who are ruled and the rulers, is that the latter are natural masters by virtue of their superior intelligence (251). This does not mean that they have the right to enslave others, "but merely that they are fitted by nature to be princes and guides" (251). Vitoria denies that it is possible to legitimately alienate Amerindian possessions but he considers the possibility of having some right of jurisdiction over them: "Hence, granting that these barbarians are as foolish and slow-witted as people say they are, it is still wrong to use this as grounds to deny their true dominion (*dominium*); nor can they be counted among the slaves. It may be, as I shall show, that these arguments can provide legal grounds for subjecting the Indians, but that is a different matter" (251). Indigenous peoples cannot be enslaved, but they *can* become subjects: if they are imperfect (foolish and slow-witted) they may be subjected (politically subordinated) to the Spanish Crown. Indigenous inferiority does not grant dominion of their possessions to the Spaniards, but it may be sufficient reason for subjecting them to the Spanish rule. They may have ownership but there are reasons for taking over their administration and government. Vitoria tries to ground colonial subjection on the principle of "natural subordination": indigenous peoples cannot be enslaved or dispossessed but they *may* be subjected. This alleged inferior nature of the "barbarians" is revisited when Vitoria discusses the eighth possible title, where he concludes that they may be governed partly as servants.

At this point, it is instructive to bring up Anthony Pagden's explanation of the category of natural servitude. "The origin of natural slavery, however, is to be found neither in the action of some purely human agent nor in the hand of God, but in the psychology of the slave himself and ultimately in the constitution of the universe. It depends on the axiom, common to much Greek thought, that there exists in all complex forms a duality in which one element naturally dominates the other" (*Fall*, 42). Pagden quotes from Aristotle: "In all things, which form a composite whole and which are made up of parts, whether continuous or discrete, a distinction comes to light. Such a duality exists in living creatures but not in them alone; it originates in the constitution of the universe; even in the things which have no life, there is a ruling principle" (Aristotle, *Politics* 1254a.28ff, in *Basic Works*, 250).

In the same conclusion Vitoria opened the door to the eighth "possible title," asserting that even if these barbarians are slow witted, it is still wrong to use this as grounds to deny their true dominion (Vitoria, *Political Writings*, 251). In the eighth "possible" title, he asks if it is possible to establish Spanish sovereignty based on the supposed lack of rationality of the Amerindians: "it may strike some as legitimate, though I myself do not dare either to affirm or condemn it out of hand" (290). Although Vitoria maintains that Amerindians are not entirely deprived of the capacity of judgment, he likewise asserts that they are "so close to being mad, that *they* [the barbarians] *are unsuited to setting up or administering a commonwealth both legitimate and ordered in human and civil terms*" (290). The Spanish theologian claims that they have neither "appropriate laws nor magistrates fitted to the task" (290). In Vitoria's mind, the other signs of civilization are likewise absent: "hence their lack of letters, of arts and crafts (not merely liberal, but even mechanical), of systematic agriculture, of manufacture, and of many other things useful, or rather indispensable, for human use" (290). Just like Columbus's descriptions of his first encounters in the New World, Vitoria describes the Amerindian in terms of absence: lacking European technology, writing, money, and steel. The Amerindians' lack of civilization is an effect of the application of technique as a standard of the capacity to master actions and things.[31]

As Pagden has clearly shown, the Spaniards' perception of Amerindian barbarism was directly linked to their supposed lack of mining, metallurgy, iron tools, and other technologies (*Fall*, 91). In Pagden's words, "What the Indians so obviously lacked in their Neolithic world—though Vitoria does not himself say so—was iron" (91). Pagden's insight is fundamental to explain the oscillation between attributing dominion to them and considering them somehow "imperfect" or incapable of governing themselves. Pagden adds "for it was the ability to mine and smelt the hard metals that made the weapons and machines of European culture possible. Without access to such metals the Indians were permanently thwarted in any natural inclination they might have to progress" (91). Iron was for many the only "true" metal. Nicolás Monardes, a natural scientist with substantial experience in the Americas, rejected the supposed merits of gold and silver as merely founded on "opinion," for none of those substances did anything.[32] Iron, on the other hand is the only "true" metal precisely because it is "so crucial in the practice of the arts" (Pagden, *Fall*, 92). Pagden adds that when describing the impressive stone building of the Inca and the Maya, the Spaniards may have been surprised by "how much had been achieved without the use of iron tools" (92).

After examining the relección on *Civil Power* and the way Vitoria derives the force of law from the first principles, it will become evident that a certain disavowed instrumentalism is the silent presupposition behind the ontology of dominion and the notion of "natural subordination" of matter to form and means to ends. The inability to rule presupposes an instrumentalist metaphysics that derives the force of law of civil power from the transposition of technique to nature and politics. Thus, the Amerindians' lack of metallurgy or technique is very close to lacking the capacity to administer or govern themselves. According to Vitoria, "It might therefore be argued that for their own benefit the princes of Spain might take over their administration" (*Political Writings*, 290). By this reasoning, the Crown could name prefects and governors of cities as well as new princes so long as this is in the interest of the barbarians themselves (290). If the Amerindians are somehow mentally incapacitated, then the Spanish king is required to protect them, "as if they were simply children" (290). The same criteria applies to both the Amerindians and the *amentes*, given that with respect to their capacity of self-government they are no better than madmen (290). According to Vitoria, their crude diet is further evidence that they are not even superior to wild beasts and must therefore be governed by the wiser men (291).[33]

To assume the role of caretaker for those who lack the capacity to self-govern is "supported by the requirements of charity, since the barbarians are our neighbors and we are obliged to take care of their goods" (291). Vitoria, however, is cautious about the consequences of this conclusion and warns against any strict interpretation of this principle: "everything is done *for the benefit and good of the barbarians, and not merely for the profit of the Spaniards*" (291). Given that the souls of the Amerindians are in danger of perdition, "In this connection, what was said earlier about some men being natural slaves might be relevant. All these barbarians appear to fall under this heading, and they might be governed partly as slaves" (291). It is important to underline that the subjection of the natives has to be done for their own benefit and not for the profit of the conquistadors.[34]

Vitoria's doubts about this title opens up a paradox, since his arguments could be used in either way. The fissure in the argument opens a space for polemics, because the same frame, metaphysical instrumentalism, can be used to justify both ways. On the one hand, he concluded that Indians could not be enslaved because they enjoy perfect dominion. On the other hand, Indians could be considered natural servants because they lack technology. This engenders what Pagden calls the "paradox of the free slaves." Pagden elegantly summarizes this paradox by explaining that "the Indian was a free and independent being; but he lost his authority over his own affairs, and in some sense his humanity too, once he had been brought into contact with civilized men. Once, that is, a society had come into being which included both natural slaves *and* natural masters, the slave had to begin to fulfil his function as a slave" (*Fall*, 55).

Once we introduce the instrumentalist presuppositions of the metaphysical division between commanding forms and obedient matter it is possible to see a paradox of the living tool. For Vitoria, the Amerindians can only be considered servile once they are not considered mere means or tools. From the perspective of metaphysical instrumentalism, servants are living tools, but they have to be governed for their own sake—and not for the profit of the ruler: they are a means, but they can only be treated as a means by paradoxically considering them an end in themselves. The Indians cannot be considered mere means, living tools, because they have to be ruled for their own benefit.

My thesis is that this irresolvable contradiction derives from the inner presuppositions of metaphysical instrumentalism. The commanding master is already understood as somebody who stands in front of the craftsman in the same way that the craftsman stands in front of the material and the final product. Since there is a transposition of technique to natural causality and political domination, the irresolvable contradictions that arise from imperial ideology are not accidental but substantial. Since metaphysical instrumentalism introduces the division between commanding masters and ruled servants, form and matter, ends and means, those who were imperfect sovereigns become inferior.

CONCLUSION

This reading of Vitoria began with an examination of the arguments in favor of subjecting the Amerindians to the metaphysical presuppositions in his political and theological works. The first valid title was given priority because it focuses on a defense of commerce and the appropriation of precious metals. In *On the American Indians*, Vitoria argues that, before the arrival of the Spaniards, the Amerindians were the legitimate sovereigns and possessors of their lands and therefore there is no legitimate reason to wage war against them unless provoked by injury or another just cause. Nevertheless, he also states that trade, commerce, and extraction of metals (in places that do not belong to anybody) are universal rights grounded on the laws of nations (*ius gentium*). If Indians deny this right to the Spaniards, they commit injury against the latter. Impeding commerce, free circulation, and evangelization provided Vitoria with the arguments needed to justify the existence of a network of extraction of precious metals and commerce. As a result, Spaniards can make just war on the Amerindians using the natural right to repel force with force.

In the relección *On the Law of War* (1539), Vitoria explains that only a "perfect" (self-sufficient) community with dominion and civil power over its subjects can declare war and repel force with force. The problems of conquest, the encomienda, mining, and commerce were historically inseparable from the problem of justifying the Spanish presence in the Indies through the consideration of natural law, the civil foundations of sovereignty, and the nature of political power as understood in the philosophy and theology of Aquinas. And Francisco de Vitoria was the first theoretician to offer a rational defense of this business enterprise. He never doubted the justice of the conquest itself but rather questioned the theoretical foundations that were applied to particular questions of colonial administration and government. Throughout his work, Vitoria deconstructs the titles inherited from medieval theocratic theories in order to elaborate a new theoretical foundation, which would apply to all future colonial ideology and the interrelations between legitimate states. He attempts to ground the empire and the emergence of a global commonwealth not only on rational arguments but also on a solid Aristotelian and Thomist metaphysics.

Vitoria provides the metaphysical basis of dominion as political mastery and the right to declare war in his lectures *On Civil Power* (1528), *On the Power of the Church* (1532), and his commentaries on Thomas Aquinas (1533–1534). In *On Civil Power*, he grounds the legitimate power of an autonomous commonwealth in the four Aristotelian causes (material, formal, efficient, and final). Political mastery over lands and subjects depends on the principle of the natural subordination of matter to form, means to an end, and the imperfect to the perfect. A close examination of this principle demonstrates that political dominion presupposes technical mastery. In other words, the principle of natural subordination presupposes artificial subordination, since God stands in front of the machine of the world in the same way that a craftsman stands in front of prime matter before making a final product.

In one and the same move, imperium borrows mastery and control from technique and labor and subordinates the transformative power of the craftsman such as the art of bridle making to a superior power such as the art of war (these examples are used by Vitoria himself). Ultimately, natural right presupposes what it pretends to demonstrate: natural causality and political mastery presuppose *technical* mastery and the subordination of inferior skills to superior skills. Despite its attempt to ground all law in a preexisting natural order that would function as universal and necessary guarantor of the civilization and Christianity, metaphysical instrumentalism ends up depending on the transposition of the realm of technique to the realm of nature. The transcendent (that is, more than material) ends are conceived against the background of merely material means. Such transposition will produce irreducible contradictions that will become visible in the material and concrete dependence of the Spanish empire on merely material means.[35]

Vitoria's whole theoretical enterprise is based on an irreducible contradiction in which the just war, a means to an end, presupposes the superior skill of the art of war and the right to repel force with force. This right is presupposed on many levels, and Vitoria keeps repeating that there is nothing more natural than self-defense. First, individuals lack the means of defense—natural weapons analogous to those of the animals. Then, he proceeds to deduce that the commonwealth can defend itself in the same way that individuals defend themselves from animals and from each other. Third, he states that disrupting commerce is an offense that can be repelled by means of conflict. Finally, the capacity to repel force with force is presupposed in the way every law derives from the eternal law. Imperium, the capacity to command, presupposes the technical subordination of every skill to the art of war. In sum, war presupposes itself.

Metaphysical instrumentalism is ideological in the sense popularized by Marx's maxim: "They do this without being aware of it" (*Capital*, 1:166–67). Marx points out a certain divorce between knowledge and practice. In examples popularized by Marx, people act as if they considered money to be an inert object, while in practice they endow it with a life of its own. In the present case of metaphysical instrumentalism, the subject of imperial ideology thinks that teleology belonged to the natural order of things, while in practice the same subject acts as if the technical means were ends in themselves.

As a consequence, while in theory it pretends to act in such a way as to benefit the Indians, treating them as ends in themselves, in the disavowed practice the subject of imperial ideology actually conflates the means and the ends. The means become the ends themselves. For instance, in Vitoria, a necessary condition of the legitimacy of subjection is to act for the benefit of the subjected barbarians beyond reducing them as mere means. As soon as the imperial subject becomes aware that the ends of the imperial system are merely rationalized ex post facto—and that the real point is to transform the ruled into mere means to that end—the ideological effect of natural law is spoiled. The ideological mystification resides in that the barbarians can only achieve a temporal profit if both the barbarians and the Spaniards believe that there is a necessary end solidly grounded in an origin.

Although the real stake of natural right is the defense of temporal profit (circulation of commodities and extraction of riches), this gain has to appear as a by-product, a secondary effect. It has to appear as if it were a necessary means subordinated to the transcendent end. As soon as the subjects of ideology perceive that the goal is to control the artificial means, the whole imperial enterprise defeats itself. The aim—the implicit end behind the explicit end of metaphysical instrumentalism—is not only the control of the artificial means but also the transformation of the newly discovered territories and peoples into means. What appears to be the end is nothing but the means of the means. Spaniards acted as if they were justifying war to pursue common good, when the act of subordination presupposed war itself.

Vitoria was silenced by Charles V, yet the apologists of the empire adopted his views and radicalized them. To prove this point, we now move to the polemics between Bartolomé de Las Casas and Ginés de Sepúlveda, who defends the concept of just war by invoking the principle of the natural subordination of imperfect matter to perfect form. Sepúlveda is the best example of modern imperialism because he draws the geopolitical consequences of the application of the principles of natural law. For him, Indians are naturally inferior and have to obey the Spanish masters in the same way that matter has to docilely obey form. Although Bartolomé de Las Casas recognizes the centrality of the principle of natural subordination, he refuses its geopolitical consequences by showing how it is not possible to impose the form of one community over the matter of the other. The debate in Valladolid can be interpreted as an impasse between two postures that derive from the inconsistencies of Aquinas and Vitoria's metaphysical instrumentalism.

2

THE IMPASSES OF INSTRUMENTALISM

Revisiting the Polemics between Sepúlveda and Las Casas

Francisco de Vitoria laid the metaphysical foundations of imperial ideology by examining the causes of just war and the possible legal titles for justifying Spanish sovereignty in the Indies. The valid titles for justifying the possession of natural resources were all grounded in the principle of the natural subordination of matter to form and means to an end. The instrumentalist presuppositions behind this principle engendered irreducible contradictions that would determine future attempts to complete Vitoria's project.

Metaphysical instrumentalism came up against several impasses in the debate staged in Valladolid in 1550–1551 between Ginés de Sepúlveda (1490–1573) and Bartolomé de Las Casas (1484–1566). In *Demócrates Segundo*, Sepúlveda affirms that the ultimate axiom of natural law is the subordination of imperfect matter to perfect form. He explicitly grounds the possibility of making war on Indians in their position in this natural order. Their position is equivalent to that of imperfect matter, which can be improved by the imposition of superior forms of government. Therefore, there is an intrinsic relation between the idea of Indians as barbarians and serfs by nature and the presuppositions of metaphysical instrumentalism.

Following Las Casas's notion of barbarism, in a strict sense it means dwelling outside law and reason and lacking mechanical arts. Barbarism is the exception to the metaphysical foundations of natural order. Las Casas momentarily shares the presuppositions of Vitoria and Sepúlveda but then takes them to the ultimate consequences by stating that, if such barbarians exist, they cannot be the majority of a continent but only a minority. Las Casas critiqued the central axiom of Sepúlveda's imperial philosophy, the subordination of the imperfect to the perfect. Las Casas dismantles this argument, saying that Spanish forms cannot be simply imposed over the Indians because they have their own self-sufficient communities.

Both irreconcilable positions derive from metaphysical instrumentalism. Sepúlveda is clearly following Aquinas, for whom the sovereign commands are to the subordinate administrators what the plan (*ratio*) of the architect or designer is to the inferior craftsmen, manual workers, and manufacturing crafts (*artificialibus*). When Sepúlveda employs the principle of the natural subordination of matter to form, he assumes this subordination of inferior crafts to superior crafts and the subordination of imperfect matter to the perfect form taken from artificial manipulation. This instrumentalist division between rulers and ruled has strong imperial connotations, since, for Aquinas, foreigners who lack self-sufficient communities and proper institutions can be serfs by nature, which he defines as living tools. Las Casas draws antithetical conclusions based on the same premises because, although he admits the importance of the hylomorphic doctrine, he affirms that Indians have the right and the duty to defend themselves because they have self-sufficient communities with the right to repel force with force. While Vitoria identifies any commonwealth with the material cause of civil power, Las Casas identifies the commonwealth with the efficient cause, which gives them the possibility of rejecting Spanish rule.

In 1512 the Spanish Crown issued the Laws of Burgos, and two years later Palacios Rubios composed his well-known Requerimiento.[1] Both documents were attempts to systematize the assertion of Spanish sovereignty over the New World. When Bartolomé de Las Casas returned to Spain in 1515 he enjoyed the support of the Dominicans in both Spain and the island of Hispaniola as he began his fervent advocacy for Amerindian rights in the Spanish court during the following six years. In 1516 Cardinal Francisco de Cisneros sent Geronymite friars to Hispaniola to defend the Amerindians' right to self-government. In 1519 Las Casas and Fray Juan Quevedo, Archbishop of Darién, debated the nature of the Amerindians before Charles V, the former defending the Amerindians' right to self-determination while the latter defended the idea of the Amerindians' natural servitude using the arguments forwarded by John Mair in 1510. In his *Secundum Librum sententiarum*, the Scottish theologian maintained that the inhabitants of the New World were barbarians and slaves by nature as described by Aristotle in the first and third books of the *Politics*. By 1520 Las Casas's indefatigable advocacy led Charles V to propose an eventual end to the *encomienda* and abrogation of the *encomenderos'* right of inheritance. Nevertheless, the Crown's official policy was that the encomienda was compatible with the freedom of its Amerindian subjects (Adorno, 103).

Throughout the following two decades, the encomenderos continued to amass power and fortune while the Amerindian population was decimated by Spanish violence and epidemics. In 1530 the Crown decreed the abolition of Amerindian slavery only later (in 1534) to authorize the enslavement of Amerindians captured in war. Between 1537 and 1539 the Dominican *catedrático* Fray Francisco de Vitoria delivered his famous lectures on the Indies in which he questioned the foundations of Spain's official imperial policies. Vitoria's influence at the University of Salamanca was such that Charles V prohibited him from further commenting on matters concerning the Crown's imperial policies.

In Peru, Manco Inca's resistance to the Spanish invaders at Cuzco collapsed, leading the the last Inca to retreat to Vilcabamba. Between 1533 and 1569 the viceroyalty of Peru was rocked by a series of internecine conflicts leading to a panorama of disorder and destruction that decimated the inhabitants of the Tawantinsuyu and destabilized any political or social structure remaining from the Incas, provoking harsh criticism from the Las Casas camp, which opposed the mining and mercantilist policies tied to the encomienda (González Casasnovas, 8). In 1540 Charles V sent Cristóbal Vaca de Castro to impose order upon the permanent state of violence and chaos in Peru. Tensions between encomenderos and royal officials continued to heighten.

In 1539 Bishop Valverde asked the Spanish monarch to intervene, and in 1541 Charles V issued a decree urging Vaca de Castro to comply with the prohibition of sending Amerindians to work in the mines against their will. The Dominican followers of Las Casas, particularly Fray Domingo de Santo Tomás, pressured local authorities and denounced the nefarious effects of the mines. These attempts to regulate mining led to the Ordenanzas de Minas, written between 1541 and 1545 by Vaca de Castro, who nevertheless allowed for the use of *yanaconas*, Indians who had left their *ayllu*, in the mines of Potosí. In 1542 the Crown sent a new viceroy, Blasco Nuñez Vela, entrusted with calling the encomenderos to heel and implementing the New Laws of 1542, which were meant to impose real limits upon the encomienda and the use of Amerindian slave labor. The New Laws decreed that Amerindians could not be sent to the mines against their will and that the encomiendas would remit to the Crown upon the death of the encomendero, thus limiting the threat of a local aristocracy. Gonzalo Pizarro led a rebellion of encomenderos in reaction to these measures, which ultimately led to his defeat and execution.

Unable to abolish the encomienda, in 1545 the Council of Indies permitted the inheritance of the institution for one generation after the death of the original encomendero. For the Amerindian inhabitants of the viceroyalty, the New Laws represented a temporary reprieve from slavery and servitude under the encomienda system, at least until the *repartimientos de labor*, the compulsory labor system imposed upon the Andeans, became a practice legitimated by the colonial state and not just practiced by individual colonial entrepreneurs. For the first generations of conquerors and encomenderos these decrees meant significant limits on what they viewed as their rights as feudal overlords. The 1530 abolition of Amerindian slavery was compensated with the arrival of the first African slaves and with a new system of labor levies requiring Amerindians to work in the mines for a nominal remuneration, a system that would later crystalize as the *mita* under the viceroy Francisco de Toledo in 1572. Thus, the mita emerged as a way of legitimizing forced labor in order to produce public and private wealth beyond the institution of the encomienda.

After Gonzalo Pizarro's defeat, the viceroy Pedro de la Gasca distributed Amerindians for use as laborers among his allies and authorized the use of Amerindian labor in the mines of Potosí. Nevertheless, in 1549 he ordered the return of the Amerindians he had sent to Potosí. He ordered the owner of the mines, Polo de Ondegardo, to conduct a survey to see if the laborers preferred to remain in Potosí. Not surprisingly, de Ondegardo's official document concluded that the Amerindians preferred to live in Potosí. This report was sent to Madrid. When an inspector was sent to Potosí to enforce De la Gasca's order, the encomenderos contributed a great amount of silver that was "returned" to their Amerindian laborers in order to pay a head tax established by the president. Furthermore, the encomenderos forced their Amerindian charges to declare their desire to remain in the mines. In November 1549, Pedro de Hinojosa, the richest encomendero of Potosí, sent a considerable sum of precious metals to Seville.

From 1548 to 1550 what would later become the practice of the mita in the mines of Peru, formally established by Toledo, began to take shape. Mining continued to be a primary target for the Dominicans, and in 1551, the same year as the second Las Casas–Sepúlveda debate in Valladolid, Fray Domingo de Santo Tomás described Potosí as "the mouth of hell" (González Casasnovas, 13). The Dominican denounced the impoverishment and disintegration of the Tawantinsuyu's once thriving communities due to the labor levies in the mines. The spread of mercantilist practices after 1545 required the radical reorganization of the precolonial economy, and Spanish authorities slowly began to take note of the process.

Las Casas's *De unico vocationis modo omnium gentium ad veram religionem* (circa 1537) was a largely theological treatise employing both Aristotle and Aquinas that defended the Amerindians' knowledge of natural law and their use of reason. His thesis was essentially that evangelization should be peaceful and cannot be preceded by violent conquest. In 1537 Pope Paul III, at the insistence of the Dominican Julián Garcés, declared in his bull *Sublimis Deus* that all Amerindians were human beings with rational souls, who should be saved and whose salvation could not come at the cost of their lands or property (Adorno, 104). Las Casas's advocacy led to the creation of the New Laws in 1542.

In reaction, Ginés de Sepúlveda entered into the subsequent debates over the legitimate possession of Amerindian lands and labor. In 1545 Sepúlveda wrote *Demócrates Segundo, o de las Justas Guerras contra los Indios* at the behest of García Jofre de Loaysa, president of the Council of Indies. Sepúlveda's work, originally written in Latin, rehashed Vitoria's arguments on just war and the right to repel injury with force. Sepúlveda was a humanist scholar specializing in ancient Greek, an intense opponent of Erasmus and Vives's pacifism, and an ardent defender of the compatibility of Christian ethics and the chivalric code of war. Sepúlveda rose to become the tutor of Prince Philip II. Sepúlveda's work generated such a controversy that the Universities of Alcalá de Henares and Salamanca both prohibited its publication. This did not stop Sepúlveda from circumventing the censorship by publishing an abbreviated version in Rome in 1550. Las Casas, who had used his influence to prohibit the first publication, nevertheless persuaded the courts to ban the importation and circulation of the work in Spain and the Indies. As Adorno expresses in a summary of the main events that led to the debate in Valladolid: "The theoretical and ideological opposition between Las Casas and Sepúlveda culminated in their personal confrontation" (120).

As a result, in 1550, at the recommendation of the Council of Indies, the emperor Charles V ordered the suspension of conquests until further deliberation by a council of jurists and theologians in Valladolid. The debate at Valladolid between Las Casas and Sepúlveda was staged in 1550–1551 before a jury of Dominican theologians. As Domingo de Soto's summary of the proceedings indicates, the purpose of this council was to determine if it was legitimate to subjugate Amerindians by violent force prior to the process of evangelization. The suspension was ordered by means of secret instructions sent to Pedro de la Gasca in Peru and Antonio de Mendoza in New Spain in April 1550. In 1551 each of the fourteen members of the council wrote out his decision: military conquest would be suspended "except for those directed by missionary priests for the purpose of Christian indoctrination and conversion" (Adorno, 83). Between 1552 and 1561, Las Casas continued his advocacy at court and undertook the writing of his monumental *Historia de Indias*, which covered the history of the New World from 1492 to 1521 using the wealth of information he had gathered from firsthand experience in Hispaniola.

The central topics of this debate were the interpretation of the papal bull of donation and the problem of the nature of the Amerindian, which was directly linked to the Aristotelian doctrine of natural slavery as outlined in Chapter 4, Book 1, of the *Politics*. One of the issues examined in the debate was whether the Amerindian could be forced to submit to the authority of the pope and the Catholic monarch through violent means. Just war and the nature of the Amerindian were the central themes of these polemics, but these issues were nevertheless inseparable from the justification of mining and the exploitation of Amerindian labor in the mines of the New World. The two sides debated the correct interpretation of papal donation, which Sepúlveda defended and Las Casas questioned. Las Casas, following the path trod by Vitoria and de Soto, was wary of conceding too much power to the pope over temporal matters. According to Adorno, the Dominican "favored Indian autonomy to the extent of advocating the full political sovereignty of Indian communities, even following their conversion to the Christian faith. Only after their conversion, and by their free will and consent, should they be placed under Spanish rule" (121).[2]

NATURAL ORDER

Vitoria had already addressed the issue of Amerindian sovereignty, not only in his lecture on the Indies but also in his lecture on civil power where he maintained that, while there are human beings incapable of self-government, this could not be used as an excuse for slavery. As both Pagden and Adorno argue, there is some continuity between Vitoria's and Sepúlveda's arguments. As Adorno states, the divergence between the two positions lies in the "degree of certainty they respectively expressed about their applicability: Vitoria equivocates and qualifies; Sepúlveda is certain" (Adorno, 113). Given this similarity in their respective positions, Adorno argues, Las Casas reacted by misrepresenting Vitoria's arguments (113). Nevertheless, there is an undeniable consensus between the two thinkers in understanding natural slavery as a "hierarchical relationship between those with the talent and training to rule and those who were better off being ruled by others" (113). In addition to the problem of empirical knowledge about the nature of the Amerindian (a key issue raised by Las Casas), the positions staked out in the Controversy of Valladolid are likewise the result of contradictory interpretations of the principle of the subordination of matter to form and means to ends already developed in Vitoria's reading of Aquinas. The same natural law that dictates that war can be declared in order to protect trade, circulation, evangelization, and exploitation of metals is grounded in a notion of mastery that presupposes technical domination, the art of war itself. Vitoria established the foundations of Iberian imperial ideology by appealing to a metaphysical instrumentalism that conceives the nature of the New World as a collection of commodities, a source of labor force, and an available raw material adaptable to the needs of the Crown. Vitoria provided a rational goal-oriented justification of colonialism that ultimately depended on the transference of technological properties to the ontological origin of natural law.

As Adorno explains, Sepúlveda's *Apología*, published in Rome in 1550, summarizes the arguments put forth in *Demócrates Segundo*, in which the theory of natural slavery is substituted for the thesis, based on Augustine's *Civitas Dei*, that slavery is divine punishment for the disobedience of an inferior to his superior (Adorno, 115). Adorno likewise dispels the common reading of Sepúlveda as justifying the slavery of Amerindians because of their supposed inhuman nature. The inferiority that Sepúlveda attributes to the Amerindians is based on a hierarchical relationship similar to the relation between any two entities "when one is more perfected than the other" (115). Against the typical mistake of attributing a certain binary logic of the human and the animal to Sepúlveda, Adorno reads him within the Aristotelian frame of the varying degrees of perfection ascending from less human (or civilized) to more human. This frame allowed Sepúlveda to understand the supposed inferiority of the Amerindian not as something immutable but, rather, as something "susceptible to improvement" (115).

I read Sepúlveda's work, as well as Las Casas's response to his doctrine, through the fundamental principles of Aristotelian physics, which are the foundation of metaphysical instrumentalism, the ontological frame of Spanish imperial ideology. Aristotle maintains that there are exactly three fundamental principles of physics: form, matter, and privation. These physical principles presuppose a metaphysical foundation, a ground that is the principle of the subordination of matter to form and means to ends. I argue that Sepúlveda explicitly employs the principle of the subordination of imperfect matter to perfecting form as the ultimate justification for the notions of just war and the natural slavery of the Amerindian. I maintain that Sepúlveda's use of this principle actually leads him to view the Amerindians neither as human nor as animal, given that they do not neatly conform to the category of "civil" human being or to the category of "wild" animal. Thus, Sepúlveda argues, it is possible to "correct" them through the use of coercion. This argument depends on both Aristotelian physics and Thomist natural rights theory, the principle of the natural subordination of matter to form and means to ends.

Both Aquinas's and Vitoria's ontological principles of natural law presuppose the transposition of technique to nature and politics. Such transposition has two consequences. First, natural law predicates that the unconditional character of the law ultimately derives from an origin (or efficient cause) that imposes order (forms, preconceived patterns, or ideas) into matter (understood as a pure, passive potency, deprived of any positive concrete determination) directing it to an end (which is also the idea preexisting in the efficient cause). From a Heideggerian perspective, this kind of mastery presupposes artificial mastery, resulting from the transposition of technique to nature. In order to illustrate how the natural hierarchical order results from God's reason, Aquinas employs the examples of an artisan who makes a statue, or the pilot who directs a ship. God stands in front of his creation in the same way that the artist stands in front of the object of art. The natural world is understood as a passive, malleable material that needs to be ordered to its final end.

The second consequence is a disavowal of technique. Since grounding natural law in technique is clearly problematic, because it robs the law of its natural, universal, and necessary character, natural law needs to subordinate technique itself. In other words, once technique fulfills its role of providing an example—a second degree copy—of the exemplified yet indemonstrable principle, technique (including the the artist's know-how, artifacts, and tools) is relegated to the status of mere prosthesis, a neutral, inert, passive device. Despite using technical examples in order to illustrate physical causality, Aristotelian Thomism reduces the artifact to an essentially inert and neutral tool with no capacity to move itself. For Aristotle and Aquinas, while nature has an inner capacity to move itself, a technical artifact does not have the interior disposition to change and cannot move itself (*Physics* 2.192–93, in *Basic Works*, 236). Thus, the inner contradiction of the ontological ground of imperial ideology—metaphysical instrumentalism—is that it transposes technique to nature only in order to subordinate technique to a superior master. As a result, metaphysical instrumentalism conceives both nature and technique as a "standing reserve," an available object of domination and transformation. Adorno was the first scholar to pay attention to the centrality of the subordination of the imperfect to the perfect in Sepúlveda's imperial frame. This chapter argues that the inevitable consequence of Sepúlveda's argument, an inherently instrumental view oriented toward the realization of an end, is inseparable from the repudiation of Amerindian technique, which is assigned to an inferior class in the hierarchy of natural physics.

Despite their spirited debate, Las Casas did not dissent with regard to this fundamental principle. For the Dominican, the barbarian (specifically, the third class of barbarism) is also like an animal without exactly being an animal, but he refuses to apply this category to the Amerindian. Las Casas's position depends less on the contrast between the empirical and the abstract than on the geopolitical limits inferred from Sepúlveda's thought and the inherent aporia of his metaphysical ground.

Demócrates Segundo is a dialogue between two fictional characters, Demócrates, who represents Sepúlveda's own defense of the just war against the Indians, and Leopoldo, who represents the pacifist opposition to the imperial perspective. Ironically, although Leopoldo is against the just war, the best synthesis of Sepúlveda's thought is found in Leopoldo's summary of Demócrates's doctrines. At the end of the work, Leopoldo, who represents the pacifist position in the dialogue, claims that it is unnecessary to debate the justice of the "war and empire" under question because this doctrine has been demonstrated by "solid reasoning, taken from principles of philosophy and theology, and is deeply rooted in the very nature of things and the eternal law of God" (Sepúlveda, *Demócrates*, 83).[3] Leopoldo adds that after hearing Demócrates's thesis he is now "free of all [the] doubts and scruples" that had beset him and is finally able to accept that war can indeed by justified by the four causes outlined in Sepúlveda's work. The first cause is the barbarians' rejection of the "dominion of the more prudent, powerful and perfect, which should allow to its own great benefit, as is proper because of natural justice" (83). This natural justice is consistent with the first principles of Thomist political theology, which dictate that "material should be subject to form, just as body should be subject to soul, appetite to reason, brutish beasts to man, that is, the imperfect should be subject to the perfect, the worst to the better" (84). Leopoldo recognizes that Demócrates appeals to the authority of not only Aristotle, "master of justice of all moral virtues and wisest interpreter of nature and natural laws," but also Aquinas, "prince of the Scholastic theologians, Aristotle's commentator and rival in explaining the laws of nature, which he demonstrated were all divine and derived from God's eternal law" (84). The second cause would be cannibalism, which "offends nature" and is associated with the problem of idolatry in the "monstrous rite of burning human victims" (84). Another argument made with "force and weight" is the divine mandate to protect those "mortal innocents the barbarians burned every year" (84). The fourth cause is the need to open the way to missionaries and defend their right to cross international boundaries. Conquest would be the only means of freeing the barbarians from their princes and priests so that they may "freely and with impunity receive the Christian religion" (84). War is the only means of eliminating "all the impediments and the cult of idols, renewing the pious and most just law of the emperor Constantine" (84–85). This argument is exemplified by the Roman Empire, whose justice is proven by Demócrates "with quotations from San Augustine and Thomas Aquinas," though this is considered the cause with "the least weight" (85). The justice of domination is nonetheless incompatible with "cruelty and avarice in its execution and dominion" for which the prince would bear the blame for the misdeeds of his soldiers (85).

Leopoldo ends his summary by imploring Demócrates's approval: "have I not summarized well and in few words your extensive dissertation in which you have expounded the justice of this war?" (85). In this brief summary of Demócrates's arguments we can already discern many important points. First, Sepúlveda acknowledges the centrality of Aquinas. Second, in his very first argument, he introduces the principle of the subordination of the imperfect to the perfect, explicitly bringing in the hylomorphic vocabulary. And third, he argues that the natural order is universal and necessary, and therefore that any offenses or transgressions must be punished.

NATURAL LAW IS IMPRESSED UPON HUMAN BEINGS

Sepúlveda begins *Demócrates* by declaring that his primary motive is to determine the justice of war in the Indies and "on what juridical reason can dominion over these peoples be founded" (1). This problem, according to Sepúlveda, is "transcendent," because it implies "consequences of the highest importance" (1). The "heated polemics" regarding this problem arose because they directly implicate the "fame and justice" of Spain's princes (1).

War is not to be "desired" for its own sake, according to Sepúlveda, but rather, it should be pursued only with "rectitude" and "piety" and only "by men of the highest virtue and piety" (*Demócrates*, 4). Consequently, princes sometimes must accept war as the lesser of evils in order to "achieve great benefits and at times out of necessity" (4). Thus, war can never be an end in itself but, rather, should be a "means to achieve peace" and should be pursued only after "mature deliberation" founded on just and necessary causes (4–5). Sepúlveda calls upon the authority of Augustine, who maintained that when war is necessary "one does not seek peace as a means for war but war as a means to achieve peace" (5). Leopoldo counters this claim by pointing out that war violates the principle of Christian charity, to which Sepúlveda retorts that while Christ certainly turned the other cheek this did not imply a complete suspension of natural law, "according to which all men are allowed to repel force with force within the limits of legitimate self-defense" (8).

For Sepúlveda, the right to repel force with force should not be a pretext for un-Christian violence but, rather, a precept that does not contradict Christian charity. The *corrective* force of the law is natural and follows from divine law, while the *corrected* force transgresses divine law. This is the fundamental axiom shared by Vitoria, Sepúlveda, and Las Casas. War, just like technique, is a means, not an end, and should therefore be subordinated to an unconditional end.

Sepúlveda's Demócrates defines natural law as "that law which in all places has the same strength without depending on circumstantial criteria" (11). Like Aquinas, he defines natural law as "participation of eternal law in all creatures blessed with reason," and like Augustine, he defines the eternal law as the will of God, "who wills the conservation of the natural order and prohibits its perturbation" (11). Human beings participate in this eternal law through the use of right reason and the inclination to duty and virtue (11). While "appetite" predisposes men to evil, "reason" reorients them toward good (11). Saint Paul echoed this law when he admitted that there were good men among the pagans who conducted themselves according to the mandates of natural reason (11). This law is "stamped" in the hearts of humanity, as the effect of "light of right reason, which is what one understands by natural law" (11–12). Aquinas defines law in the following way: "hence, since all things subject to Divine Providence are ruled and measured by things participate to some degree in the eternal law: that is, in so far as they derive from its being imprinted upon them their inclination to the activities and ends proper to them" (Article 2, *Summa Theologica* IaIIae.91, in *Political Writings*, 86). Imprinting is an essential part of law, since "just as man, by such pronouncement, impresses a kind of inward principle of action upon a man subject to him, so God imprints on the whole of nature the principles of their proper actions" (Article 5, *Summa Theologica* IaIae.93, in *Political Writings*, 110). Finally, this act of imprinting is also an act of promulgation: "The impression of an inward active principle is to natural things what the promulgation of law is to men: because law, by being promulgated, imprints on man a directive principle of human action" (110).

As natural law is universal, it is not subject to exceptions, nor does it depend on circumstances. In maintaining that natural law "participates" in divine law, Sepúlveda offers a consummately Aristotelian-Thomist definition of natural law. He employs a Neoplatonic theory of participation between things and ideas by means of an analogy linking being and entities. This analogy implies that the essence of entities is in some sense the same and in some sense distinct from being as the ground of the totality of all entities. Right reason is the means by which human beings can participate analogically in divine law, transcending carnal and animal appetites. Most importantly, this law is inscribed or stamped onto the hearts of all humanity through the light of right reason in the same way that the form of a statue is molded from a malleable substance. The light of reason impresses the law on our hearts, subjecting them to the *imperium* of the divine mandate.

Sepúlveda's Demócrates, moreover, explains that the rational character of natural law is found not only in Christian authors but also in pagan philosophers versed in "natural and moral philosophy and all kinds of politics" (*Demócrates*, 12). Demócrates makes an analogy between the Christian interpretation of the pagan philosophers and finding precious metals, since miners did not create the gold and silver found in the New World "but took [it] from mines, as Divine Providence, which fills all things" (12). The principle and ultimate end of right reason is like the gold provided by the hand of Providence that guides history according to a divine order. Sepúlveda mounts a rhetorical defense of philosophical principles, and he will end up opposing these philosophical principles to legal principles.

THERE CAN BE ONLY ONE WAY

In order to justify war, it is first imperative to define the "good end" and "right intention," without which any martial action would be a sinful transgression of Christian charity. While there are diverse and sinuous paths to sin, there can be only one straight road to virtuous action. Just as there is only one straight line between two points, "the archer only has one way of hitting the target, but many ways of missing it" (14). When choosing the straight line, "the final reason is the principal" (14). Just as in mathematics, the end is the foundation of any proposition, "and it is just to denominate things by their end" (14). Sepúlveda coincides with Aquinas who thinks that, "whatever exists for the sake of an end must necessarily be adapted to that end" (Article 1, *Summa Theologica* 96, in *Political Writings*, 138). In Aquinas we read that "just in the case of speculative reasoning nothing is validly established other than by being inferred from indemonstrable first principles, so too in the case of practical reasoning nothing is validly established other than by being ordered to the final end, and whatever is grounded in reason in this latter of sense has the character of law" (Article 2, *Summa Theologica* IaIae.90, in *Political Writings*, 80). The final end is the commanding principle of practical reason since, as Vitoria put it, "everything must be as it is because of that purpose or final cause, which is the true and necessary cause of all things" (*Political Writings*, 6).

As observed in Sepúlveda's thought, the principle of the subordination of means to ends presupposed Aristotle's principle of unity: "being is spoken of in many ways, but always relative to one term" (Aristotle, *Metaphysics* 4.2.1003 a33, in *Basic Works*, 732). There are many ways but only one end. And only right reason can serve as its own guarantee when searching for the right way for achieving the unconditional end. The end is the commanding principle and pretends only to presuppose itself. Not all means can be justified by the end. Only those means that are completely subordinated to the end are just. For instance, if in pursuing a licit end, such as earning money, somebody employs illicit means, such as adultery, then that person should be deemed an adulterer (*Demócrates*, 14). Consequently, in order to justify war it is necessary to always keep in mind the justice of the end in the election of means (14). Therefore, to fight is not a crime, but it is a sin if it is done for the sake of plundering; "to govern the republic is not a crime, but it is a crime to do it to augment its riches" (14). War is a neutral means that is good when pursued for virtuous ends and evil when solely subordinated to the accumulation of wealth or power.

In order for a war to be just, its causes must be just, and the "most important and natural is to repel force with force when there is no other choice" (16). Demócrates maintains that natural and human law permits anyone to "repel force with force in self-defense and in defense of its interests" (16). It is not necessary to respond to violence at the moment of receiving injury, "since this can be done at the earliest possible opportunity" (16). Moreover, in order for war to be legitimate it must be declared and initiated on the authority of a prince, otherwise "there would be legal consequences for the violence and damage inflicted" (16). Although

Sepúlveda does not cite Vitoria's lectures *On the American Indians*, his doctrine nevertheless fits squarely within the philosophical framework he used. Vitoria also argues that Spanish expeditions can be backed by military force, since "It is lawful to meet force with force" (Vitoria, *Political Writings*, 282). Vitoria alludes to the famous principle according to which it is possible to repel force with force (Watson, *The Digest of Justinian*, 98), which will become the only true and solid foundation for declaring war on the natives in *On the Laws of War*. Sepúlveda makes explicit Vitoria's use of the principle of self-defense.

TO REPEL FORCE WITH FORCE IS TO CORRECT VIOLENCE

Sepúlveda distinguishes on the one hand between legitimate defense and illegitimate violence, the just war exemplified when a "republic or its supreme authority is injured or attacked with a war that must be repelled with war" (*Demócrates*, 16). The animal body parts and human weapons are means for achieving natural ends that are part of a divine order: "And so, it was precisely for this purpose that Nature armed the other animals with claws, horns, teeth, shells, and other defenses, and equipped man for every kind of war by giving him hands, which substitute for claws, horns, and shells, as well as the sword and spear, and all sorts of weapons that can be manipulated by hand. Additionally, it endowed him, as the Philosopher says, with ingenuity and other natural faculties of the soul (called prudence and virtue)" (16).

Like Vitoria and Aquinas, Sepúlveda states that nature equipped animals for conflict, while men have only reason and virtue (Aquinas, *Political Writings*, 5; Vitoria, *Political Writings*, 7). Their purpose is to demonstrate that individuals lack the means for being self-sufficient and need to live in association with other human beings. The three authors presuppose an instrumentalist vision of bodily extensions, since these extensions serve a further purpose, which is self-defense. The three authors also presuppose that humans supplement their natural lack in self-sufficiency with reason and the product of their hands, weapons. The ability and right to repel force with force is the result of an instrumentalist view of the world in which means must be subordinated to an end or product. Even skills and natural dispositions, such as prudence and virtue, are means for preserving political power. An autonomous political and social pact can only exist with the help of techniques that supplement or replace the natural defensive prosthesis, as Vitoria maintained in his theory of political autonomy. Herein lies the foundational aporia hidden under the guise of naturality: the dependence on artificial means, the original technique, which substitues an essential lack in human beings when compared to the animal world. Nevertheless, Sepúlveda even more clearly brings into focus the tautological relation between his martial instrumentalism and the metaphysics of perfection and autonomy. While Aquinas and Vitoria appeal to the similarity between human weapons and animals' natural defenses in order to argue for the need to live in society, Sepúlveda is arguing for the need to go to war, making explicit the belligerent world view of Vitoria himself.

THE NATURAL CONDITION OF BARBARIANS IS OBEDIENCE

Sepúlveda adduces another cause for war based on divine and natural law, "which is the most applicable to those barbarians commonly called indians" who are of such "natural condition" that "they should obey others" (*Demócrates*, 19). For Sepúlveda, there is a categorical obligation to obey for those who are incapable of self-rule: "if they refuse the dominion [of those qualified to command] and there remains no other recourse, they should be dominated by force of arms since that kind of war is just according to the most eminent philosophers" (19). Aquinas also naturalizes the difference between the rulers and the ruled:

> The soul, by nature, rules over the body, and human beings by nature over irrational animals. Therefore, all human beings who differ from others as much as the soul does from the body, and as human beings do from irrational animals, are because of the eminence of reason in them and the deficiency in others, by nature masters of the others. In this regard, Solomon also says in the Proverbs 11:29: "The stupid will serve the wise." And those human beings whose chief function is to perform manual tasks are disposed in this way, namely, as irrational animals are to human beings, or as the body is to the soul. (Aquinas, *Commentary on Aristotle's* Politics, 30)

After Demócrates's affirmation of the natural condition of those destined to obedience, Leopoldo replies, "You have just finished explaining a strange doctrine, Demócrates, one quite different from the common opinion of men" (*Demócrates*, 19). Demócrates retorts "strange perhaps, but only for those who have only greeted philosophy from the doorway and therefore I am shocked that a man so learned as you can't take as a new dogma an ancient doctrine held by philosophers and very much in conformity with natural law" (19). Sepúlveda maintained that the concept of natural slavery could only seem strange or novel to those who are ignorant of the principles of philosophy. Sepúlveda is not making a new point, since the division between those destined to obey and those capable of commanding was central to both Aristotle's and Aquinas's politics. Sepúlveda attempts to separate philosophical principles from canon law because those who study the law do not understand that philosophy is its foundation.

Pagden argues that the theologians of the School of Salamanca rejected Sepúlveda because his humanism represented a rhetorical and literary defense of philosophy's primacy. Sepúlveda trangresses the disciplinary boundaries observed by Vitoria's followers (Pagden, *Fall*, 109–18). The objectional "rhetorical" element of Sepúlveda's work is his literary defense of philosophy instead of theology or canon law. Adorno also points out that, while Las Casas uses canon law as his main source of inspiration, Sepúlveda uses philosophical discourse. My own thesis is that, by radicalizing Vitoria's imperial metaphysical instrumentalism, Sepúlveda somehow obliges theologians to face the disavowed consequences that necessarily fall out of certain philosophical foundations and principles. Although Sepúlveda somehow steps out of philosophy in order to defend his ontological ground by rhetorical means, he confronts philosophy with its own imperial presuppositions and consequences. In short, Sepúlveda draws the imperial consequences of Aquinas's hylomorphic doctrine of the preeminence of form over matter.

Demócrates continues, arguing that it is necessary to distinguish between the jurists' "concept of servitude" and the philosophical definition (*Demócrates*, 20). The former defines the law as "a certain adventitious condition that has its origin in the power of men, in common law, and at times in civil law," while the latter defines "servitude" as "congenital incapacity" affecting the intellect and resulting in "inhuman and barbarous customs" (20). Sepúlveda makes the same distinction as Vitoria makes in his lecture *On American Indians*. On the one hand, there is civil slavery, which involves the ownership of the slave. On the other hand, there is a more philosophical notion of slavery, which consists of men with "natural deficiency, because they need others to govern and direct them" (Vitoria, *Political Writings*, 251). Vitoria says that Aristotle did not think that masters "had a legal right to arrogate power to themselves over others on the grounds of their superior intelligence, but merely that they are fitted by nature to be princes and guides" (251). Aristotle bases mastery in necessity and expediency:

> But is there any one thus intended by nature to be a slave, and for whom such a condition is expedient and right, or rather is not all slavery a violation of nature? There is no difficulty in answering this question, on grounds both of reason and of fact. For that some should rule and others be ruled is a thing not only necessary, but expedient; from the hour of their birth, some are marked out for subjection, others for rule. And there are many kinds both of rulers and subjects (and that rule is better which is exercised over better subjects—for example, to rule over men is better than to rule over wild beasts; for the work is better which is executed by better workmen, and where one man rules and another is ruled, they may be said to have a work). (Aristotle, *Politics* 1254a17–28, in *Basic Works*, 1132)

Aquinas comments on this passage, saying "we perceive that there is a distinction regarding human beings from their very birth, such that some are fit to be subjects and others fit to rule" (*Commentary on Aristotle's* Politics, 27). When, in *On the American Indians*, Vitoria opens the possibility of subjecting the Indians to Spanish dominion, he opens the door to Sepúlveda's argument. Vitoria did not think the inferiority of the Indians could justify their enslavement and dispossession, but he opened the possibility of subjecting them to the jurisdiction of the Spaniards, thus opening the door to Sepúlveda's analogical notion of dominion. While, for Vitoria, natural slavery is only a possible and therefore doubtful title, Sepúlveda has no doubts about the natural inferiority of the Indians.

THE PERFECTION OF FORM AND THE DOCILITY OF MATTER

Moreover, Sepúlveda, like Aquinas, argues that power (*dominio*) is analogical, not univocal: it "is not always exercised in the same way but in very diverse ways" (*Demócrates*, 20).[4] This implies that different legal domains have different "foundations": "the dominion of father over son, husband over wife, master over his servant, magistrates over citizens, the king over the people and individuals subject to his empire are different and have different juridical foundations" (20). This diversity in the exercise of power is nevertheless based on a common foundation: "and even though these dominions are so diverse, nevertheless when they are based on proper reason they all have their basis in natural law, which, within its variety, derives, as wise men have taught, *from a single principle*, that of *natural dogma*, the empire and dominion of the perfect over the imperfect, strength over weakness, sublime virtue over vice" (20; my emphasis). Sepúlveda reached the ultimate presupposition of imperial reason, the "only principle and natural dogma," the dominion of the perfect over the imperfect. Aquinas's ultimate philosophical presupposition is inherent in the idea of law as such. When discussing whether law is always directed to a common good, Aquinas responds to those who think that law is not always directed to the end:

> Law belongs to [reason, which is] the guiding principle of human acts because it is their rule and measure. Now just as reason is the first principle of human acts, so reason itself must be guided by something which is the first principle of everything it does; and it is to this guiding principle that law must chiefly and mainly be directed. Now the first principle in practical matters, which are the object of practical reasoning, is the final end; Law must therefore attend especially to the ordering of things toward blessedness. Moreover, since every part of something is ordered in relation to the whole as the imperfect to the perfect, and since one man is part of a perfect community, law must attend to the ordering of individual things in such a way as to secure the common happiness. (Article 2, *Summa Theologica* IaIae.90, in *Political Writings*, 79)

Aquinas also states that when "the part is included in the whole, it is advantageous for the part and the whole. Likewise, the same thing, namely, the soul ruling the body, is advantageous for the body and the soul. And he shows that the slave is related to his master as the body is to the soul, but also as a part of his master, as if he were a living instrument that is a separated part of his master's body. For this separateness distinguishes the slave from the master" (Aquinas, *Commentary on Aristotle's Politics*, 38). As we can see, expediency and usefulness form an essential aspect of the subordination of the imperfect to the perfect. Aquinas's text emphasizes that slaves by nature are living tools, making evident the instrumentalist presuppositions of his metaphysics.

For both Sepúlveda and Vitoria, there are many different forms of domination, but all of them depend on the same principle that acts as a causal and explicative ontological foundation: the hylomorphic "dogma" that subordinates imperfect to perfect.[5] Moreover, the subordination of the imperfect to the perfect is the only principle that unites the different analogical conceptions of dominion.[6]

The mandate to subordinate imperfect matter to the perfection of form applies to the totality of objects, whether they are "unbroken or separated," and is demonstrated whenever we "observe that one of them, that is, the most important has dominion over the others, as the philosophers teach, and those others are subordinated to it" (*Demócrates*, 20). Sepúlveda adds that one of the parts of this totality, "the most important," is destined to exercise control and authority over the rest. One entity is called upon to embody the principle of subordination and become the foundation of all other entities.[7] Again, on this view Sepúlveda is consistent with Aquinas, who asserts that there is also a sort of political subordination to harmonic unity "in a mixed material substance, in which one of the elements always predominates" (*Commentary on Aristotle's* Politics, 28). Moreover, there is always an exceptional element that provides unity to multiplicity: "The Philosopher says that all things belonging to one genus are measured by the one primary member of that genus, for if there were as many rules or measures as there are things measured or ruled, the rules or measures would cease to be of any use, since the usefulness of the rule or measure lies precisely in the fact that it is a single standard that applies to many instances" (Article 1, *Summa Theologica* IaIIae.96, in *Political Writings*, 139). As we can observe from this last statement, exceptional political unity is ultimately based on usefulness, since it is impossible to have a rule or measure for each ruled or measured thing. The usefulness of having one single standard is the ability to direct things to their proper end.

Sepúlveda goes further than Vitoria in explaining the genesis of the imperial *subjection*. One particular element arises from amongst a multitude, attaining an organizing power, a hegemonic structuring role, as a universal subject that proceeds to subordinate to itself the multitude of particular predicates. A particular thing becomes universal. Sepúlveda does not see that the paradox resulting from this concerns the mixture of particularity and universality: one dominating *universal* grounded in and directly embodied by one of its own subordinated *particulars*. This paradoxical unity is the result of one exceptional particularity that orders other particulars, creating the illusion of transcendence, which is the illusion that this exceptional particularity preexists separately and above the rest. The elevation of the exceptional one is the product of the violent struggle for structural dominance in which particulars combat other particulars in order to enact themselves as commanding universals.

Sepúlveda illustrates the principle of the subordination of the imperfect to the perfect and the need of a single unity that precedes and regulates entities by means of the hylomorphic doctrine. This doctrine divides entities between a perfect ruling form and an imperfect ruled matter: "all inanimate objects composed of matter and form, the latter, as more perfect, presides over and seems to dominate; matter, on the other hand, is subordinated to it and seems its handsmaiden" (*Demócrates*, 20). The dominant form "presides" because it is an *end*, a perfect and complete object representing the substantial unity of a harmonious composite of servitude and mastery. Form presides, it is presupposed in the same way that the end itself is given. As Aquinas maintains, the form is anterior because it is itself an end: it is given and as such is preeminent in the order of being: "For form is the end of matter. Therefore, for matter to seek form is nothing other than matter being ordered to form as potency to act" (*Commentary on* Physics, 72). The simultaneously identical and mutable matter presupposed by this theory is subjected to the mandate of form: "and since the matter of all things that are born and dies is identical, we see in its variations that matter follows form just as the slave follows her mistress, wherever she directs her" (*Demócrates*, 21).

Matter is imperfect but undetermined, while form is perfect and determining. A "substantial variation" occurs when an element such as earth is converted into fire, and "the matter that first gravitates to a lower condition now, influenced by form, gravitates toward a higher condition" (21). This process is even more evident in the animal kingdom, "the soul then has dominion and is like a mistress; the body is subordinated and is like a slave" (21). According to this framework of analogical thinking, "even in the soul, the part endowed with reason presides and administers a nevertheless civil control; and the other part, bereft of reason, is subordinated to its dominion and obeys it" (21). Once more, the metaphor links the human being's "substantial unity" with that of civil society. Deprived of determination and reduced to a plastic and amorphous raw material at the disposal of an alien will, matter must be subjected to form.

From the foregoing passage it becomes clear that imperial reason is grounded in the physical principle of subordination of the imperfect to the perfect. Deprived of strength, virtue, and perfection, matter must be subjected to the "empire and dominion" of form, which lends it the superior strength, prudence, and perfection that it lacks. The analogy of participation consists of subordinating the multiple to the singular, or in showing that the part is subordinated to the whole, that its being is an effect of the whole. Moreover, Sepúlveda maintains that matter is identical because difference can only proceed from form. Matter is the element that is deprived of perfection and that cannot become if not through form, which creates difference and degrees of perfection among entities. Matter is identical because it is deprived, or emptied, of determination, while form imposes a hierarchy among entities through governance and the assignment of difference. The empire of the master over the slave by nature is analogous to the empire of form over matter.

What, then, is the *origin* of form—that is, the origin of the origin? The principle of the subordination of the imperfect or deficient to the perfect unity is based on a "decision and divine and natural law according to which the more perfect and the best things predominate over the imperfect and mutable things" (21). If this mandate is based on the natural dogma of the principle as the ground of the ground, then this underlying cause is a kind of decision prior to even the preexisting underpinnings of "natural" subordination itself. Sepúlveda emphasizes that the rational order imprinted upon reality by this principle is based upon a natural and divine mandate. To reduce this affirmation to simply an outlying expression of theocratic militarism within the broader context of a multicultural empire would imply ignoring the ontological construction of Sepúlveda's ideology and the subsequent appropriations of this doctrine. The principle rules and commands because it is founded on the natural and divine fiat. The tautology that grounds the mandate (or imperium) on the origin understood as final cause (or telos) is the ontological foundation of imperial expansion. By explicitly attributing the principle behind the principle to a divine mandate, Sepúlveda exposes the ground of the ground, that is, the manner in which the metaphysics of origin emerge from an arbitrary decision to attribute this decision to a preexisting master. As explained above, the genesis of the exceptional one that precedes and rules over the multitude of particulars is the result of the struggle between particulars. Such struggle, the preceding antagonism of society, is disavowed in the name of a preexisting and encompassing unity without fissures. We are in a better position to understand Adorno's statement about how "Vitoria equivocates and qualifies" while "Sepúlveda is certain" (113). If Vitoria vacillates on the principle of the divine mandate it is because of his belief in the wise counsel as the good conscience of the prince. Humanist rhetoric is the best artifice for dissimulating and naturalizing this decision, the action of laying the foundation of his own principles, thus creating the illusion of consistency in the ontological edifice of natural law.

For Sepúlveda there are men, "sound in body and soul," who preserve their nature as "whole" and "uncorrupted," but there are also "vice ridden and depraved" men for whom "the body frequently dominates the soul, and appetite dominates reason, an evil and unnatural thing" (*Demócrates*, 21). Similarly, Demócrates adds, within an individual man one can observe the "inherent dominion" of the soul over the flesh as well as "civil and royal dominion" that the mind exercises over the appetites (21). Hylomorphic doctrine clarifies the logic of natural law based on the repulsion of force with force. The force that is to be controlled is characterized as "a bad and unnatural thing," a depraved body that controls the soul, a carnal appetite that clouds reason and judgment. Founded upon a natural and divine mandate, true reason should emend and rectify precisely that anti-natural, deviant force. Insubordinate passion is seen as pure violence against the natural order.

Sepúlveda argues that the supremacy of "reason over appetite" is "natural and beneficial" because it occasions "parity," while on the other hand "the mutability of dominion is destructive for all" (21). This "same measure and law" reigns over both human societies and the animal kingdom (21). Domesticated animals are even superior to savages because "submission to the dominion of man is better and more beneficial for them for the simple reason that they are able in that way and in no other way to survive" (21). In a phrase almost identical to that used by Vitoria, Demócrates declares "for the same reason, husbands have dominion over wives, adults over children, the father over the son; in a word, the superior and more perfect have dominion over the inferior and the more imperfect" (21).[8] The reason that dictates the subordination of women and children to the authority of their husbands and fathers "is valid for the rest of mankind in their mutual relationships, since among them there is one class in which some are by nature masters and others by nature servants" (21–22). According to the natural order, which dictates the preeminence of the rational soul over the body's appetites, some people are born to command and others are born to obey.[9]

The "masters by nature" should exceed in "prudence and talent" but not in physical force, while those who are "physically vigorous [but] slow and inept of understanding" are necessarily destined to be "servants by nature" (22). The philosophers maintain that, for the slave by nature, "it is not only just but also useful" to obey and serve their lords as the natural order dictates (22). Those who have the capacity to rule possess "political prudence," a virtue or habit that functions through the intellect in concert with wisdom and technique, while the slave lacks this gift. While the ostensible purpose of society is the common good, only those who possess the virtue of prudence will understand how to order people, actions, and matter to this end. Consequently, the slave benefits from the mastery of the lord because only through prudence, the intellectual virtue par excellence of the master, can one see the form and end to which undetermined matter must be conducted. Recall how Vitoria grounded civil power in a top-down vertical division between the formal cause (force of law) and the material cause (the commonwealth).

Sepúlveda likewise cites the Book of Proverbs, saying "that the fool shall serve the wise" and adding that "it is a widely held belief that this also describes barbarous and inhuman savages distanced from civil life, a law-abiding conduct and the practice of virtue" (22). For slaves, it is "beneficial and more in conformity with natural law that they be subordinated to nations or princes who are more human and virtuous" (22). The purpose of the slave's subordination to the wise man is to leave behind his barbarous state and embrace a more human existence by practicing virtue, obeying the dictates of natural law, and following the example of prudence exercised by the rulers (22). If the slave by nature should disobey the "principle" and "natural dogma" upon which this authority is founded, "they may be obliged to do so by force of arms, and the philosophers teach us that this war is by nature just" (22). The legality of riches acquired through war "springs in a certain sense from nature," given "that one part of [nature] is the faculty of hunting, which is not only to be used against animals but also against those men who, having been born to obey, refuse dominion, for such war is just by nature" (22). Sepúlveda echoes Aristotle who, in Chapter 8, Book 1, *Politics* 1256b, writes that "the art of war is a natural art of acquisition, for the art of acquisition includes hunting, an art which we ought to practice against wild beasts, and against men who, though intended by nature to go governed, will not submit" (*Basic Works*, 22–25). Disobeying the principle of the subordination of imperfect matter to perfecting form is a just cause of war. Sepúlveda's bellicosity is grounded on a metaphysical instrumentalism that lays the foundations for imperial ideology and violent conquest of the "servile barbarians."

This examination of the underpinnings of Sepúlveda's argument demonstrates that his imperial ideology is marked not only by the humanist role of grammar, rhetoric, and the Greco-Roman tradition but also by Aquinas's metaphysical framework based on the subordination of means to ends. Sepúlveda's rhetoric appeals to Aquinas's authority in order to explain the natural order. The division between humanist rhetoric and Scholasticism can only be applied to Sepúlveda at the cost of rending his use of language from the metaphysical foundation upon which it is constructed.[10] The principle of the subordination of matter to form and of means to ends is the key to Sepúlveda's justification of the Amerindian's natural inclination to servitude, which he considers a natural consequence of matter's resistance to form and the right to repel force with force. Sepúlveda employs the same philosophical underpinnings as are found in Vitoria's work in order to cement his case against the Amerindian and to support an imperial rationale that lends legal and moral force to the actions of Spanish conquistadors. In short, Sepúlveda uses his rhetorical skill to radicalize Vitoria's frame.

Vitoria sanctioned military violence in defense of commerce, and commerce is grounded in natural law, which presupposes eternal law. The proof that Vitoria offered for this claim is that inferior crafts are subordinated to superior crafts. Yet he had doubts about the servility of the Indians. Sepúlveda, nevertheless, employs the same strategy with the topic of slaves by nature. Sepúlveda unequivocally grounds natural servitude in what Vitoria understood as the subordination of the Indians to the Spaniards in the principle of the natural subordination of matter to form. As a result, Sepúlveda identifies Indians with the imperfect matter that needs to be corrected and guided to its proper end by the imposition of form. If imperfect matter disobeys the perfect form, then its disobedience can be corrected by force. Sepúlveda's hawkish stance is not contrary to Aquinas's doctrine, either; in fact, it is the natural consequence of a fundamental political and theoretical decision that fills the gap left open by both Aquinas and Vitoria in the attempt to apply universal, necessary, and indemonstrable principles to concrete and contingent historical circumstances.

PRUDENCE AND VIOLENCE

To ground the subjection of the Indians, Sepúlveda discusses how the rule of the most prudent people has a universal and necessary scope. Demócrates asserts that it is "naturally just and beneficial for both parties, that good men, excellent because of their virtue, intelligence, and prudence, should rule over their inferiors" (*Demócrates*, 23). This dynamic of filial relations reflects a doctrine from natural law and the law of nations that is likewise valid for the world order, being authorized by "universal consent and general practice among peoples" (23).[11] The philosophers demonstrated that this "consent and practice" is the natural law observed by "all nations governed by proper policies" as well as "just monarchs" when appointing others for public positions, that is, "those persons who, according to their judgment, will guard the interests of the nation" (23). Those at the helm of the republic should possess "virtue or prudence, because they judge that only in such a manner will the republic be saved and maintain a just and moderate empire" (23). This is the only means of assuring that the people of the republic are ruled by the "good and wise," lest they are left to be guided by "passion, vice, injustice," or "lack of prudence" (23). The subordination of vice to virtue is the "foundation" upon which "stands the entire political doctrine of the philosopher" (23). Demócrates argues that "the rationality of the natural order" mandates that the government should always be "in the hands of the best and most prudent," the only ones who can determine "the good of the community" (24). In the absence of the "best" and most "prudent" governors, the kingdom lacks "valor" (24). In Aquinas's political philosophy, judgment is a "right decision" that can be "either speculative or practical" (Article 1, *Summa Theologica* IIaIIae.60, in *Political Writings*, 193). In his commentary to Aristotle's *Ethics,* he states that prudence, which is the practical deliberation about how to direct means to an end, is the most important virtue, "for inquiry is ordered to judgment as to an end, and judgment to command" (392).

Sepúlveda illustrates the political prudence of the ruler with the example of a physician. The physicians determine that the "healthy humors" must reign over the human body for the "preservation of its natural state and good health" (*Demócrates*, 24). When these are offset by the corrupting "humors," it is necessary to use any means necessary to "nip that imbalance in the bud and minimize its pernicious influence" (24). At the same time, if in attempting to "exterminate" these noxious humors the patient's life is put in danger, then the prudent physician will abstain from action, "not because they are ignorant that such perversion of humors is prejudicial and unnatural, but because it is better to preserve a sick life than to destroy the entire body" (24). Those "who govern with foresight imitate the physicians' prudence when sick kingdoms, "as if insane," abide "evil princes," not in contravention of the natural principle of governance by the "best" but, rather, to "avoid the outbreak of uprisings and civil wars" (24). Thus, "a lesser evil, according to the doctrine of the philosophers, sometimes substitutes for the good" (24). Augustine recommends tolerating the "wicked" in the name of peace, and Aquinas likewise argues that the prince's sin cannot be punished if it would result in harm to the multitude (24–25). Consequently, Demócrates advocates tolerating an evil if in doing so a greater harm to the spiritual and temporal welfare of the republic is avoided (25). Sepúlveda is taking for granted Aquinas's corporatist notion of society, which sees the political bond in terms of a hierarchical and harmonious totality that disavows its own internal antagonism by projecting it onto the figure of an intruder that can be amputated for the greater benefit of the common good.

Again, Sepúlveda's position is entirely consistent with Aquinas's metaphysical instrumentalism. Reaffirming that it is permitted to kill beasts given that they are destined for use by man "as imperfect is ordained to perfect," Aquinas takes his inquiry a step further and asks if it is licit to kill sinners (Article 2, *Summa Theologica* IIaIae, in *Political Writings*, 253). First he repeats that "every part is directed to the whole as imperfect to perfect; and so every part naturally exists for the sake of the whole" (253). Based on this need to direct the imperfect to the perfect, Aquinas argues, "we see that if the health of the whole body requires the removal of some member, perhaps because it is diseased or causing corruption of other members, it will be both praiseworthy and wholesome for it to be cut away" (253). Since individuals are part of the whole, if an individual is "dangerous to the community, causing its corruption because of some sin, it is praiseworthy and wholesome that he be slain in order to preserve common good" (254).

Moreover, in his answers to the objections to the main thesis, Aquinas provides more information about this individual who is to be amputated from the social body. He asserts that the sinner loses the dignity of his humanity: "By sinning, man withdraws himself from reason, and in so doing falls away from the dignity of his humanity, by which he is naturally free and exists for himself, and descends instead toward the slavish condition of the beasts: becoming liable, that is, to be disposed of in whatever way is useful to others" (253). Aquinas cites Psalms 49:20 and Proverbs 11:29: "'Man, when he was in honour, did not understand; he hath been compared to senseless beasts, and made like to them': and it is said at Proverbs 11:29: 'The fool shall serve the wise'" (253). Aquinas concludes with a lapidary decree: "And so although it is evil in itself to slay a man while he remains in his dignity, it can nonetheless be good to slay a man who is a sinner, just as it can be slay a beast. For a wicked man is worse than a beast, and does more harm, as the Philosopher says at *Politics I* and *Ethics VII*" (254). Aquinas is quoting Aristotle's passage in *Politics* 1253a32 where he writes: "For man when perfected, is the best of animals, but, when separated from law and justice, he is the worst of all" (Aristotle, *Basic Works*, 1130).

BOTH MEANS AND ENDS SHOULD BE LAWFUL

Leopoldo maintains that in just war it is not only necessary to have good intentions but also a "proper method" in pursuing them (*Demócrates*, 27). He disputes that the wars against the Indians were waged with the purpose of accumulating "great quantities of gold and silver" by either lawful or unlawful means (28). Therefore, those who bear the "hidden desire" to obtain wealth in the persecution of the Indians are committing a crime. Given this, the Spaniards' actions are neither "just nor reasonable" but, instead, "injurious and cruel," and they could justly be obliged to "make restitution of their depredations to the savages, as would thieves to the travelers they've stripped of their property" (28). Yet good intentions, or the will to pursue noble ends, are not enough. According to Leopoldo, one must also be able to evaluate both the end as well as the means to achieve it. Thus, just war is not merely part of a cynical discourse meant to justify unjust means by the invocation of noble ends, but more importantly it leads to the belief that there is *only one manner* of achieving any end. Prudence, understood as the strategic calculation of means, must submit entirely to the preexisting and transcendent end. Leopoldo, just like Las Casas, argues that ill-gotten wealth must be restored to the *"bárbaros."*

Demócrates responds to his interlocutor's objections by affirming that the authority of the prince cannot be limited by the crimes of his ministers and subjects (28). Acts of wanton "avarice and cruelty" do not invalidate the cause pursued by the prince unless he is consciously complicit in these crimes or negligent in preventing them (28). Alluding to sources such as Cortés's letters to Charles V, Demócrates outright rejects the veracity of "certain accounts of the conquest of New Spain" according to the sources he uses (29). Moreover, he redirects the topic of the debate from the question of the "moderation or cruelty" implicit in Spanish soldiers' actions to the nature of the war with regard to the king and his ministers (29). The nature of the war is determined by the "greater benefit" to the "conquered *bárbaros vencidos*" rendered unto the Indians, which cancels the moral stain of "whatever benefit" be gained by the Spaniards (29). Sepúlveda appeals to Vitoria's idea that the Indians could only be subjected for their own benefit and not for the benefit of the Spaniards.

The supposedly logical process behind the declaration of a just war consists of a series of clearly defined stages. First of all, the "barbarian" nation must receive notice of the "great benefits" of submitting to the law, customs, and religion of Spain (29). While the recipients of this message should be given time to deliberate their response, Demócrates cautions: "but it is improper to give them an excessive amount of time," given that the time required to truly understand the content of these laws, customs, and religion could continue indefinitely (29). Notwithstanding the appearance of a straightforward procedure, this first step implies a fundamental contradiction, given that it is impossible to understand the "great benefits" until after accepting Spanish "domination, with continuous contact with our men and with the doctrine imparted by teachers of morality and religion" (30). Thus, the addressee is presented with a classic double bind: accept the rationality of a culture that you cannot by nature comprehend; alternatively, your incomprehension or resistance is a sign of your barbarity and will be justifiably met with violence.

Once informed of their duty to comply, the addressees must be brought within the fold of Christianity and must be subject to conditions consistent with their "nature" in order to serve under a tributary system (30). If they refuse to listen or if they offer resistance then "both they and their property will fall into the hands of the conquering prince so he may dispose of them as he sees fit, providing that prudence and the rationality of peace and the public good control his will with rules which must be applied whenever the enemy is punished after a victory" (30). A war that is not justified by natural law or within "the will and decree of a just prince" is unjust, and Demócrates maintains that it can only be classified as "theft," given that the category of enemies only befell those upon whom the Roman Empire had openly declared war (30). Any goods obtained outside of a legitimately declared just war following the foregoing procedure would thus be the spoils of simple *latrocinio* and should consequently be restituted (30). Basing the procedure of just war upon a double bind, Sepúlveda thus justifies not only empire but also domination, or the appropriation of the property of the conquered. As we can observe, Sepúlveda follows the same steps as Vitoria in his explanation of the first and second valid title: the defense of trade, communication, and evangelization by military force.

LIKE MONKEYS TO MEN

Sepúlveda attempts to demonstrate that the principle of subordination to the "prudent, good, and humane" justified the Roman Empire, and this principle was subsequently upheld by authorites such as Aquinas in *De regimine principum* and Augustine in the *City of God*. According to Sepúlveda, God conferred an empire upon the pagan Romans with the aim of impeding the vices and avarice of less civilized nations who had not achieved the relative political virtues of their Roman overlords (32). The Romans did not only seek ephemeral glories "but, rather, solid glory by following a rational course and legitimate methods" (32).

After a digression upon the distinction between the pursuit of glory and common vices, Sepúlveda returns to the purpose of his treatise: "whether it is licit and just that the best and those who stand out by nature, customs, and laws rule over those inferior to them" (33). According to Sepúlveda, it can be demonstrated that Spain's dominion over the nations of the New World is also based on an examination of the "nature and morality" of both sides (33). The barbarians exhibit "prudence, intelligence, and all virtues and human sentiments," which make them "as inferior to the Spaniards as children are to adults, men to men, the cruel and inhuman to whose who are extremely gentle, the exaggeratedly intemperate to those who are moderate and self restrained, and, I am tempted to say, between men and monkeys" (33).

At first sight, it would look as if the Indians themselves were animals, and yet things are more complicated. The result of the application of the principle of natural subordination to the Indians is the establishment of a certain parallelism wherein the Indians are to the Spaniards what animals are to humans. The relation between the superior Spaniard and the inferior Indian is like the relation between the human and the inhuman. What matters is not so much identifying the Indians with animals as establishing a certain relation of parallelism, that is, of establishing an analogy. This does not mean that Sepúlveda is directly identifying Indians with women or children but that there is a relation of analogy, according to which the slave is to the master as the woman is to the man, the child to the adult, the animal to the human, that is, as matter is to form.

Adorno has demystified the image of Sepúlveda as defender of the idea that the Indians are "half-human" by demonstrating how he "speaks of the progress that had been made in bringing the natives to the practice of European customs" (116). For Adorno, Sepúlveda's notion of natural servitude is a mixture of Aristotelian philosophy and Roman jurisprudence: "Although almost always overlooked, Sepúlveda's proposal is novel. It is not a direct throwback to Aristotle's theory of natural slavery, but rather a new position that finds inadequate both [the] philosophical notion of natural slavery and the juridical institution of civil slavery when it comes to defining the hierarchical relationship that, in his view, ought to be obtained between the European colonizers and the Amerindians" (117). The problem, as I see it, is how to evaluate theoretically the novelty of Sepúlveda's assertion about the inferior character of the Indians.

Aristotle clearly anchors the idea of the slave by nature in human passions as a series of qualities that can be described. Sepúlveda appeals to an inhuman and inferior character of the presumedly servile Indians, but this inhumanity cannot be properly pinned down or described. Such a notion of inhumanity cannot be exhausted by describing the qualities of the Indians because it consists of an analogy based on a structural relationship, where A is to B as C is to D. Ultimately, the Indians are to Spaniards as imperfect matter is to perfect form. It is a structural relation in which the Indians occupy the inferior position in a hierarchical structure that is taken for granted by the principle of natural subordination. Indians do not belong to a different genus or species, they are still humans, but their inhumanity emerges as a consequence of being deprived of the characteristics of their own species, just as matter is deprived of form and children are deprived of the proper use of reason. This "lack" within the Amerindian is an undefined variable that eludes identification: it is *like* the difference between men and animals without being a positive feature reducible to animals. Directly identifying them with animals would mean that they were external to humanity. For Sepúlveda, Indians are not external to humanity since they lack something proper to humans while still being human. The novelty of Sepúlveda's position resides in the affirmation of the inhumanity of the Indians as something that does not make them animals while still making them inferior: they occupy the same place as imperfect matter in relation to perfect form.

In order to argue for the Indians' deficit of humanity, Sepúlveda examines the conquest of Mexico, affirming that the native Mexicans were cowardly while the Spaniards were brave and prudent (*Demócrates*, 36). If Demócrates's primary aim is to demonstrate that the Amerindian is naturally inferior, that is, imperfect matter to be subordinated to the perfection of form, "can there be any greater or clearer proof of the advantage some men have over others in terms of intelligence, ability, strength of character, and virtue?" (36). Sepúlveda adds: "in any case, the fact that some of them seem to have a certain facility for artisanal work is no argument in favor of their having more human prudence, since we see that certain insects, like bees and spiders, create works that no human ability can manage to imitate" (36). Leopoldo retorts that inhabitants of New Spain and Mexico—Tenochtitlan—may well have possessed the most civilized institutions of the Indies, "as if it were not sufficient proof of their industriousness and civilization that they had rationally constructed cities and kings designated not by hereditary law and age but by popular suffrage and that they engaged in trade like other civilized nations" (36–37). Yet, Demócrates replies, herein lies the problem: it "resides precisely in their public institutions, since almost all are servile and barbarous" (37). Demócrates's argument culminates in an attack on the supposed prudence of the Mexicas: "in any case, the fact of their having houses and some sort of rational life style in common as well as the commerce brought about by natural need, what does that prove except that they are not bears or monkeys completely lacking in reason?" (37).[12]

Positive facts, empirical features like the undeniable fact that Indians have institutions and artifacts, cannot disprove their inhumane nature. Sepúlveda's ideology succeeds because the arguments against it start to function in favor of it: the obstacle is transformed into its enabling condition of possibility. The empirical arguments against the inferiority of the Indians such as having institutions and artifacts are transformed into arguments against them: they are not monkeys, which transforms them into something even worse, inhumane beings that occupy a middle space between the animal and the human. Sepúlveda's power, his novelty, consists in establishing a correspondence between the Indians and the metaphysical notion of imperfect matter, which works as the inherent presupposition of the imperial ideology. This presupposition cannot be refuted by a neutral examination of reality, because it serves to give a sense of direction to reality itself.

It is possible to draw some consequences from Sepúlveda's notion of Indians as servile. First, attributing inhumanity and imperfection to the Indians means attributing a certain privation identified with their nonhumanity as a defining feature. Their lack of perfection is a nonidentifiable feature irreducible to a series of empirical qualities. The undefined lack that marks the Amerindian as inferior has no empirical, positive demonstration, because it is the objectivation of an absence cast upon the matter in question by the always already presupposed end form. It could be pointed out that Sepúlveda gives long lists of virtues lacked by the Indians, and yet Sepúlveda answers every attempt to prove the humanity of the Indians by saying these show that the Indians are not animals but they do not prove that Indians are humans or political animals. Having houses and some rational way of life such as practicing commerce merely proves that they are not bears or monkeys totally lacking in reason. In sum, being inhuman is not a property of animals. Animals cannot be inhuman, because only humans can lose their humanity.

Second, their inhumanity derives from metaphysical instrumentalism, which, with its two central compotents teleology and hylomorphism, appeals to a relation of subordination of imperfect matter to a final form. The relation of natural subordination results from the transposition of artificial subordination to the natural order. In Aquinas's worlds, "now God is the Creator of all things by His wisdom, and He stands in the same relation to them as a craftsman does to the products of his art" (*Political Writings*, 102) Therefore, the Spaniards stand in the same relation to the

Indians as the Divine Artifice stands in relation to their creation. The Indians are a malleable and imperfect matter that has to be improved by the imposition of new cultural forms. The consequence is that a culturally inferior being must become subordinated to a culturally superior one, "for subordinate arts are directed to the end of a superior art, as the art of horsemanship to the end of military art" (Aquinas, *Commentary on Aristotle's* Metaphysics, 14).

Third, the attribution of inhumanity/imperfection to the Indians is not only metaphysical but also ideological and intrinsically fetishistic. The fetishism here consists of attributing the characteristics of the social network to one particular element. For instance, a particular thing is elevated to the character of the universal, such as in the case analyzed above where, according to Sepúlveda, there is always an exceptional particular that becomes universal by imposing its unity over the other particulars. The illusion of the operation consists in making believe that the universal idea or form was an exceptional thing that existed in a transcendent plane. The illusion of the inhumanity of the Indians is the illusion that they are inherently inferior, when their inferiority is the result of a tautological and performative operation that retroactively posits their inferiority. The Indian is the exact correlate of the superior exceptional element. Just as there is always the exceptional one that imposes an order, there is always another exceptionally inhuman element that needs to be tamed and corrected. The servile barbarian condenses all the violence intrinsic to the contradictory civilizing program into what appears to be some inhuman excess external to civilization. If Vitoria laid the foundations of imperial ideology, then Sepúlveda is *the* ideologist of modern colonialism: all the contradictions of civilization are condensed and projected into the figure of the Indian, which is both inhuman yet not animal.

IRON IS MORE VALUABLE THAN SILVER

Sepúlveda thinks of the relation between those who command and those who obey as a kind of exchange done in benefit of the ruled people. Sepúlveda argues that the public good includes Spanish subjects as well as the "well-being and salvation of the barbarians" (*Demócrates*, 78). It is necessary to compensate collateral damages incurred in war with "much greater benefits" (78). According to Sepúlveda, the sum of the goods produced by the conquest are greater than the incidental evils, "since if we were to make a detailed accounting of the evils and benefits this war occasions with regard to the barbarians, the evils would doubtlessly disappear completely in the face of the number and weight of the benefits" (78). While Sepúlveda justified the amputation of a limb in order to preserve the health of the body, he here demonstrates the benefits to the totality as a result of the sacrificed part.

Among the lesser sum of evils, Sepúlveda includes the few Amerindian princes that were "to a great extent despoiled of moveable goods, like gold and silver and metals that among them have little value" (78). The Amerindians not only did not use precious metals as currency, but furthermore "in compensation they received iron from the Spaniards which has more utility in life for an infinite number of reasons" (78). While gold and silver are inessential to the level of civilization attained by a nation, this is not so for materials with more practical applications. Given that precious metals "constitute the material of money, it is not their nature that gives them value but law and the voluntary agreement of men, and therefore using the same logic, iron and bronze would be preferred to gold and silver" (78). Sepúlveda privileges the use value of iron and metallurgy over the exchange value of gold, which is based on the mutual accord of all nations.

In his *Diálogo del hierro de sus grandezas y excelencias* (1574), Nicolás Monardes, an early scientist that worked in the Spanish Casa de Contratación and lived in Spanish America, states that iron constituted a true metal precisely because of its practical, mechanical, and even medicinal applications.[13] Gold was judged merely an object of social value and opinion. As a fundamental condition for the development of technique, without metal there could be no arts or professions. Lacking iron, Amerindian nations were deemed lacking in the necessary techniques for forming a civilized society. Of course, the great architectural accomplishments of Amerindian societies, so often the object of Spanish chroniclers' awe, were constructed from stone without the aid of iron. Once more, Sepúlveda echoes Vitoria's eighth possible title, which appeals to the Indians' lack of metallurgy, a sign of their incapacity to govern themselves. This makes visible Sepúlveda's instrumentalist presuppositions, which ultimately imply that civilization is attached to the capacity to invent useful things.

According to Sepúlveda, iron has a particular value for human life because, without it, "many necessary instruments will disappear" (78). Citing Pliny, Sepúlveda declares, "with iron we plow the earth, plant bushes, fruit trees, force vines to rejuvenate themselves every year by pruning them; with iron we build houses and split stones, and it is useful to us in many other applications in life" (78). The purpose of this argument is to show that "with iron alone" the Spaniards have sufficiently compensated the Amerindians for the other ill-gotten precious metals and that "the benefit is restored to them with interest" (78). In addition to iron, Sepúlveda enumerates an open list of goods unknown to the Amerindian, including wheat, barley, cereals, legumes, horses, mules, asses, oxen, and numerous plants and trees (78). Sepúlveda concludes that "the benefit of each of these particular things is greater than the benefit the barbarians derived from gold and silver" (79). From this Sepúlveda concludes, "since the Spanish monarchs have invented so many highly useful things, both necessary and unknown in those regions, with what gifts, with what favors, with what honors could the barbarians compensate them in equal terms for so many and such immense benefits?" (79). Indeed, any impediment to Spanish dominion would represent an obstacle to the humanization of the Amerindian (79). Sepúlveda then moves his instrumentalist presuppositions to the next level of the kings of Spain, who have the right to tribute based on the fact that they benefited the inhabitants of the New World by providing them with humanity and civilization.

Unlike Vitoria, Sepúlveda thinks that Indians were not in perfect possession of their lands and possessions because they were all inherently servile. He describes the Indians as barbarian communities "in which those who are servants by nature are controlled by other, slightly more intelligent servants, who with their own impetus could, each one in his own realm, expel foreigners and forbid them to exploit gold and silver mines along with pearl fishing" (83). He employs the same arguments from Vitoria's first valid title, the defense of trade and communication by military force, grounded in a natural law that prohibits the barbarians from impeding trade with foreigners or the extraction of precious metals in places that do not belong to anybody. The jurist theologians argue that, while objects have an owner who exercises dominion (*dominium*), a territory belongs to a republic or prince, "although for certain uses they are held in common" (83). The legality of the transfer of dominion to Spanish princes is not derived from an absence of dominion but, rather, from the fact that "those mortals who occupied them were completely outside the dominion of Christians and civilized nations" (83). Sepúlveda adds that the papal bull issued by "the Pontifex Maximus and the Vicar of Christ, who has the power and the obligation to seek out opportunities for making enmities between Christian princes disappear and to bring to the fore for those he deems proper the mission of propagating the Christian religion, whenever a rational and juridical opportunity presents itself" (83). While here Sepúlveda demonstrates an affinity with the assertion of the Amerindian princes' dominion, he nevertheless asserts that the inferior nature of the Amerindian, who lacked the essentials of a republic guided by natural law, was sufficient grounds for the transfer of sovereignty to the Spanish monarch. In short, while the Amerindians were natural lords before Spanish discovery and conquest, after receiving the "light" of Spanish Christian civilization, it was only natural that they become subject to superior lords.

To conclude, a close reading of Sepúlveda's *Demócrates Segundo* shows not only how he falls squarely within metaphysical instrumentalism but also how he represents the most radical ideological consequence of such an ontological frame. By framing the problem of the barbarians' servitude with the principle of the natural subordination of matter to form and means to an end, Sepúlveda invents a new category, that of the inhuman/imperfect barbarian who needs to be improved by civilization. The novelty of this category resides in how he expands the principle of natural subordination from within the community to a relation between communities. Sepúlveda confronts Vitoria and the Salamanca School with the disavowed consequences of their thinking by showing that, once natural subordination is introduced, it ends up necessarily justifying indigenous subjection since it is not possible for the inferior to reject the superior just as it is not possible for matter to reject the form imposed by the Divine Artisan. Metaphysical instrumentalism inevitably leads to identifying the Amerindian as a slave by nature, not as a result of verifiable empirical characteristics but, rather, as a nonempirical absence—of a pure lack, privation, or indetermination—that emerges as soon as we compare the imperfect part with the perfect whole to which it must be subordinated. According to this imperial ideology, the Amerindian becomes synonymous with this lack at the center of a system of traits (sloth, alcoholism, lust, sodomy, polygamy, etc.) that are the negative shadow of imperial reason, that is, the unreason upon which this empire constructs itself.

Up to this point I have been analyzing Sepúlveda to make explicit the instrumentalist presuppositions behind his theoretical decisions. I will now analyze Las Casas's attacks on Sepúlveda in order to show how, although both thinkers depart from the same theoretical point of departure (which is Aristotle's and Aquinas's teleological hylomorphism), they inherit the potential contradictions in Vitoria's treatment of the problem of the legitimacy of the Spanish Empire. With this purpose in mind, I now move to the analysis of the problem of barbarism in Las Casas.

LAS CASAS ON BARBARISM

In his *Apología*, Las Casas did not reproduce the arguments from *Demócrates Segundo*, which he confesses he did not read.[14] Nonetheless, Las Casas opposes his own quadripartite theory of "bárbaros" to Sepúlveda's identification of the Amerindian with the slave by nature. Las Casas maintains that to the first category of "bárbaros" belong those who behave cruelly and inhumanly as if they were beasts. Then, the second category is that of barbarians in a relative sense since they are defined as barbarians because they do not speak the language of another. The third category are the barbarians "in the strict sense," which I will analyze at length below. The fourth category of barbarians is those who are not Christian. I will focus largely on the third kind of barbarians because this category is based on the philosophical presupposition of teleology. This third class of barbarians is the exception to the principle of reason according to which everything has a purpose and according to which nature rarely fails. Teleology presupposes instrumentalism, and part of the characterization of this kind of barbarian is that these barbarians lack the proper use of artifacts. Therefore, the philosophical presuppositions behind this kind of barbarian go back to Aquinas's metaphysical instrumentalism, a perspective shared by both Sepúlveda and Las Casas. Las Casas dismantles this argument by pushing it to its end and producing an impasse internal to imperial ideology.

Las Casas does not doubt the existence of beings who are "bereft of reason" and who live in a state of barbarism: "Taking this term in its proper and strict sense, which is that of those men who, out of impious and terrible instincts, or because of the terrible conditions of the region they inhabit, are cruel, ferocious, squalid, and stupid, bereft of reason, which men are not governed by laws or constitutions, who do not cultivate friendship, who have not constituted a republic or city in an organized way; who moreover lack a prince, laws, and institutions"

(127). These barbarians are barbarians in the "proper and strict sense." This category univocally designates its own object independently of its historical circumstances. Las Casas explains that this precise and rigid definition is, succinctly stated, alien to reason. Las Casas here employs the *vía negativa* in order to explain what it means to be exterior to reason and thereby foreign to the human species. The barbarians in this category lack basic institutions such as monogamy and commerce, "they neither buy nor sell, they do not rent, they do not compose societies, they are ignorant of juridical institutions of deposit, rent or bailment" (127–28). They have no knowledge of natural rights or the laws of nations, nor do they enjoy the unity of political life given that they "live dissipated and scattered, inhabiting forests and woods, contenting themselves only with their women, like animals, both tame but also wild" (128).

These barbarians "in the simple and proper sense" are "lacking in reason" and any "moral orientation corresponding to men and all things that are proper to men by the law of custom" (128). According to Las Casas, these are the category of barbarians to which Aristotle referred when he spoke of "natural slaves, for they lack a natural principate and political institutions, for there is no order among them and they are subject to no one, nor do they have a prince" (128). They lack princes, given that "nobody among them is distinguished by prudence," nor do they possess "such skill and rectitude in judgment," because they "lack laws that inspire fear or guide all their actions" (128). Thus nobody can promote "good works," "foment virtue," or "rein in vice through punishment" (128). To this point Las Casas shares common ground with both Vitoria and Sepúlveda in his belief that the "bárbaros" lack laws, *both those of princes and of principles*. The inherent quality of this class of barbarians is their exteriority in relation to right reason, as evidenced in their lack of civilized institutions.

The barbarian is thus the positive materialization of an absence, that is, the sum of negative characteristics viewed from the perspective of perfect rule. Just as in Sepúlveda, these beings are *like* animals without being reduced to animality *strictu sensu*; they are inhuman beings who live outside civilization. The ideological force of this construction lies in the fact that these barbarians represent the exception to civilization. The descriptive force of this category does not depend on positive, observable empirical data but, rather, derives from a positional, performative act pronounced from the center of the empire. These barbarians are thus the negative shadow that outlines the profile of imperial metaphysics: they assume a commonality behind a series of negative properties that serve to bridge the abyss between the imperial reason and empirical reality.

Despite this affinity with Sepúlveda's interpretation of Aristotle, Las Casas hastens to add that the Amerindian does not necessarily correspond to this category of barbarian. In fact, the Dominican friar maintains that barbarians of this class, "or savage peoples, are very rare in any part of the world, and are few in number when compared to the whole of humanity, just as Aristotle affirmed, in the same manner that those men endowed with heroic virtue, who we consider to be heroes or demigods, are likewise very rare" (128). Las Casas again cites Aristotle who asserts "Of all the things that exist, nature always produces the best"; "Nature takes special care to produce the most noble of all things"; "Nature produces the best possible of all things" (128). Las Casas employs an Aristotelian teleological frame to demonstrate that this sort of barbarian must be the exception and not the rule. Las Casas thinks that there cannot be entire populations of people that are so inhuman or cruel, because he takes literally Aristotle's definition of humanity as a social animal and the principle of sufficient reason according to which all ends tend to actualize themselves.

In a world in which the natural order reigns supreme, all things tend toward perfection, and the final cause is eventually realized. There is a continuity between Sepúlveda's "principal dogma" and the principle of natural perfection in Las Casas whereby "those things that are in the realm of nature are determined by a cause for which they exist, which they always, or almost always, realize" (129). Las Casas further states that nature usually "engenders and produces that which is best and perfect," which implies that it is aberrant that natural causes should cease to produce the effects that are proper to their nature (129). Some examples of these aberrations would be men who are born lame, or blind, or missing extremities, or "with the soles of their feet pointing upward" (129). The general rule, Las Casas continues, "is that fire engenders fire, the olive tree olive trees, and men other men" (129). As entities participate analogically in being, a species always reproduces its likeness. Las Casas appeals to the ultimate, foundational principles of nature, which are predetermined and which impose the exception as the limit to its universality. Each member of the species "perfectly" engenders its likeness, and thus "every man by nature understands and accepts these first principles" (129). The unifying cause by which all things are perfectly engendered by principles that guide them toward their natural end is found in the fact "that all works of nature are by the hand of the Great Intelligence that is God," as affirmed by Thomas Aquinas (129). Divine Providence and generosity cause nature to "produce" things that are "better and more perfect" in greater number than those things that are "imperfect and inferior" (129). Nature is thereby understood as a process of reproduction wherein the good, the finished, the perfect and sufficient, far outnumber the exceptional cases of incompleteness.

Las Casas asserts that "savage, imperfect and inferior men" are aberrations of nature, or "monsters within the rational order" (129). Averroës had likewise compared any "error" in reason or falsity in opinion with monstrous deformities in corporeal matter (129). Monsters are thus accidents of nature, mutilated beings deprived of their inherent end as perfecting form. The rational nature of men is determined by Divine Providence, by which human beings are differentiated from other creatures not only as a species "but also as each individual is determined" (129). Therefore, "it is clear that it is impossible to find such monstrosity or natural sins within a rational creature, that is, to find the absence of a common human reason, except in very rare cases" (129). The rarity of this sort of barbarian shows how imperial ideology depends on a basic tautology that reveals the performativity of colonial reason: the exceptionality of this absence of natural reason is a function of the limit drawn by reason itself. If these are universal, necessary, and self-evident principles, then the aberration or deviation must remain singular as not to be confused with the universal. There can be exceptions to natural law as long as they remain exceptions. Monstrosities are accidents and aberrations, that is, effects of a *privation* characteristic of indeterminate matter that has not been corrected by a superior end form.

Las Casas further writes that the "good and omnipotent" God created all things in the universe "for the utility of human beings" and providentially directs all human actions, "enlightening the understanding of each individual and inclining him towards virtue with the necessary spiritual capacities" (129). Rational nature derives its "force exclusively from God" (129). Thus, nature produces "perfect" men endowed with intelligence and rarely gives birth to barbarous or imperfect beings (129). God endowed men with understanding because they were chosen to be superior to the rest of the animals in his creation (129). Understanding is then the specific mark of difference between animals and humans. The reason that true barbarians are so rare is that rational nature and intelligence are given to men directly from God. Understanding as a characteristic common to all men is what holds together the hierarchical order of nature decreed by God.

According to Las Casas, Sepúlveda's assertion that there exists a multitude of "beastly, uncivilized, and stupid" barbarians in the Americas that exceeds the rest of humanity in number is an impious affirmation in contradiction with the divine order (129). Aquinas argues that good is always given in the same proportion to all communities within the natural order; consequently, the absence of good is observed only in very exceptional cases (130).[15] Human beings endowed with knowledge for pursuing a properly human life are always in the majority with respect to those who lack such knowledge (130).

Las Casas concludes therefore that those barbarians who lack the benefit of understanding such that they cannot be instructed and converted to the Catholic faith are "extremely rare in number" (130). Although Las Casas uses the logic of probability, this does not mean that his criteria are purely quantitative. On the contrary, he employs a qualitative category. The number of human beings inhabiting an entire continent represents a significant percentage of the species, which is by definition a being whose essence, as willed by God, is in its superiority of understanding above the rest of the animal kingdom. To believe that there is an immense number of barbarians lacking in the "natural light common to all peoples" is tantamount to believing that God failed in the execution of his will by producing a "great detraction to the perfection of the universe, which would be a great embarrassment and unthinkable for any Christian" (130). Such a detraction of perfection amounts to a fissure, an absence, in the metaphysical foundation of the totality of the universe.

Las Casas insists that he had already expanded on this topic in his *De único modo*, in which he concluded that "it would be impossible to find anywhere in the world an entire race, nation, or province stupid or insensitive and that, as a general rule, would lack sufficient reason or natural ability to rule or govern themselves" (131). According to Aristotle, this species of barbarian should be governed by the Greeks, given that their "stupidity and wildness of intelligence" made them incapable of self-government and dependent on those who were educated in "political life" (131). Furthermore, given the "robustness" of these barbarians, they must also contribute to the city with their labor (131). These barbarians could be "seized" or "hunted" like beasts in order to be brought within the fold of the "proper rule of life" (131). Nevertheless, the power to thus coerce the barbarian is limited to "princes and heads of state" (131).

Las Casas distances himself from this Aristotelian argument by claiming that these barbarians cannot be coerced but, rather, must be "peacefully attracted and kindly won over to adopting better customs" (132). Citing Paul's Letter to the Romans, in which the disciple exhorts the strong to lift up the weak instead of seeking to profit from their weakness, Las Casas asserts that the divine mandate to love one's neighbor applies even to barbarians (132). The Pauline doctrine, according to Las Casas, requires the Christian to love the barbarians, even if they are "dull and stupid," given that they are made in the image of God and participate in divine grace (132). Las Casas, just as Vitoria before him, asserts that even the pagan Greek philosopher who affirmed the prerogative to gather barbarians like beasts did not authorize that "barbarians should be killed or subjected to harsh, hard, cruel, and rigid labor, like beasts of burden, and to that end may be sought out and captured by those who are wiser" (132). Las Casas thus abandons Aristotle categorically: "let us set aside Aristotle, since from Christ, who is eternal truth, we have the following commandment: 'love thy neighbor as thyself'" (132). Las Casas invokes St. Paul's universalism in opposition to the Aristotelian category of slave by nature.

Moreover, Las Casas argues, Christ does not look for riches or power: "He, who is the only immortal King of kings, looks for neither idleness nor pleasure of human kind for which he offered his life nailed to the tree of the cross. Whoever wishes to have numerous subjects so he can (following the doctrines of Aristotle) behave toward them like a cruel butcher and oppress them with slavery and in that way enrich himself, is a tyrant, not a Christian; a son of satan and not of God, a thief, not a shepherd, inspired by the spirit of the devil, not the heavenly spirit" (133). Las Casas insists that the apostolic enterprise should be based on teaching Christian doctrine "gently, moderately, smoothly and humanely" to the Amerindian catechumen (133). The brutality suffered by these "extremely innocent peoples, well-disposed to abstain from evil and to receiving the word of God," had condemned Spanish perpetrators as "sons of the devil and the most cruel of all thieves" (133). He ends up affirming that

Aristotle thought it was possible to hunt barbarians because he lacked the necessary knowledge of true faith (133).

Las Casas concludes that the barbarians who appear in the first book of Aristotle's *Politics* are "barbarians in the strictest sense" who lack "sufficient reason to govern themselves" and live "without law" (134). He adds that, in the third book, Aristotle admitted there are barbarians who possess a "legitimate, just, and natural government even if they lack the art of writing" (134). They do not lack "prudence and ability to rule or govern themselves," and they possess kingdoms and communities that are governed "prudently with laws and customary institutions" (134). Therefore, "their government is legitimate and natural, even if it somewhat resembles tyranny" (134). The princes of these kingdoms are bestowed with the light of reason, and "they do not lack justice and peace" in their realms (134). Las Casas asserts that in the Indies there are multitudes who live in great cities "in conformity with a political and social regime" and with "organization of commerce, trade and rent" and "contracts inherent to the laws of nation" (134). Las Casas concludes that Sepúlveda had "viciously and in blameworthy fashion" misinterpreted not only Aristotelian doctrine but also the reality of Amerindian polities (134). The governments of these nations, according to Las Casas, was in many ways superior to Spanish government and "could astonish the wise men of Athens" (135).

LAS CASAS ARGUES AGAINST SEPÚLVEDA'S NATURAL SUBORDINATION

Las Casas goes on to argue that lack of education in liberal arts does not justify the domination of another people (135). Turning his gaze back to Spanish history, he reminds Sepúlveda that Trogo Pompeyo considered the inhabitants of Hispania to be "fierce and barbarous" until Augustus Caesar bestowed the form of Roman province upon them and required them to accept "a more civilized way of life" (135). Las Casas asks Sepúlveda if the war waged by the Romans upon Hispania was just, which would imply that resistance to Roman domination was unjust (135). Las Casas, who calls the Spaniards "thieves and torturers" of the Amerindian, asks if Sepúlveda would have approved of the Romans enslaving the "fierce and barbarous people" of Hispania and forcing them to work in "gold and silver mines to extract and purify the metals" (135).

Referring to the distant ancestors of his adversary's *patria*, Las Casas asks: "would you, Sepúlveda, allow Saint James to evangelize your fellow Cordovans in such a way?" (135). From here, Las Casas then precedes to unravel the geopolitical application of the principle of subordination in relation to the justification of mining.

Las Casas exposes Sepúlvedas's primary motivation for justifying natural slavery, which was, in his view, a pretext for exploiting Amerindian labor in gold and silver mines. Las Casas does not simply contrast the category of barbarian to the empirical facts of Amerindian societies, for he furthermore appeals to the historically contingent application of the category by appealing to Roman imperialism in the Iberian Peninsula. While Las Casas maintains the validity of the principle of the subordination of the imperfect to the perfect, he likewise asserts that the Spaniards are by no means perfect themselves, as they are descendants of ancestors considered barbarian by the Romans. This is not simply a moral condemnation or rhetorical strategy. By applying the argument of just war as subordination of the imperfect to the perfect in his reading of Spanish history, Las Casas reveals that behind the divine mandate (the "natural and divine decision") that grounds the principle of just war there is nothing other than an arbitrary, human decision made by "thieves," "torturers," and a "fierce and barbarous people."

Las Casas maintains that the Amerindians have the potential to cultivate the liberal arts and are inclined to embrace Christianity and renounce their sins. He says, "this is taught to us by experience" (135). These nations likewise possess an aptitude for mechanical arts and create beautiful works of architecture and clothing (135). Las Casas accuses Sepúlveda of disdaining manual labor and ignoring the fact that such work is an "operative act of understanding which is usually defined as proper reasoning of real things, derived from the acts of reason, by means of which the artist acts in an orderly and fluid way and without error in the proper act of reason according to the authority of the philosopher" (135). While Sepúlveda employs the analogy of political prudence and mechanical arts to justify domination, he nevertheless repudiates any sign of Amerindian aptitude for such labor. Thus, where Sepúlveda uses the example of medical technique to illustrate the necessity of violent correction, Las Casas adduces Amerindian technique as an example of an "operative act" of right reason through applied arts.

Las Casas furthermore claims that the Amerindian is capable of learning liberal arts such as grammar, logic, and music, which he explains in the second part of his *Apología* (136). He opposes the "lies" of Sepúlveda's informants to his own direct experience with Amerindians (136). Las Casas casts doubt upon the whole corpus of empirical evidence adduced by Sepúlveda to prove the Amerindian's inferior intellectual capacity, partiularly the work of Fernández de Oviedo, whom he accuses of keeping Amerindians "subjected to slavery like animals" (137). Las Casas further maintains that, even if one admitted to the Amerindian's lack of intelligence or "industriousness," this would in no way justify waging war against him or dispossessing him of his sovereignty and wealth (137). While natural law dictates that one is required to accept Christianity as long as it is preached by rational and legal means, it is nevertheless prohibited to coerce another into accepting the faith (137).

Accepting natural law as complying with eternal law implies the recognition that "there exist no motives for one people, under the pretext of culture, to try to dominate another or to destroy foreign kingdoms" (138). Employing the reductio ad absurdum, Las Casas maintains that if one accepts that a superior nation can subjugate an inferior with natural law, then one could likewise use the same argument to defend the Turks and the Moors, given that "some think that such Turks or Moors are superior to us in terms of their political constitution" (138). Such an argument, Las Casas concludes, would be tantamount to accepting "a true anarchy in all things both divine and human" (138).

Thus, Las Casas demonstrates the "true anarchy" at the foundation of Sepúlveda's geopolitical application of the principle of subordination, which is based on an arbitrary and contingent decision outside any law. In order to justify a fundamental transgression of divine law it would be enough for any enemy of the faith to consider himself more "perfect" than the Spaniards. This divine law mandates unconditional love for one's neighbor, the end of which is universal good. Las Casas concludes that there is no motive or reason that justifies the destruction or domination of one nation by another. The ability to limit the geopolitical consequences of the principle of subordination largely depends on the ability to demonstrate the limits of the subordination of matter to form. Las Casas demonstrates that the Amerindian does not represent an intrusion or violent exception to the natural order, but rather, the arbitrary and violent excesses are attributable to the Spaniards themselves. Thus, the limits placed on the geopolitical application of the principle of subordination are twofold: on the one hand, the conquest of the Americas is "anarchic"; on the other hand, the Amerindian possesses political prudence and the capacity for self-determination.

LAS CASAS ATTACKS SEPÚLVEDA'S APPLICATION OF HYLOMORPHISM

Las Casas's assault on Sepúlveda's doctrine focuses on the latter's use of hylomorphism to justify natural slavery (139). Las Casas summarizes the essence of Sepúlveda's argument: "that the more imperfect things should cede, by nature, upon encountering more perfect things, just as matter cedes before form, the body before the soul, and sentiment before reason, which is certainly something that I do not deny" (139). As both Las Casas and Sepúlveda share the same fundamental metaphysical principles, the Dominican friar Las Casas does not refute the law by which imperfect matter must submit to perfecting form. Nevertheless, he limits the scope of Sepúlveda's reasoning by declaring, "this is true when the two things come together by nature *in actu primo*, as when matter and form concur in a single thing" (139). An example of this, Las Casas states, is when the body and soul "combine and form an animal or when sense and reason exist in the same subject" (139). Therefore, if the perfecting element and the imperfect matter are separate entities belonging to distinct subjects, then the imperfect cannot submit to the perfect because these are not united *in actu primo* (139). This absence of a unity *in actu primo*, or original contiguity, attacks Sepúlveda's assumption that whenever there is a multiplicity of particulars, one particular imposes itself over the rest, elevating itself to the status of the universal. Las Casas argues that each particular substance and subject has *its own* originary form. An example of an original unity is the "substantial form" given to indistinct matter, as pure passive potentiality, by being. This substantial form is the original unity of form and matter, just as prime matter would be an original potentiality. Being itself emerges from this original unity of form and matter. The first act is this bestowing of unity and being upon an entity, therefore form and matter (along with privation, which is elided by both Las Casas and Sepúlveda) are the principles of this substantial unity. If the entities in question are societies or any geopolitical organization separate and distinct from others, then form cannot be imposed from without because each entity proceeds from an original unity.

According to this logic, imposing the form of one perfect community over another equally perfect community would be just as absurd as imposing the soul of one human being upon another. Briefly put, Las Casas attacks the logic of exception as developed by Sepúlveda when the latter says "observe that one of them, that is, the most important has dominion over the others, as the philosophers teach, and those others are subordinated to it" (*Demócrates*, 20). Las Casas concludes: "If cultivated [perfect] or uncultivated [imperfect] people live constituting the same political body or *under the same prince or governor*, in such a case those lacking culture should subordinate themselves to the judgment of the more cultured governors of the republic, that is, to their king, their laws or governors, and those that refuse to do so could properly be forced to do so and punished according to natural law" (*Apología*, 139; my emphasis).

No person or free nation is required to submit to another, even if the latter exceeds in "prudence" (139). Las Casas declares, "no free people can be forced to submit themselves to another, more cultured people, even if that submission brings with it great advantages for that first people" (139). Las Casas argues that, when Aristotle "adduces the argument that matter gives way to form, he means only that nature produced some men apt, out of their innate prudence, to be able to govern others who are not endowed with such great intellectual capacity" (139.) Thus, Las Casas maintains that Aristotelian doctrine "teaches that such persons, more prudent, should be drawn to manage the tiller of the republic for the conservation and usefulness of the republic" (139). Las Casas is repeating Vitoria's distinction between slavery as a civil institution and the natural subordination of inferior people to wise people within a given community. Las Casas adds, "to such persons it is necessary that the rest be subordinated as matter is to form, and the soul to the body" (139). Thus Sepúlveda's argument is based on a false syllogism: "Sepúlveda's last argument, that all men may be forced, even against their will to do that which is useful if considered in absolute terms, turns out to be an extremely false proposition" (139).

Las Casas denies that the hylomorphic dogma can be a pretext to declare war. He accuses Sepúlveda of giving a geopolitical leap, transgressing the limits of the subordination of the imperfect to the perfect. He admits the principle of the natural subordination of the imperfect to the perfect but he thinks it applies only within a perfect community. While Las Casas insists that the principle only works within one community, Sepúlveda believes it also works between communities. In my understanding, this impasse results from the internal contradictions of metaphysical instrumentalism within the imperial ideology of Vitoria.[16] This is to say that this discussion is far from being a mere dispute over abstractions since central to the discussion is the nature and treatment of the Amerindians.

In both *De thesauris* (*On the Tomb Treasures of Peru*) (1564) and in the *Tratado de las doce dudas* (*The Treatise of Twelve Doubts*) (1564), Las Casas radicalized his attack on the encomienda system. As Adorno states, "there he expounded the same principles of secular dominions and the rights of all peoples to sovereignty in their own lands on the basis of canon law principles he had espoused most of his activist life" (Adorno, 85). In 1556, Adorno continues, the Crown's economic crisis led Philip II to accept the Peruvian encomenderos' offer to pay for the rights to use land and indigenous labor (85). Las Casas attempted to dissuade the king by proposing a counteroffer that consisted of restoring property to the *kurakas* (native lords) of Peru who would in turn pay the regular tribute owed to the sovereign by his subjects (85). In the two years it took the Council of Indies to create a commission to consider the encomenderos' proposal, Las Casas organized the *kurakas* and in 1560 proposed the sum of one thousand ducados on top of the encomenderos' offer for the purchase of their freedom (85). The *kurakas* would additionally pay one-third of all the treasure discovered in hidden tombs or from mining (86). The discovery of treasures hidden in the tombs of Peru reignited the conflicts over the legitimacy of the conquest and the questions regarding the right of ownership of these riches. In *De thesauris*, Las Casas unequivocally maintained that the inhabitants of the New World were legitimate sovereign subjects, and given that the pope had only granted Spain the right to evangelize, the only just solution was the restoration of sovereignty to the Amerindians of Peru (87). The obvious question was what would become of the Spanish Crown's role in America. Adorno summarizes Las Casas's response: "Philip would serve in a merely symbolic role as 'universal lord,' being acknowledged for this universal hegemony with an annual token of payment from the Inca of a single jewel, such as the king of Tunis gave each year to the emperor" (88). As Adorno asserts, Las Casas's position became ever more radical as he realized the impossibility of defending the interests of both the Amerindians and the Spanish Crown (88).

In *De thesauris*, Las Casas asserts that the Incas were the legitimate lords of Peru, and given that the consent of the ruled is the metaphysical and legal foundation of sovereignty, the abrogation of their rights was an act of tyranny and transgression of natural law. While both Las Casas and Vitoria share the same metaphysics grounded in the subordination of matter to form and means to ends, they follow divergent paths at the moment of taking a theoretical stand with respect to efficient cause and the principle of privation. In *De thesauris,* Las Casas sustains that the Crown had right to the Indies but not over the Indies:

> As the peoples and inhabitants of that world of the Indies, with its kings and princes, neither freely consent to the institution regarding them issued by the Papal Bull, nor accept it as juridically valid, nor surrender possession to our illustrious kings of Spain, who have nothing more than a title, a cause for acquiring the supreme principality of that world and a "right to the thing," that is, a right to the kingdoms and supremacy or universal dominion over them, which originates in the title; however, they lack the "right over the thing," that is, over the kingdoms. (Las Casas, *Los tesoros,* 279)

Las Casas asserts that free consent of the Indians to subject themselves to the Crown is the only valid title for the Spanish right over them: "Therefore, as the aforementioned peoples and inhabitants, along with their kings, do not freely consent, etc., our kings will only have a right to those kingdoms, not a right over those kingdoms, that is, they have no power to exercise jurisdiction or to act as supreme princes" (281). Las Casas offers several proofs borrowed from both the legal tradition and Thomist causality. First he argues that:

> Because our kings lack the ground and efficient cause, that is, the consent of the peoples and their kings to exercise every class of jurisdiction, legal authority, or power, as well as to acquire the principality of that world, and thus Baldus [de Ubaldis] states that consent is required as a natural ground; it is meant to come into possession of a post or magistracy fairly and legally; and this consent should be simple, that is, without the least shadow of violence or fear, of the sort that, if violence or the outcry of the enraged people restricts the electors, it is said that such election, by the force of law itself, is null, since liberty is not considered to exist where there is any kind of restriction. (281)

While Vitoria thinks that God was the efficient cause of the civil power, Las Casas thinks that the efficient cause was the people.[17] Las Casas shifts the efficient causality of sovereignty from God (as maintained by Vitoria) to the Amerindian peoples. The continual radicalization of Las Casas's thought toward the affirmation of Amerindian agency and autonomy is without precedent in Western political thought. While Vitoria thinks that the commonwealth is a material cause that cannot administer itself without the mediation of magistrates (form), Las Casas thinks that the Amerindian communities were perfect communities and their commonwealth was the efficient cause of civil power. The efficient cause is nothing but the political agency itself, since it commands and dictates the law. Only the free consent of the efficient cause can submit itself to a papal bull, which remains merely potential until it is actualized by the acceptance of the efficient cause. Sepúlveda's imperfect matter is Las Casas's efficient cause. Las Casas's geopolitical thesis performs an inversion of Vitoria's political physics on a fundamental level.

Another important reversal takes place at the level of the final cause: "This is proven in the second place because then the final cause would cease to exercise its temporary jurisdiction over them. Then the final cause through which the Vicar of Christ can intervene in temporary affairs, especially regarding that which concerns those peoples, is Faith's prediction of them and their conversion, which one must ordain and refer to as this business is carried out" (*Los tesoros*, 283). The final cause is the well-being and conversion of the indigenous peoples, and there cannot be such a thing without their free consent: "However, as the conversion of those nations and peoples and even their temporary utility were the final cause from which such an institution originates, promotion, donation, or concession (call it what you will) as demonstrated without consent, does not endow right over those realms, which is the exercise of legal authority according to which the aforementioned is derived; consent is, therefore, the law and the principal efficient cause of the principality of our kings" (285). The consequence of this apparently abstract theoretical move is the radical questioning of the colonial enterprise: mining or the appropriation of indigenous riches is no longer legal or compatible with the transcendent ends of the empire:

> All the gold, silver, precious minerals, pearls, jewels, gems, and all other metals and precious objects beneath the earth or the water or the surface

that the Spanish have taken, from the time that they discovered it until today—except those that the Indians, from the beginning of the aforementioned discovery until they were mistreated by the Spanish, conceded freely or as gifts or in some places through voluntary exchange—were all stolen, unjustly usurped, and perversely snatched away, and consequently, the Spanish commited a theft or robbery that was and remains subject to restitution. (357)

The precious metals under the ground no longer belong to the Spaniards because they have been simply taken away from the Indians. The veins of metals that grow inside the mines are objects of the Indians' common good:

> However, our people, without permission from those kings or peoples, even against their will, violently robbing, killing, cutting throats, devastating, with mortalities and fires, despoiled the proper owners, which is to say, those people, of their dominions, jurisdictions, dignity, lands, forests, mountains, fields, rivers, seas, mines or veins of gold and silver, and all the other things that come from there and which nature has engendered for common necessity, the sufficient adornment and the utility of the common good of each kingdom or community, and the solace of the Indians. As we have seen, they usurped and appropriated infinite quantities of gold, silver, pearls, gems, and other precious objects that belonged to them; therefore, all gold, silver, and other precious objects that our people took during that time were stolen, unjustly usurped, and viciously appropriated; consequently, they committed theft, robbery, and acts of the worst tyranny. (360–61)

Las Casas clearly operates within the frame of Aquinas's metaphysics by employing the notions of efficient and final cause, but he draws anti--imperial conclusions from it by concluding that it is imperative to restitute to the Indians their sovereignty and riches.

In the first chapter I concluded that Vitoria's metaphysical instrumentalism was the ideology of the Spanish Empire because it disavowed its instrumentalist consequences. Now, I argue that Las Casas's anti-imperial politics consists of taking the metaphysical premises of natural law more seriously than Sepúlveda and Vitoria did themselves. Las Casas takes imperial ideology seriously by addressing its implicit presuppositions and consequences. On the one hand, the explicit imperial ideology is sustained by the silent metaphysical presuppositions. Vitoria and Sepúlveda's explicit ideology of imperial dominium was sustained by a silent presupposition that the indigenous were not a perfect community but an imperfect matter because they were technologically inferior. On the other hand, Las Casas's strategy of dissidence in *De thesauris* took the metaphysical instrumentalism of Vitoria more seriously than Vitoria did himself by making explicit its silent presupposition in two ways. First, he takes literally Vitoria's conclusion from the first part of *On the American Indians*: "before arrival of the Spaniards these barbarians possessed true dominion, both in public and private affairs" (Vitoria, *Political Writings*, 251). If they possessed true dominion, then only their free consent, which Las Casas interprets in terms of the efficient cause, can submit to the jurisdiction of the Spaniards. Second, if the final cause of their subjection is the common good of the Indians, as Vitoria states in the eight possible titles, then again, it is not possible to subject them without their consent. The radical anti-imperial consequences drawn by Las Casas are potentially present in Vitoria's own contradictory oscillations when evaluating the problem of the dominion of the Indians.

THE IMPASSES OF METAPHYSICAL INSTRUMENTALISM

In order to show how both irreconcilable positions derive from metaphysical instrumentalism let us examine Article 3, *Summa Theologica* Ia IIae.93, titled "Whether every law is derived from eternal law," where Aquinas responds to those who do not think that human law derives from eternal law arguing that law rationally directs acts toward an end (*Political Writings*, 106). The secondary governor derives its power of directing acts to an end from the first governor, in the same way that the lower craftsman who works with his hands derives some practical knowledge from the designer who has a better conceptual grasping of the end (106). The core of the argument is that, since every directing to an end presupposes a preexisting overaching origin, then all human laws and plans of government derive from the presupposed valid source (106). A law that is not grounded on a divine law is not a valid law but an arbitrary caprice (106).

According to this argument, the sovereign commands are to the subordinate administrators what the plan (*ratio*) of the architect or designer is to the inferior craftsmen, manual workers, and manufacturing crafts (*artificialibus*). When Sepúlveda employs the principle of the natural subordination of matter to form he assumes this subordination of inferior crafts to superior crafts. This instrumentalist division between rulers and ruled has strong geopolitical implications, since Aquinas himself writes in his commentary to the *Politics* of Aristotle: "But most foreigners are physically strong and mentally weak. And so there cannot be among them a natural order of ruling and being ruled. And because there is by nature rule among those in whom reason abounds and not among foreigners, the poet says that it is fitting that Greeks, who were endowed with wisdom, rule over the foreigners. That is as if to say that being a foreigner is the same as being a slave" (*Commentary on Aristotle's* Politics, 12). At the same time, Aquinas defines slaves by nature as living instruments: "The rudder is an inanimate instrument of the ship's pilot, and the lookout (i.e., the sailor who guards the front part of the ship, which we call the prow) is a living instrument. For assistants in crafts have the nature of instruments, since as master craftsmen move their tools, so also they move their assistants by their commands" (23).

According to what we might call Sepúlveda's imperial thesis, the principle of natural subordination derives from a superior unity in which a community can be called to subordinate foreigners, turning them into means, living instruments, for their own sake. Sepúlveda's imperial thesis is entirely consistent not only with Aquinas's instrumentalism but also with Vitoria's imperial ideology. For Vitoria, the authority of the ruler is not limited to the community in question, since it has the power to punish foreigners who contravene natural law:

> The prince has the authority not only over his own people but also over foreigners to force them to abstain from harming others; this is his right by the law of nations and the authority of the whole world. Indeed, it seems he has this right by natural law: the whole world could not exist unless some men had the power and authority to deter the wicked by force from doing harm to the good and the innocent. . . . If the commonwealth has these powers against its own members, there can be no doubt that the whole world has the same power against any harmful and evil men." (*Political Writings*, 305–6)

Let us examine Las Casas's antithesis to show how it is also grounded in Aquinas's instrumentalism and Vitoria's political ideology. According to Vitoria, the proof that every law derives from a superior law is the fact that "inferior crafts are subordinate to superior ones, as bridle-making is subordinate to the art of war. God is the supreme legislator" (Vitoria, *Political Writings*, 168). Political subordination is exemplified by appealing to technical mastery, yet in the case of Vitoria, the specific technique transposed into natural law is the art of war as a *self-defense technique*, since the "proof" that law is subordinate to eternal law "is that inferior crafts are subordinate to superior ones, as bridle-making is subordinate to the art of war" (168). The proof behind the way all laws derive from eternal law is the manual to the art of war, whose inherent function is to repel force with force. Vitoria thinks that there is nothing more natural than repelling force with force. When explaining how the material cause of civil power is the commonwealth, he states that "every man has the power and right of self-defense by natural law, since nothing can be more natural than to repel force with force" (*Political Writings*, 11). Since communities cannot lack what individuals have, they must also be able to "command single limbs for the convenience and use of the whole" (11). The right of self-defense is an attribute of individuals before the elective human franchise, and "there is no reason why one man should have power more than another" (11). The commonwealth cannot give up the power to defend itself: "If a man cannot give up his right and ability of self-defense and of using his own body for his own convenience because his power belongs to him by natural and divine law, by the same token the commonwealth also cannot by any means be deprived of the right and power to guard and administer its affairs against violent attack from enemies, either from within or without" (18–19).

As Las Casas would say, there is no reason a free community has to renounce its power to self-defense, since it is protected and mandated by natural law itself. As a result, metaphysical instrumentalism reached an impasse. The principle of natural subordination of matter to form and means to an end presupposes the subordination of inferior crafts to superior crafts, "as bridle-making is subordinate to the art of war." The power to command and dominate can be used both to defend autonomy and to superimpose the authority of one perfect community over another one.

CONCLUSION

Vitoria performed a critique of imperial reason whose purpose was to strenghen the validity of the empire by examining the problem within the limits of reason alone. Since imperial reason was its own guarantee and presupposed certain instrumentalism, it fell into a circular auto-referentiality that transformed the means into ends in themselves. Las Casas exhausts this critique by operating within the legal and metaphsycial European frame but with the purpose of pursuing an anti-imperial politics of radical restitution. His anti-imperial politics is clearly compatible with the rational limits of the empire as demarcated by Vitoria. In Las Casas, imperial reason is turned against itself, and instrumentalism is put into the service of indigenous restitution.

Sepúlveda also operates within Vitoria's frame, radicalizing his presupposed war-oriented instrumentalism. Sepúlveda's modernity consists of conceiving the New World in terms of a passive and imperfect matter that has to be shaped and molded by directing it to a superior end. The impasse between considering the Indians as autonomous subjects in control of their own ends and considering the indigenous labor force and riches to be means to different ends arises from within imperial metaphysical instrumentalism itself. Both views are irreconcilable and produce a fissure that was present in Vitoria's metaphysical instrumentalism since its very inception. These impasses will become visible to the Spaniards when they find themselves no longer able to hide the destructive consequences of mining and the impossibility of reconciling this with the transcendent ends of the empire. The work of José de Acosta will take Vitoria's frame as a point of departure and find a solution to the inconsistencies that materialized in the discussions around the Spanish dominion of the natural resources of Peru from a pragmatic and utilitarian position. Acosta cuts through the impasse generated by the inner contradictions of metaphysical instrumentalism both by acknowledging that the Spanish Empire was grounded in illicit violence and by appealing to the expediency of the means in order to justify the end, namely, the salvation and civilization of the Indians. As a result, Acosta displaces the principle of the natural subordination of matter to form into the sphere of nature by providing a causal and providentialist frame in his natural history, which is designed to justify the exploitation of the mines of Potosí.

3

MASTERING NATURE

José de Acosta's Pragmatic Instrumentalism

The debate between Bartolomé de Las Casas and Ginés de Sepúlveda materialized the inner inconsistencies of the metaphysical frame conceived by Francisco de Vitoria. The thinkers' positions were irreconcilable, yet they both derived from the same metaphysical frame, the natural subordination of imperfect matter to perfect form. In José de Acosta's two main works, *De procuranda Indorum salute* and the *Historia natural y moral de las Indias*, on the other hand, he finds a solution to the impasses of instrumentalism in a rational program of the control of nature and in his justification of mining. First, *De procuranda Indorum* revisits the arguments of Vitoria, Sepúlveda, and Las Casas about the conquest and colonization of the West Indies. Although Acosta accepts that the Spanish Empire is founded on unjust conquest and violence, he maintains, along with Vitoria, that circulation and evangelization are sufficient reasons for subjecting the Indians to Spanish rule. He uses Thomist teleological arguments in order to justify the *mita*, which was a compulsory system of indigenous labor in the mines of Potosí. Acosta also analyzes the subject of mining in the *Historia natural*, where he discusses the composition of metals. He uses the Scholastic principle of the subordination of matter to form and means to an end in order to explain the extraction and appropriation of metals in the New World. Their role and function is assigned through their subordination to the ultimate ends dictated by Divine Providence.

I argue that, while in Aquinas, Vitoria, Sepúlveda, and Las Casas, the principle of natural subordination underlies all discussions about political mastery and dominion, Acosta applies it directly to the task of justifying the mining practices of the Spanish Empire. Precious metals also fulfill the role of money by being not only an instrument of exchange but also something with intrinsic value. Acosta's corpus represents the most perfect example of metaphysical instrumentalism, since it not only justifies mining but also describes it using Aquinas's principle of the natural subordination of matter to form and means to an end. Finally, Acosta's description of metals as an intrinsically valuable money form is fetishistic and engenders aporias and contradictions that result from the inherent problems of metaphysical instrumentalism. I interpret Acosta's descriptions of the inhumane working conditions inside the mines as a materialization of the inner contradiction of metaphysical instrumentalism that nevertheless supports the entire imperial enterprise.

According to José de Acosta's modern Jesuit biographer, the young Spaniard from Medina del Campo ran away from home at the tender age of twelve in order to join the Society of Jesus and pursue a career as a missionary in the Indies or Africa (Burgaleta, 10). As a student at the University of Alcalá de Henares from 1559 to 1567, Acosta would follow the standard curriculum of philosophy and theology, absorbing the influence of the "open Thomism" of the First School of Salamanca and studying Aristotle's physics and metaphysics through the commentaries of Francisco de Soto (Burgaleta, 18). Just as all the Jesuit students at Alcalá, Acosta would have focused on positive theology, which John Major and the church fathers had defined as the study of Scripture for the purposes of its practical application through ministries, missions, and conversion. While humanists employed positive theology as a weapon against the vacuous speculation of the Scholastics, the Jesuits carried on the reform of Thomism initiated by Vitoria and Domingo de Soto at Salamanca with the goal of balancing the study of theology with missionary practice. While missions among the infidels were considered the culmination of the Jesuit vocation, the Jesuits' parallel pursuits in theology likewise made them the heirs of the Dominicans with the contributions of Luis de Molina and Francisco Suárez (Pagden, *Fall*, 147).

With this solid foundation in positive theology, but plagued by ill health, Acosta embarked for missionary work in the Indies in 1571. The distinguished Jesuit scholar arrived in Peru just as Francisco de Toledo was carrying out reforms such as the use of mercury in the amalgamation of silver and the implementation of the mita system of labor levies in the Potosí mines, which would radically alter the social and economic landscape of the viceroyalty and produce unprecedented wealth for the Spanish Crown. While Toledo administered the stabilization of Peru after prolonged civil conflict between Spanish factions, Acosta brought the influence of the First School of Salamanca to the Jesuit college in Lima and the University of San Marcos, where he would later participate as a senior theologian in the Third Council of Lima in 1581 (Brading, 185). Initially Acosta was an important advisor in Toledo's inner circle, but their relationship soured as the viceroy perceived the Ignatian as acting in the interest of the Society of Jesus over and above the authority of the king's proxy (185). After sixteen years in the Americas, including extensive travel within the viceroyalty of Peru and a yearlong sojourn in New Spain, Acosta returned to Spain in 1587 where he became a close advisor to Philip II and later acted as the king's emissary to Rome in delicate negotiations between the pope and the Jesuit general (185). In 1588 Acosta obtained permission to publish revised versions of his works on missiology and natural history, which had been published as a single work in Peru in 1581. The work on missiology, titled *De procuranda Indorum salute*, appeared in Latin in 1588, and the *Historia natural y moral de las Indias*, a work that would define New World natural history until the eighteenth century, was published in 1590.[1]

Acosta's work and legacy have become an important subfield of colonial studies in which scholars with diverse methodological approaches offer divergent interpretations of the Jesuit's place within early modern Spanish America. For example, while Claudio M. Burgaleta describes Acosta's work as reflecting a "Jesuit theological humanism" (73) that reconciles Scholasticism and Renaissance humanism, Ivonne del Valle sees the Jesuit's *De procuranda* as a proto-baroque political treatise emerging from the violence of Spanish colonialism (Del Valle).[2] Del Valle reads Acosta against the background of Machiavellian politics, showing how "Acosta turns this policing into reason of state because in the name of evangelization he accepts the *need* to continue domination of the Indies despite the illegitimacy of its basis" ("Violence and Rhetoric," 59). Del Valle's approach is unique in demonstrating how Acosta assumed the task of elaborating a multi-programmatic work that presents a baroque synthesis of the often antagonistic interests of the colonial world (ecclesiastics, *encomenderos*, merchants, bureaucrats, caciques, and Amerindians), "thereby freeing Spain from the obligation to accept either the Machiavellian or the Lascasian positions" (48). Del Valle also shows how Acosta justifies the infernal violence materialized in the mining business in the name of the administration and evangelization of the Indies, providing a baroque synthesis of the positions of Las Casas and Sepúlveda. Thus, Acosta has variously come to represent a Scholastic humanist, a baroque political theorist, the father of modern ethnography, and a precursor to modern empirical science.

Acosta also enjoys a place of privilege in Cañizares-Esguerra's revisionist history of Iberian science as a precursor to the empirical, pragmatic tradition inquiring into "how things work and how colonial peoples thought, so as to use and manipulate the former and to convert and govern the latter" (25).[3] Cañizares-Esguerra criticizes a predominant academic interest in Acosta's ethnographic production that ignores his contributions to the empirical understanding of natural laws, a view likewise shared by Antonio Barrera-Osorio, who argues for Acosta's emphasis on empirical observation and his development of a theory that restructured empirical information "within the hierarchy of cognitive functions" (118).[4] By portraying Acosta as a pioneer of modern science within the context of Spanish American colonialism and commerce, Cañizares-Esguerra and Barrera-Osorio seem to equate the Jesuit's contributions as an observer of nature with those of imperial bureaucrats and transatlantic merchants, thus dismissing the metaphysical foundations of his works. While recognizing the merits of these historiographical contributions to an understanding of Acosta's place in the development of modern scientific methods and networks, in this chapter I will uncover the onto-theological underpinnings that predicate the Jesuit's observations and collection of data in the New World as a material means to transcendent, more than material, ends.

While these revisionist histories of modern science emphasize observation of the "natural" in Acosta's works against the tendency to highlight his contributions to the "moral," or political, understanding of the New World, I maintain that this arbitrary division between nature and politics is itself a product of certain unacknowledged instrumentalism inseparable from the Scholastic principle of "natural" subordination. By analyzing the onto-theological foundation of both *De procuranda* (1588) and the *Historia natural y moral* (1590), I hope to reveal the underlying contradictions of an instrumental logic that continues to define our inquiries into the nature and history of the Americas. The principles of political life and the laws of nature are inseparable within the Scholastic philosophy developed by Acosta, who transposes technique to the sphere of physics while repudiating its technical foundation. Thus, any approach to Acosta's work is incomplete without an understanding of the principle of the subordination of imperfect matter to perfecting form.

This Aristotelian-Thomist principle was the foundation for the School of Salamanca's definition of civil power (*potestas*), Sepúlveda's argument for just war, and Las Casas's defense of Amerindian sovereignty. In Acosta, the metaphysical instrumentalism inherent in the principle of subordination reveals its artificial, technical, and disciplinary foundations.[5] The irreducible contradictions that undermined Vitoria's ontological foundation of empire become manifest in Acosta's instrumentalist use of the principle of the subordination of matter to form and means to ends, which is itself based on the ungrounded decision to transpose technique to physics and politics, thus hiding the tautological foundations of onto-theology. Such an ungrounded decision posits a preexisting origin-end after the fact while disavowing its own technical, contingent, and antagonistic origin.

The first part of the chapter explains how Acosta uses metaphysical instrumentalism in order to advance his idea of colonial administration as business, grounding the need to discipline the Indians. In the second part, I examine his radicalization of Vitoria's arguments in favor of the free circulation of merchants and missionaries as well as Vitoria's teleological defense of the institutions that make possible the extraction of precious metals. In the third part, I analyze Acosta's examination of the role and function of the precious metals using the principle of the subordination of imperfect matter to the perfect form, showing the mutual correlation between Acosta's pragmatic and empiricist approach and his Thomist metaphysical instrumentalism. In the last section, I examine the connection between money and mining and the spiritual economy of the New World, showing how Acosta jumps from the instrumentalist notion of money as *medium*, to the mercantilist idea of money as value in itself, value that commands everything else. I conclude that José de Acosta's metaphysical instrumentalism represents a pragmatic program for the technical manipulation of nature as well as the transformation of the New World into an object of inquiry and source of riches.

Acosta's onto-theological defense of Spanish power in the Americas, as developed in both *De procuranda* and *Historia natural*, dismissed the arguments made by Sarmiento de Gamboa (1530–1592), who, under the auspices of Viceroy Toledo, maintained that the Incas were themselves the usurpers and tyrants of Peru. Acosta declares that such titles are incomprehensible and purely motivated by the "desire to expand royal power" (*De procuranda*, 462).[6] According to Acosta, to invoke such titles to justify the usurpation of Amerindian sovereignty and precious metals is tantamount to justifying the possession of stolen property because it was taken from a thief: "what reason or justice can deprive the tyrants (assuming that's what they are) of their power?" (463). Instead of attempting to justify the origins of Spanish power in the Americas, Acosta marks the temporal distance from this violent foundation in a defense of the status quo that will later be cited by Juan de Solórzano y Pereira in his *Política indiana*: "although they have been usurped by violence, they have the validation of many years, and they enjoy their sovereignty, which we must admit is the basis of empires if we do not wish to disrupt the established institutions" (463). God orders his servants only to *correct* what is given, not *undo* it, asking "what kingdom or empire does not have violence as its origin?" (463). Acosta presents his reader with a clear choice: on the one hand, the established order based on a dubious and violent origin or, on the other hand, the threat of disorder and chaos in the absence of colonial institutions. As Del Valle points out in "Violence and Rhetoric," for Acosta, the violent origin of an empire is therefore not an argument against its legitimacy. This seemingly arbitrary and artificial decision to exclude the violent origins of Spanish power from his considerations of colonial administration constitutes a fissure that Acosta is never able to resolve within his particular political theology.

While he distances himself from the foundational acts of wanton violence and greed that established Spanish power in Peru, Acosta nevertheless defends the necessity of violence, both direct and indirect, throughout *De procuranda*. This violence is defended as the means of subordinating the barbarous and intransigent inhabitants of the New World to a natural order organized toward the realization of the common good. In an attempt to formulate an onto-theological foundation for Spanish power, Acosta follows Vitoria's careful distinction between a just war as retaliation against "injuries from the barbarians" or other "offenses committed against our holy faith" and the "general and certain title" that would justify the administrative and disciplinary regime of colonial government, which, according to the Jesuit, has its finality in the common good (*De procuranda*, 463). Even in the absence of legitimate titles, Acosta argues that the Spanish monarchs are required to pursue the salvation of the Amerindians, which constitutes the ultimate end to which human actions are directed.

Acosta seems to have recognized that the attempts of Toledo's inner circle to establish legitimate titles in the origin of Spanish power in the New World would lead to the risk that these very titles could be used subsequently in favor of the Amerindians and thus undermine the administrative and economic structure of the viceroyalty. Acosta's writings dramatize the increasingly unstable onto-theological foundations (first visible in Vitoria) of Spain's imperial power in the New World. Acosta's admission that violence is always present at the origin of empires is an indirect reference to Augustine's comparison of empires and piracy: "Justice removed, then, what are kingdoms but great bands of robbers? What are bands of robbers themselves but little kingdoms?" (*City of God*, 147). For Acosta, violence becomes the exception of the law that is also the law's foundation. The origin of an empire, even the universal Christian monarchy, is an event, that is, it is an ultimately unjustifiable and violent decision that lacks any further justification in natural or divine law. To search for an origin of the law is to deprive the law of a principle that determines the law as necessary: the law becomes historical and contingent.

Acosta's mission was to formulate a way out of this labyrinth and reestablish the providential framework for Spanish power in the New World. He inherits the contradictions of Vitoria's thought that later materialized in the positions of Sepúlveda and Las Casas, and his work centers on the task of suturing the fissure that threatens the unity of imperial reason. Ultimately, the Jesuit must opt for a position closer to Sepúlveda. For Acosta, the principle of natural slavery becomes a tool for defending the disciplinary regime of the viceroyalty while also distancing the foundation of Spanish power from its violent origins. At first viewed as an impediment to the legitimate title of Spanish power, the absence of valid reasons for justifying the conquest becomes the justification of the necessity of colonial administration. In Acosta, the lack of valid reasons requires a violent and arbitrary decision that is justified by the pursuit of the common good (*utilidad común*). Briefly stated, in the absence of a preexisting solid ground or origin, Acosta argues in favor of an instrumental end that will confer retroactive necessity to the contingent and violent origin.

COLONIAL ADMINISTRATION AS *NEG-OTIUM*

In Chapter 7 of *De procuranda*, Acosta outlines the methods of dealing with the Indians in order to bring them to the Catholic faith. The Jesuit recognizes the difficulties inherent in this evangelical enterprise in which the "bestial customs" of the Amerindians are a constant trial for the virtues of the missionary (409–10). Acosta furthermore compares the opinion that Spaniards have of the Amerindians with the disdain that the Greeks showed for barbarians, and the opinion of both nations with regard to the inferiority of beasts of burden to men (410). This division between the animal and the human is ultimately, according to Acosta, a difference in the degree of rationality evidenced in a providential scheme of salvation wherein "Rational men are saved by His justice, irrational men by His mercy. The former are provided for, the latter are ruled over" (410).

Notwithstanding this hierarchy of being, Acosta insists on the Christian duty to bring these irrational, near-brutes within the fold of full-fledged humanity. The focus on supposedly "bestial customs" is similar to Sepúlveda's assertion that lack of technique is evidence of the Amerindians' imperfect humanity. Yet, as Pagden argues, Acosta understands the Amerindians as brutish only insofar as they have been degraded by unnatural customs, habits, and laws.[7] While both Vitoria and Sepúlveda cite the principle of subordination's bellicose origin in their deliberations on the subject of natural slavery, for Acosta this principle grounds the instrumental, technical, and administrative ends of colonial rule.

Acosta denounces the "impiety" of those who assert the Amerindian's incapacity to understand the message of Scripture and willingly convert to Christianity, though he likewise describes the arduous task of inculcating this message as that of disciplining a beast of burden with the whip (410). Just as the rider must spur on the beast, so too must the missionary lead forth the Amerindian: "tighten the reins and bit on the donkey's jaw, burden him with a proper load, use the whip if necessary, and if he should buck, don't lose your temper or abandon him. Punish him with moderation; rein him in little by little until he is accustomed to obedience" (411). Acosta asserts that, just as the owner of the ass cannot abandon his investment, neither can the Christian reject the souls paid for by Christ's sacrifice (411). Sepúlveda's justification of violent coercion as a means of correcting the Amerindian as imperfect and unyielding matter becomes in Acosta an obligation to undertake the arduous task of disciplining and domesticating the Amerindian in order to ensure a return on investment.

Of course, Acosta recognizes that, despite the self-evidence of the Christian message, obedience is not guaranteed, which begs the question, "What to do?" (411). The Jesuit advocates restraint in the use of direct force while suggesting other coercive measures: "one must attempt a more cautious and attentive approach with them, one must use the lash only as a Christian" (411). These measures are similar to those used to discipline children, asserts Acosta, who cites Proverbs 29:15: "The rod and the reproof give wisdom, but a mother is disgraced by a neglected child" (411). The principal means that Acosta suggests for disciplining the Amerindian within this colonial pedagogy is constant labor, for the Amerindian, "when burdened with work, thinks of leisure. What if he were free and at ease? He would think of fleeing" (411). Whereas Sepúlveda's doctrine amounted to a program for the expropriation of the Amerindian, Acosta's work outlines a pedagogical regime for the surveillance and discipline of a subject to be brought into the spiritual, political, and economic fold of imperial reason. While the mandate to subordinate imperfect matter to perfecting form remained an abstract law eclipsed by the problem of war in Vitoria's and Sepúlveda's writings, for Acosta imperial reason is retroactively grounded through practical, administrative measures. As the principle of subordination assumes an instrumental character, it likewise loses its capacity to ground, thus revealing the fundamental aporia of imperial reason, which is in the last instance always dependent on the matter that it must discipline but without which it cannot exist: Amerindian labor.

Acosta appeals to the lessons of experience in the Indies, which have confirmed that both Africans and Amerindians "are in principle free men, yet in their customs and character are like slaves" (411). Citing Augustine's diatribes against the "Donatists and Circumcellions, a wicked caste of men," Acosta argues that there are men who are dominated more by their animal nature than their rational or spiritual potential and with whom one must turn to severity instead of charity (411). Again, these men are not excluded from salvation, but they must nevertheless be instructed by bending their wills through "work and continuous activity" and "the check of fear" (411–12). The Jesuit unequivocally asserts that experience in the Indies dictates that "in this manner they forced them into salvation even against their will" (412). The quasi-animal nature of the Amerindian—and the coercive means used to discipline him—is not "experienced" but, rather, interpreted, constructed, and adapted to the mandate to subordinate matter to form and means to ends. This is the foundation of Acosta's program to educate, or humanize, the barbarians, a principle that ultimately brings him closer to Sepúlveda than to Las Casas.

Despite his affinity with Sepúlveda, Acosta nevertheless cautions against the excessive use of force, "which is contrary to the spirit of Christ" (412). Although the education, discipline, and administration of the Amerindians may strain the patience of the colonial administrator or missionary, Acosta asserts that severity must yield to the Christian principle of charity. This balance of severity and charity, "the force of healthy fear," is not only proof of the colonial agent's capacity for rule but also an example to the Amerindian who must learn to suffer patiently: As Saint Paul declares "Love suffers all, resists all, awaits all; it is patient and kind" (412). Like Sepúlveda, Acosta employs the metaphor of medical technique in order to illustrate the means of correcting or improving a deviant nature, yet he emphasizes that the doctor responsible for curing the animal nature of the Amerindian must take care to temper his treatment with charity lest the patient turn away from the medicine of Christian doctrine. Unlike Sepúlveda, however, the Jesuit rejects expropriation or just war as a consequence of the mandate to subordinate imperfect matter to perfecting form; instead of a doctrine of war, Acosta develops a guide to colonial pedagogy and administration.

Acosta clearly distinguishes between the slave by nature and a slavish condition determined by barbarous and bestial habits. The Amerindian is not destined to be a slave because his nature is essentially imperfect, as Sepúlveda argued, but rather, his bestial nature is a product of a deeply flawed education. Acosta declares that "the Indians' mental incapacity and untamed habits are not so much a product of birth or race, or of the climate but, rather, stem from a deep-seated upbringing and lifestyle that is not too different from that of animals" (412). As the life of the Amerindian cannot be reduced to a purely animal nature, Acosta condemns outright violence without a rationalized regime of discipline and conditioning of the will. Where Sepúlveda's description of the Amerindian as indeterminate matter projected an absence of humanity upon the cultures of the New World, Acosta fills this absence with empirical determinations that trace the Amerindians' supposed inhumanity—or excess of animal nature—to degenerate customs. It is on this point that the Jesuit undermines the hylomorphic foundation of Sepúlveda's justification of war and dispossession.

Without directly naming Sepúlveda, Acosta revisits the arguments outlined in *Demócrates Segundo* with regard to the question of whether it is justified "to force them to abandon their idolatry and abominable rites, their dealings with the devil, the unspeakable sin between men, incest with sisters and mothers, and other such crimes" (432). Acosta begins by insisting on the primacy of experience when treating any matter concerning the Indies, given that many histories purporting to faithfully represent the nature and inhabitants of the New World are "vastly inferior to reality" (433). While insisting that "the customs of the majority of Indians are those of animals," the Jesuit nevertheless discredits as folkloric superstition the belief that the Amerindian is literally a hybrid creature with animal and human parts (433). This brings into relief Acosta's earlier comparison of the Amerindian with a beast of burden, which is a purely rhetorical strategy for explaining the unity between the immaterial faculties of the soul and the material appetites of the flesh. From Acosta's Thomist perspective, in which man is essentially a rational animal, popular fantasies describing the Amerindian as human-animal hybrid can only ever lead to unchecked violence. Acosta's "rational animal" is *one unified substance* that is a composite of animal matter and rational form and cannot be subjected to obedience by mere physical coercion.

Alluding to the debate between Las Casas and Sepúlveda, Acosta maintains the need to overcome the unilateralism of polemics, which have enshrouded the truth with a "dense fog" (435). Only a thorough examination of both positions can lift this fog and clearly reveal that "it is just as sinful to commit an injustice as it is to pursue justice unjustly" (435). Echoing Las Casas's position, Acosta asserts that no republic or prince has the right to violate the sovereignty of another republic or to dispossess it of its property, a power that "within a brief span would upset the entire planet and fill the world with discord and death" (435).

According to Acosta, a prince or republic can only dictate laws and punish crimes within its own borders and has no more authority over another republic than a citizen over his peers. The only case in which one republic may punish another is in order to repel force with force, that is, in self-defense or as retribution for injuries received, a right that is moreover reserved only for princes and not extended to citizens. Thus, other ends, such as "honor," "the accumulation of wealth," or "the extension of dominion" (even in the name of the Christian faith) are not proper titles for waging a just war (435). Citing the authority of Augustine, Acosta declares that any actions based on these titles would be more akin to the conduct of "bandits" than the army of a republic (436).

Like Las Casas, Acosta imposes limits on the geopolitical extension of hylomorphism by arguing that, if one entity (or republic) were to arrogate the right to subject another already formed entity, then this entity would be acting illegitimately on behalf of the totality. Yet there is no common superior court that could legitimately and impartially judge the actions of both, which means that the dispossession of Amerindian sovereignty and property is here exposed as an arbitrary and artificial act of violence. The political consequences of this ontological problem are clear: the dispossession of the sovereignty and property of the Amerindian cannot be justified by appeal to the natural order inherent in things themselves but only by a metaphysical instrumentalism, the principle of the subordination of imperfect matter to perfecting form. While Las Casas questioned the existence of a prior unity that would validate the imposition of one form over another, Acosta takes this argument a step further and declares that this supposed unity is always already undermined by the absence of a principle (or prince) that could rule over the totality.

Confronted with this impasse, Acosta returns to Vitoria's defense of the rights to circulate freely and to preach the Gospel, which can be legitimately defended by force if necessary. Thus, as in Vitoria's lectures, Acosta dismisses imperial violence through the door only to later let it back in through the window. Acosta invalidates Sepúlveda's ontological justification of war, yet it nonetheless serves as the foundation for the Jesuit's defense of the *encomienda* and mita. In *De procuranda* we witness a transition in imperial reason from bellicose hylomorphism to an instrumentalism that justifies the dispossession of natural resources and Amerindian labor as disciplinary means.

Although Acosta rejected Sepúlveda's justification of war based on the principle of natural slavery, he nevertheless admits that there is a natural order that determines those fit to rule and those destined to obey within a given republic. The Jesuit asserts that this principle was evident in antiquity where the "Greeks were born to rule because of their wisdom, and the barbarians were destined to serve due to their ignorance and brutishness" (437). This principle, however, does not justify the dispossession of "barbarian" republics, just as it does not justify the dispossession of a kingdom ruled by a minor or a woman, or to "depose an ignorant prelate of pontiff. All of which is an obvious contradiction of divine and human laws" (437).

Acosta's reevaluation of the ontological and physical underpinnings of the principle of natural slavery adds a new dimension to the problem: "Because one question is what can be done in conformity with reason and nature, *and another question is what, once done, can or cannot be undone*. It is rational that the wisest and noblest rule, but if an ignorant or barbarian ruler should come to power, it is not just but, rather, a crime to depose him" (437; my emphasis). Therefore, it is impossible to invalidate the sovereignty of the "barbarians" without undermining the very foundation of the law. The mandate to subordinate means to ends and to correct imperfect matter with perfecting form also dictates that that which is executed cannot be undone lest the whole order fall into anarchy. Perhaps as a consequence of his own religious vocation, the Jesuit shows a reverential respect for any constituted authority, even that which emerges from illegal, tyrannical, and violent origins. Acosta thus leads us to the foundational contradiction of his defense of imperial reason, which establishes Amerindian sovereignty at the same time that it maintains the impossibility of undoing the "barbarous" origins of Spanish power in the Americas. Restitution and withdrawal, as Las Casas proposed, are not an option, thus leaving only the Spaniards' duty to discipline, order, and educate their inferiors.

Just like Las Casas, Acosta steps back from the brink of extending hylomorphism and teleocracy to the geopolitical sphere lest this lead to a justification of unchecked physical violence against the nations of the New World. The Jesuit strikes a disdainful, professorial tone in criticizing Sepúlveda's position, which he describes as "obscure and one suspects that it comes not from philosophical reason but, rather, from popular reason" (437). Yet this is not necessarily a defense of Las Casas's pacifist doctrine, as Acosta seems more concerned that a bellicose, popular doxography may contaminate the purity of his Aristotelian-Thomist framework. No one can doubt, Acosta declares, that it is just to declare war against those "barbarians" who lack republics, magistrates, laws, and who "like beasts wander about without settlements or stable government" (438). Unlike Las Casas, Acosta is convinced of the abundance of such men living outside the law both in antiquity and modern times, yet he cautions that any war against nations outside the law governing the totality should employ only "moderate force with which they may be convinced to live like men and not beasts" (438).

Acosta here reaches the point of inflexion at which it is difficult to determine the difference between the duty to civilize (or evangelize) and the naked ambition to dominate that is found at the origin of all empires. He explains that the Romans ruled their subjects with more just laws than those of the barbarians, yet even Augustus Caesar confessed that many of their conquests had been tyrannical in nature: "It was not the law that provided arms, rather the arms provided the law" (438). Acosta's doctrine of "moderate force," exercised through a pedagogical regime and backed by fear of punishment is, despite the Jesuit's efforts to distinguish his prescriptions from imperial domination, likewise impossible without the foundational, violent contradiction of empire.

While Roman historians celebrated their wars as pursuing "the Faith or public well-being," Acosta reminds us that Augustine denounced the Roman Empire's violent origins and its conquests as "honorable thefts" (438). Acosta maintains that the church fathers attributed the wars of the Israelites and Canaanites to the will of God, who used these nations as vehicles for his divine wrath. Both Abraham and Joshua, therefore, were simply executors of an unfathomable divine will, which does not imply "that a father may kill his son or that a prince may take up the sword against another nation, howsoever impious it may be" (438). Although the church's authorities defend the wars of the Hebrews against the Amorites, they nevertheless clarify that idolatry and crimes against nature were not "sufficient cause for the Hebrews to exile those people from their lands and destroy them, given that much diligence was spent in finding other causes" (438). In the event, the "*bárbaros*" of the New World are guilty of lesser crimes than those attributed by Scripture to the nations stricken down by God's wrath. Thus, the teachings of Scripture and the church fathers do not justify conquest; on the contrary, God mandates that his followers sacrifice themselves in defense of the faith and in the rejection of idolatry (439).

Unlike Sepúlveda, Acosta maintains that a barbarian prince or magistrate who betrays his office "has God as his judge, not another republic or foreign prince" (439). Acosta, just as Vitoria and Las Casas before him, uses reductio ad absurdum to expose the paradox inherent in the argument put forth by Sepúlveda, for if one recognizes the right of nations to punish others "when our own princes or magistrates commit grievous crimes, the French, the English, or the Italians would have the right to punish the crimes of the Spanish Republic and the princes of these nations would jointly share this power [over Spain]" (439). It is therefore an act of tyranny to punish the "madness or immaturity" of foreign rulers even with the consent of their subjects (439). In extreme cases, it may be possible to free innocents from the tyranny of their barbarian princes, yet if the "corruption of custom" has led these subjects to voluntarily adhere to these laws, "they cannot be compelled to virtue through violence by a foreign nation" (439). In emphasizing the nature of the Amerindians as conditioned under a barbarous culture Acosta departs from the hylomorphic doctrine developed by Sepúlveda: "thus, in short, the important question here is that the barbarians are not so by nature but, rather, by taste and habit. They are children and madmen by choice, not by nature. Therefore, it is nobody's place to punish them, regardless of their crimes" (439–40). In his rejection of the violent origins of empire, the Jesuit likewise rejects the notion of a metaphysical ground that would justify the submission of one part to another within the totality. While bracketing the violent origins of the Spanish Empire as the design of a divine, inscrutable will, Acosta nevertheless admits this violence as an irreducible remnant that continues to ground the execution of the law: while there is no nonviolent principle grounding imperial domination, there are nonetheless practical ends that continue to justify coercive means.

Acosta therefore maintains that the violent origins of Spanish power in the Americas cannot be undone, which frees the Spaniards to embrace the duty to civilize and evangelize their Amerindian subjects through the disciplinary means of education and fear. In Chapter 9 of *De procuranda*, Acosta addresses the question of the possibility of imposing tributes on the Indians in order to turn them away from leisure. According to Acosta, those versed in such questions maintain that is necessary to impose "onerous tribute payments" on the Amerindians, "because, being a lazy and indolent people, if they are not forced to work and apply themselves in order to pay tribute, then they will continue in a slothful and bestial state" (471). Such tribute payments and constant labor are proposed as means to direct the activities of indolent Amerindian subjects who would otherwise surrender to shameful and irrational appetites without planning for the future (472). Experience has proved to Acosta and others that constant activity is not only beneficial but "completely necessary in order to properly constitute their republic" (472).

Just like Juan de Matienzo (1520–1579), author of *Gobierno del Perú* and an important precursor to Toledo's policies in Peru, Acosta maintains that the concept of profit is lost on Amerindians who do not work for money. Thus, just like Toledo, Acosta argues that the best way to govern the Amerindian subjects of Peru is to employ the same methods used by the Incas. The vice of sloth was the reason that the Incas put so much effort into administrating their society "so that it would remain just and lasting by making their citizens labor as much as possible without leaving them a moment of idleness " (472). Such was the Incas' concern with this vice that they would invent superfluous tasks just to keep their subjects busy, ordering some to "collect a certain quantity of insects, or others to move rocks from one spot to another" (472). In justifying this regime of forced labor and tribute payment, Acosta turns to the principle of natural slavery adduced by Sepúlveda in his defense of just war: the Amerindian must be constantly occupied "because the barbarians are all of a servile condition, and it was proverbial among the Ancients, as Aristotle claims, to never leave a slave unoccupied, because idleness made them insolent" (472).

Acosta defends the restricted use of forced labor, but he cautions that it is first necessary to ask: "For whom do they work? For whom do they cultivate? For whose benefit should they serve?" (472). The difference between the just ruler and the tyrant, in his opinion, is to be found precisely in this distinction between those who seek their own profit and those who seek to benefit their subjects. Acosta again returns to metaphors from the animal kingdom to illustrate the difference between the just and unjust imposition of labor upon the Amerindian subject. He condemns the *encomendero* who behaves like the beekeeper who leaves the bare minimum of honey to keep his bees working or like the shepherd who shears his flock leaving only enough root to continue exploiting its wool. A "prudent charity" should mediate the administration of the Amerindian in order to avoid such exploitation.

In Chapter 11, Acosta examines the three causes of the encomienda, featuring three senses of the word *encomienda*. The first is the "payment" for having called the conquerors who discovered "at their expense" much of the New World, "suffering incredible work and overcoming great difficulties" (475). Thus, the Crown granted them the perpetual right to receive annual payments from the indigenous peoples who were placed under the king's tutelage, in the same way that the Roman emperors granted an income as a reward for some field soldiers" (475). A second and more important cause was the distance that existed between the king and the New World, which required the service of the trustees to govern the barbarians, "restraining the freedom of the barbarian, defending borders from enemy incursions" (475). The third and most important cause was imparting Catholic doctrine to the neophytes so they could "grow accustomed to the discipline and Christian morals" (476). To summarize, the three causes are (1) the remuneration of the conquistadors for winning these kingdoms for the empire, (2) the geographical distance between the monarchy and the New World, and (3) the need to discipline and evangelize the natives.

In the next chapter, which explains how a just tribute should be calculated, Acosta explicitly turns to the mandate to subordinate means to ends. Citing Aquinas, he asserts that in decreeing onerous laws "one must bear in mind the end to be pursued and only impose those duties required to achieve said end" (473); thus, the ruler must seek a proper balance in the exchange between the dictates of Christian reason and the means of labor. A just proportion rules over the balance between the subordination of the part and the order of the whole, that is, between the means employed and the end defined as the common good. Thus, according to Acosta, the encomienda should be an exchange of just proportion within the providential design of the Spanish Empire. In order to achieve this just proportion, Acosta proposes two measures: first, restrictions on the amount of tribute or labor demanded of the Amerindian, and second, a just remuneration for their services. This conjunction between justice and payment of a fair wage is in accordance with "what the theologians—among them Saint Thomas, the glory of all theologians—conclude when they say that it is licit to impose the tributes that are necessary for the end that is pursued, and they may be increased insofar as the end demands" (473).

Acosta denounces the encomenderos who treat their Amerindian charges as property to be extenuated to the limits of human resilience in order to realize a profit (474). A just exchange between encomenderos and Amerindians, according to Acosta, should be analogous to the relationship between a doctor and a patient, or between a lawyer and a client, in which a fair price is charged for services rendered, because the encomendero receives Amerindians' labor as a payment for the "benefit that he does them" (474). This logic of the doctor-patient analogy should determine the just measure of Amerindian tribute and labor, thus limiting the encomendero to charge the strictly "necessary for the *end to which they are directed*" (474; my emphasis). The natural reason impressed on the colonial order through divine will dictates that the Amerindian should be governed for his own spiritual and political health: "And *reason dictates that end*, which is to provide the ministers necessary for the health and good government of the barbarians, and so that these ministers should not be lacking and should enjoy honor and a decent reputation" (474; my emphasis). The most important goal is, of course, the spiritual—and the spiritual shepherds provide the nourishment of the divine words and sacraments (474). To conclude, Acosta provides an explicitly Thomist and teleological defense of the encomienda, grounding it in the principle of subordination of matter to form and means to an end. Like Vitoria, Acosta grounds imperial reason in metaphysical instrumentalism, where the aim of the unconditional end (common good) is to justify the technical means (encomienda).

The secular administrator doctor, who is responsible for the political life of the Indian Republic, is likewise an integral part of this just exchange within the spiritual economy of the New World. According to Acosta, these secular officials can be divided into two classes: judges and magistrates on the one hand, and on the other the "patrons of the Indians, commonly referred to as encomenderos, who may receive tribute payments in return for the care and attention that they must show with those who are entrusted to their faith and tutelage" (474). The right conferred upon the universal Catholic monarchy to exact tribute from recently discovered and converted nations, whose spiritual well-being is the responsibility of the Spanish king, can be transferred to these encomenderos under certain conditions (474). According to Acosta, the Spanish monarchy essentially has the right to subcontract the spiritual and political administration of the New World to private enterprises who are allowed to usufruct the labor and tribute owed to the Crown.

This right, however, leads to the question of determining the limits to be imposed upon the pursuit of material wealth as a means to the spiritual benefit of the Amerindians, that is, the question of the equivalence of temporal and spiritual goods. For Acosta, this is a problem of just proportion, or avoiding any excess or want that may cause the colonial order to stray from the common good. Yet, Acosta occludes the central question of appetite and desire for precious metals, always based on surplus or scarcity, as the source of the spiritual and political crises in the New World. Confronted with this destabilizing element, he seems to fall back on Aquinas, who maintained that justice consists of each receiving what he deserves according to his place in the hierarchy of forms. Justice, within Acosta's vision of imperial reason, implies that there are always some parts naturally subordinated to others, the former charged with manual labor and the latter with the intellectual-administrative work, but all must work toward the balance of the common good.

Acosta recognizes that the problem of the encomienda is as "difficult and arduous, and as necessary as any other" (485). Indeed, the encomienda and the mita—both based on the extraction of tribute and Amerindian labor by private enterprises in the name of the civilizing and evangelizing project of the universal Catholic monarchy—not only undermine any metaphysical ground that would justify Spanish dominion but also endanger the pragmatic defense of the status quo as proposed by Acosta. The Jesuit distinguishes the "natural servitude" resulting from the curse cast upon the descendants of Ham and a "civil, legitimate" servitude imposed by the victors upon the vanquished, under which it is unjust "to deprive them of the natural fruit of their toil" (485). The manual labor exacted from the Amerindians, whether it be in the construction of houses, the service of messengers (*chasquis*) or the herding of cattle, must be justly remunerated, "and whosoever denies this is to be condemned as a murderer" (485). Acosta turns to Las Casas's censorious rhetoric in denouncing the encomendero who defrauds his charges of a just wage: "And if to burden another's mule or ride a horse one must pay a fair price to the owner, how much more sinful is it to subject the body of a free man to labors and burdens without paying him a fair price?" (486).

Scripture condemns those who delay, diminish, or withhold a just salary, and those who think they have fulfilled their Christian duty by paying with rotten food, threadbare clothing, or a miserable plot of land must be mindful that "God sees all, he is a fair judge who acts as a ready witness in favor of the poor" (486). Like Vitoria, Acosta maintains that civil servitude does not justify the dispossession of Amerindian lands, property, or labor. Servitude is based on the difference between matter and form and the hierarchy of being implied by this difference, which determines the degree of perfection of a particular entity and the duty to subordinate the less perfect to the more perfect. The perfection of the end or the final form of imperial reason (the common good) requires just remuneration for the labor exacted from the Amerindian.

The mita, or forced labor levies in the silver mines, represents a particular challenge to Acosta's doctrine of imperial reason. On the one hand, it is unjust to force anyone to work, even for a wage, "if the free man is coerced, he is not free" (487). On the other hand, if the Spaniards do not adapt their methods to the condition and ill disposition of the barbarians, nothing will get done (487). According to Acosta, the Amerindian is a free vassal of the Crown, yet due to his inclination to idleness he will refuse to work. Acosta performs a classical ideological move by displacing the weakness internal to the system, the lack of unity and consistency, into an external figure that is presented as the true cause of such weakness. Essentially, Acosta maintains that the Amerindian subject is free—but not free to refuse to work for the Spaniard. How, then, does Acosta justify this "forced" freedom? The answer is found in the necessary unity in which both Spaniards and Amerindians participate.

Acosta's justification of the mita is based on the duty shared by every citizen to contribute to public works and the maintenance of a healthy, functioning republic. The problem arises in the New World when Spaniards refuse to undertake certain labors such as shoveling earth, brick making, or herding livestock because these jobs have low prestige, and even if the Spaniards wanted to engage in them, they would not be able to do all of them (487). The central question for Spanish administrators, then, is if the Amerindians refuse to willingly perform this work "is it licit to oblige them through violence and fear?" (487). Acosta answers Las Casas's vehement negative response to this query with a defense of the status quo. The Jesuit describes as "pure nonsense and fraught with difficulties" the opinion that those Spaniards unwilling to do the work they force upon Amerindians should return to Spain (487). Even if Spaniards are willing to do the work allotted to the Amerindians, Acosta asks, "in what proportion are these few men to the multitude of tasks at hand?" (487). The suspension of labor levies and the return of Spaniards to their homeland would, according to Acosta, amount to the abandonment of the evangelical enterprise in the New World; thus, the encomienda and mita are ultimately defended for their utility in maintaining an order whose justice is unquestionable. Acosta mitigates these coercive means with an appeal for just remuneration and consideration for workers' health, yet it is the exploitative system itself that permits the spiritual and temporal order to exist. Ultimately, this order is based on a division of labor between those destined to rule and those destined to obey.

Acosta posits three causes, or conditions, that justify this combination of forced labor and remuneration. The first is "the common opinion of learned and virtuous men, who from the beginning believed a republic should be organized in this manner" (487). The second is the "very ancient custom of the Indians themselves, observed during many centuries since the origin of their Incan kings (487). For the third, Acosta cites the natural law that dictates the subordination of parts, assigning different functions to each, in order to collaborate toward the good of the whole (whether the family or the city). Likewise, social unity is the principle that holds together the two elements of colonial society, given that, "of course, the multitude of Indians and Spaniards form a single republic, not two separate republics" (487).

Amerindians and Spaniards have the same king and are subject to the same laws, "and there cannot be one law for the former and another for the latter, rather it is the same for all" (487). Consequently, the services that the Amerindians perform for the Spaniards are performed for their "fellow citizens" and not for "foreigners" (487). Acosta is clearly employing here Aristotle's metaphor of the social body as a statue.[8] Despite the failure of the statue to join the parts and the body, both the Spaniards and the Indians constitute the body of the statue (487). It comes as no surprise that the iron body of the statue should sometimes oppress the head, but "what is desirable is that the statue not fall to pieces" (488). The fundamental lesson comes from Aristotle, "the great teacher of public matters," who argues that it is natural for those who exceed in reason to rule and those who are adapted to labor to serve, "as long as they mutually help each other, and the former lends his eyes to see while the latter his feet to walk" (488).

Acosta finds a sufficient reason for this combination of forced labor and wage in three conditions. The first is the common sense of the learned and virtuous men who believed that it was necessary to organize the republic this way (487). The second is the ancient custom of the Indians, preserved for many centuries since the earliest Inca kings (487). And finally, natural law mandates the conservation of the whole (the body of the family or the city), providing that necessary parties can be forced to work, assigning different roles or functions to each part (487). Also, the social unit is the principle that holds together and unifies the two elements of society, because multitudes of Indians and Spaniards form a single republic and not two separate republics (487). Both have the same king and are subject to the same laws (487). For this reason, the services provided by the Indians to their fellow citizens are not provided to strangers (487). Even if the head sticks poorly to the iron body, according to the prophetic prediction, both the Indians and the Spaniards constitute the feet of the statue (487). Even if the iron of the body of the statue puts some pressure on the foundations, what matters is that it does not destroy or shatter the statue (488). The fundamental lesson of Aristotle, "the grand master of public affairs," is that it is natural to divide society into those who command because of their capacity for reason and those who obey because of their capacity for work (488).

Acosta justifies the encomienda and labor levies using teleology—and the mita using the need to preserve social unity or the *"utilidad común,"* of Spaniards and Amerindians, who are bound by the basic division of intellectual and manual labor. This comes to be the prior unity that mandates and justifies all subordination of manual labor to the needs of the totality. Despite this unity under a common goal, there remains one part of the whole that has achieved a greater degree of perfection and thus must undertake the intellectual labor of administration. In the realization of the perfect political community, a part of the whole may be subordinated to another, because the unity of the republic is more important than the utility of its parts. In using the statue metaphor to illustrate this subordination of the parts to the unity of the form, Acosta transposes the production of an artificial end to the supposedly natural sphere of the republic. The artificial nature of this metaphor, which marks the instrumentalism of Acosta's doctrine, provides retroactive justification to the inherent violence or lack of ground of the Spanish power. Just as the supposedly transcendent and absolute natural order can only ever be given an artificial form, it is likewise impossible to escape the foundational violence of Spanish imperialism, which continues to undermine Acosta's onto-theological justifications.

Acosta first proposes this division of labor as a practical means of inculcating civic and Christian virtues in the Crown's Amerindian subjects, but with the use of the statue metaphor he turns this practical measure into the ontological ground of political unity. The negation of Amerindian *otium* thus becomes the ontological foundation of Spanish colonial administration. The institution of the mita is furthermore justified by its pre-Hispanic origin: it is "an institution that we took from the ancient regime of the Incas, as so many other things full of wisdom" (488). While the mita and encomienda are ultimately justified by the principle of subordination, Acosta nevertheless condemns the excesses committed by Spanish officials, "things that are wholly unjust and born more from the wickedness of men than from any true right; and they have not been implemented in good faith but, rather, through the wickedness of the times or of men" (488). In line with typical Jesuit prudence in their criticism of colonial policies, Acosta limits his condemnation to the abusive implementation of these institutions by Spaniards who refuse to fairly remunerate the labor of their Amerindian charges or unjustly prolong the duration of labor tribute. While skeptical with regard to the justifications for the harsh treatment of Amerindians, Acosta nevertheless concedes that the mita and encomienda are necessary, given the Amerindians' demonically inspired refusal to work for the Spaniards, an argument later taken up by Juan de Solórzano Pereira in *Política indiana* (488).

Having established the justice and necessity of this division of labor, Acosta moves on to the pressing question of how a dutiful Christian must act with regard to an Amerindian who refuses to observe the natural hierarchy. The Jesuit first recommends appealing to a competent official with the authority to command the disobedient subject, but in the absence of such an authority he concedes the use of "some adequate force and violence, and it is not wrong to frighten them as children who do not obey the just and necessary, although they often cross the boundaries of justice and equity!" (488). Acosta, like many of his Jesuit confrères, prefers the coercion of symbolic and exemplary violence over the well-documented excesses of many of his countrymen in the Indies: "What is the purpose of these lashes if words suffice? Why inflict wounds if fear is enough to guide toward reason?" (489). This preference for the violence of the letter is likewise found in Aquinas, who links the disciplining of children to the authority of the law in Article 1, *Summa Theologica* IaIIae.95:

> Now for those young people who are inclined to acts of virtue by a good natural disposition, or by custom, or rather by Divine gift, paternal discipline, which is administered through admonitions, suffices. But because some are found who are headstrong and prone to vice, and who cannot easily be moved by words, it was necessary for these to be restrained from evil by force or fear, so that they might at any rate desist from evildoing and leave others in peace, and that they themselves, by having habits formed in them in this way, might be brought to do willing what formerly they did from fear, and so be made into virtuous men. And this kind of discipline, which compels through fear of punishment, is the discipline of the laws. (*Political Writings*, 127–28)

Thus, the division of labor and the use of symbolic violence justify even practices decried by some "patrons of the Indians" as inhumane, such as the use of Amerindians as beasts of burden. Acosta claims that the Amerindians are accustomed to carrying loads on their backs, as this was a common practice in Inca times, and therefore "they suffer no injury or harm, as long as one complies with what has been said in reference to weight, labor, and salary" (*De procuranda*, 489). Thus, if the Amerindians are justly remunerated, then there are no grounds for accusations of illegitimate violence. In Acosta's thought, the doctrine of natural servitude is displaced by the subordination of imperfect, or disobedient, matter to perfecting, administrative form in exchange for a just wage. This shift constitutes a kind of administrative, disciplinary hylomorphism in contrast to Sepúlveda's war-oriented ontology.

Like Vitoria and Sepúlveda, Acosta maintains that civil power is founded exclusively on the principle of the subordination of imperfect matter to perfecting form. In the case of Vitoria, this principle grounded the civil power's capacity to repel force with force, protecting commerce and evangelization. Sepúlveda, on the other hand, considered that the rebellion of matter could be justly remedied through the subordinating violence of a just war. Acosta maintains that the privation of matter should be corrected through the secular administration and the coercive discipline of symbolic violence. While Acosta, like Vitoria, tacitly recognizes that the principle of subordination does not provide a direct justification for the violence of war, he nevertheless provides a retroactive justification invoking the goal-oriented frame of natural law.

In the lack of a preexisting ground for the empire, Acosta provides a retroactive justification that borrows its power of persuasion from technique. He justifies the administration and discipline of the Amerindian through a tautological method that invokes the necessary character of the disciplinary techniques, based on a necessary end, employing the same hylomorphic and teleological language as Vitoria and Sepúlveda. This reduces the Amerindian to a docile body to be purified through labor and transforms nature into a standing reserve to be exploited by technical means. The privation of matter is simultaneously assumed and imposed by a tautological, performative logic that fills the gap opened by the violent origins of the empire, a crack in the natural order of being that will return in Acosta's *Historia natural*, where the depths of the silver mines present an uncanny symbol of the material, artificial means that sustain the imperial order. As Acosta recognizes the dependence of the metropolis on the subordinate periphery, the fissure of imperial unity, which had remained speculative in Vitoria's thought, materializes in the paradoxical dependence that perfecting form has on imperfect matter.

CIRCULATION, PRECIOUS METALS, AND PROVIDENTIAL DESIGN

Acosta employs Vitoria's defense of the rights of circulation, commerce, and mining: he first establishes the justice of the end, ultimately based on technical presuppositions of the providential design impressed within imperial reason, and then he argues for balance and measure in the implementation of administrative means to achieve this end within the given structure of colonial rule. Acosta's pragmatism or utilitarianism, as extolled in the work of Cañizares-Esguerra, is inseparable from the Jesuit's onto-theological doctrine, in which the distinctions between theology, administration, and technique become indiscernible. Just as with the exploitation of Amerindian labor, Acosta is also faced with justifying the exploitation of precious metals as an activity within the natural order of Spanish rule. Like Vitoria, Acosta appeals to the natural right of circulation or travel, which is intimately linked to commerce and the mining of precious metals. In her analysis of Scholastic and imaginative literature in the Spanish American colonial context, Elvira Vilches demonstrates that the chaotic flows of precious metals eroded the moral foundations of empire, thus inspiring a body of works, such as Acosta's, that sought to reconcile the ideal and material interests inherent in the imperial enterprise (Vilches, 52).[9] In *De procuranda*, Acosta attempts to attenuate this chaos at the heart of the Spanish Empire.

Like Vitoria before him, Acosta argues that the right to circulate and to spread the Gospel was sufficient pretext for Spanish presence in the New World. This right is "common in nature," meaning it applies to all men and must be respected by all nations (*De procuranda*, 450). For example, the Chinese laws that decree capital punishment for any foreigner entering the territory without official permission are a blatant transgression of the natural right "to the land, which God made for the use of all, held by the peaceful guest who does no harm and gives no reason for suspicion" (450). For Acosta, this fundamental right to pursue knowledge grounds the natural right to circulate freely across the borders of sovereign nations, "which is no small aid in the discovery of human matters and the physical sciences" (450). The onto-theological ground for Acosta's defense of the natural right of circulation is found in Aristotelian epistemology, in which the desire to discover and increase human knowledge is always directed toward a just end (*Nicomachean Ethics* 1.1.1094a, in *Basic Works*, 935). In his commentaries on Aristotle's *Physics*, Aquinas states, "natural appetite is nothing but the ordination of things to their end in accordance with their proper natures" (72). The end, again, *commands*. Acosta invokes the right to empirically study the nature of the New World in order to justify free circulation. The justification of an imperial network of knowledge-power and management of American natural and human resources was possible only by invoking these basic Thomist onto-theological principles.

In citing the pursuit of knowledge as the foundation of the Spanish right to circulate freely in American territories, Acosta appeals to the most basic ontological ground: all human actions are ultimately guided toward ends that are intrinsically just. In their focus on Acosta's contributions to modern empirical science, neither Cañizares-Esguerra nor Barrera-Osorio account for this onto-theological foundation of Acosta's thought, which maintains the necessity of colonial violence as means to enforce the natural right to circulate and pursue knowledge of the natural world. Acosta's teleocratic doctrine not only defends the right to circulate but in the last instance *mandates* the free circulation of Spaniards in the New World at the same time that it restricts the circulation of Amerindian subjects as a subordination of means to ends. Acosta thus provides the onto-theological ground of his *Historia natural* and radicalizes Vitoria's defense of the inalienable right to circulate and pursue knowledge as technical means directed to the common good of the international commonwealth. In the Spanish American colonial context, this inalienable right likewise presupposes the use of violent force in the face of any resistance to the free circulation of Spaniards.

For Acosta, the right to freely circulate is not only bound to the right to pursue knowledge but also to the right of commerce, which is based on the free exchange of goods across borders: "it corresponds to the exercise of trade to take abroad that which abounds in one's country, and at the same time bring back what is lacking at home" (*De procuranda*, 450). Commerce, according to Acosta, is the means that God chose to "associate all mortals and maintain them united in mutual communication for their mutual benefit and utility" (450). Thus, Providence determines the distribution of tasks and resources among nations in order to bring them under a kind of confederation: "Metals abound in one part, in another precious stones, in another lumber, spices, herbal medicines, silk, manufactures, and a thousand other things" (451). Thus, the distribution of precious metals is central to a Providential division of labor in which some nations are destined to provide raw materials and others the perfection of form. Peru's place in this international distribution of labor is clear: "I do not know if there is another region richer in silver than this Peru, which in almost every other good is lacking" (451). Like Vitoria, Acosta attributes a common end to the world community: the "common good" justifies the Spaniards' right to circulate, conduct trade, and pursue knowledge as means to realize the "mutual communication" foreordained by Providence. The excess of silver in Peru and the asymmetrical absence of the metal in Spain is sufficient reason for Spaniards to navigate the Atlantic and circulate freely in the Indies. While Spanish merchants doubtless seek their own profit, Acosta maintains that the nations of the New World likewise benefit from the exchange generated by this "excellent work" (451).

The pursuit of knowledge, commerce, and conversion are guaranteed by the natural right of circulation, which not only justifies the Spanish presence in the New World but also grounds the human and physical sciences. Acosta thus places the techno-scientific and commercial expansion of the Spanish Empire at the core of natural right. Not only the evangelical mission, but more importantly the pursuit of knowledge, is covered by the umbrella of a universal and necessary force of the law that subordinates the inhabitants of the Americas. The Spanish are thus authorized to impose this natural right through force and to respond violently to any injury provoked by the Amerindians in its pursuit. Acosta remains within the framework established by Vitoria, wherein he attempts to remedy the absence of an ontological ground for Spanish power with an appeal to practical reason. This practical reason is in turn based on the absence or excess of precious metals, which becomes the motor of global circulation. In the physics of Aristotle and Aquinas, this absence/excess is understood in terms of privation, an accidental principle of physics: the desire or appetite is deprived of its object, which then appears under the sign of excess in Peru. In the last instance, the impulse to pursue knowledge, circulate, and trade is moved by the common desire to obtain the basic object and natural end of human desire, which is the common good. While Acosta justifies the division of intellectual and manual labor within Peru with an administrative, technical hylomorphism, the providential distribution of precious metals guarantees an international division of labor that grounds the very presence of Spaniards in the New World. God placed abundant deposits of silver in the New World for the common good of the totality, and the inhabitants of the Americas are required by natural law to share their resources with the Spaniards by the common right of circulation and commerce.

Acosta takes great pains to distinguish the pursuit of the common good from the "greed and excessive avarice" that leads to "harmful curiosity or vain ostentation" (451). The Jesuit concedes that there are those who are solely motivated by a vain curiosity and base greed instead of the pursuit of transcendental ends: "Who would deny this? Yet bear in mind that we are not dealing with the vices of men, but rather with that which is conceded by the common good" (451). Just as he will later argue with respect to the encomienda and the mita, Acosta here maintains that the perverse application of the principle does not invalidate the principle itself; thus, Amerindians' resistance to the circulation of Spaniards in the New World, even when predicated on the past actions of the Spanish conquerors, is a transgression of the natural order and action against the common good. This common good is further realized through "the very special cause" that guides Christians to spread the Gospel through the divine mandate transmitted through the New Testament: "Go therefore and make disciples of all nations" (Matt. 28:19).

Although the first apostles traveled unshod and without arms, Acosta argues that modern times require different measures, including the protection of soldiers. Thus, even the project of evangelization is predicated on the use of violent force in order to subordinate matter to form, given the "savage condition of the barbarians" and the "immense distances of the regions" (451). True to his Scholastic roots, Acosta subsumes the project of universal salvation within the concept of common good, which is the immediate end of the perfect community and the totality of the world order. In an attempt to distinguish the simple desire for riches, power, or territorial expansion (all illegitimate titles for conquest) from the true desire for the common good, Acosta appeals to the end of all ends, universal salvation, which again implies the necessity of military force given the nature of the Amerindian. Whereas Vitoria remained undecided on the foundation of empire upon the indemonstrable first principles of reason, and Sepúlveda opted for the infallible certainty that those very same principles provided, Acosta weighs the arguments for and against the justice of Spanish power under the tacit presence of Las Casas and ultimately returns to Sepúlveda's doctrine. The exercise of violent force is a means of protecting the project of evangelization, which is in turn guaranteed by the right of circulation as means of realizing the common, universal good. Again, the violent foundations of Spanish power in the New World is thrown out the door only to later sneak in through the window: the rights of circulation, knowledge, commerce, and evangelization are in the last instance guaranteed by the right to repel force with force.

Acosta seems to recognize a certain fissure in the foundations of the natural right to circulate, which is extended to the Spanish at the same time as it is restricted among the Amerindians. The Jesuit cites Aristotle's prejudice against travelers and foreigners, the source of continuous sedition in Athens, but argues that the situation is very different in the New World, where "they need foreigners to properly organize their republic" (451). Given their supposed lack of civil order and violation of natural laws, Acosta argues that it is just to "coerce them to a certain extent—excluding, of course, slavery and death—which is in the end a kind of benefit" (452). The Jesuit, therefore, maintains that Aristotle was correct in asserting slaves by nature can be subordinated justly through violent force, though he pleads for moderation in the implementation of this principle, excluding the possibility of slavery and death as just punishment. Acosta's hylomorphic geopolitics establishes the justice of the administrative, pedagogical regime implemented in Peru: "but we allow that they [Amerindians] be generously entrusted to those who are superior and wiser in order to rule and teach them with a view to their salvation" (452).

Once Acosta establishes the geopolitical order underlying the division of labor in the New World, he then addresses the contingent and problematic concern that motivates the Spanish presence in Peru: the "extraction of gold and silver in the mines, *which is what our people care about most*" (453; my emphasis). Given that the Amerindians do not value the precious metals under their feet, Acosta claims they must therefore cede this resource to the Spaniards, who possess the technique "to dig for metals or search for gold in the sands of rivers, or to dive for pearls on the sea floor, or to extract precious stones, or finally, to search for whatever is rare and unknown or undervalued by the Indians, [and] it is not unjust that those who appreciate these things should acquire them through diligence and industry" (453). Acosta cautions that these activities should be restricted within the limits of good government: "one should take care that our compatriots do not take by force or deceit whatever is being used or held in esteem by the natives, or that the natives are not forced to work for our benefit rather than their own, both of which present many dangers" (453). While Acosta justifies military force as a just means of defending the right to circulate and to seek the "rare and unknown" as grounded in the common good, he is likewise careful to distinguish between the enslavement of the Amerindians through war or the unjust exploitation of natural resources and Amerindian labor as the realization of the perfect community directed toward transcendental ends. The principles underlying Acosta's defense of circulation, ultimately founded on providential design, do not allow for the use of war or enslavement as punishment for crimes against nature, but they nevertheless mandate the direction, regulation, and subordination of beings in the pursuit of the end of all ends, the common good.

In establishing the right of circulation as the cornerstone of Spanish *potestas* in the New World, as well as the justification for the administrative and mercantilist regime of the colony, Acosta attempts to distance the structure the Spanish American colonial government from its violent origins. Nevertheless, in establishing the Spanish presence in Peru as a providential means to achieve the common good, Acosta returns to the just titles for war adduced by both Vitoria and Sepúlveda. The constant threat of violence and the implementation of exemplary and symbolic acts of terror are, according to Acosta, justified by the "barbarians'" frequent and unprovoked destruction of the signs of the Spaniards' free circulation: fortresses, farms, and ships (453). Acosta thus defends the right to avenge such injuries by "the force of arms" (453). This significant concession to the exercise of military force in order to defend the right of circulation and the fruits reaped therein is not, however, extended to the Amerindians: "As the Indians are immature and lacking intelligence it is better to treat them like children or women, or better yet like beasts" (453). The place of the Amerindians within the natural hierarchy of intellectual and manual labor restricts the use of force in response to injuries, which should only be employed in order to "punish and intimidate them, and rather than sharpening one's sword one should employ the whip, so that once chastened they learn to fear and obey" (453). Instead of indiscriminate killing, enslavement, and pillaging, Acosta recommends the prudent calculation of force required to subordinate the Amerindian subject: "Yet [the question of] to what extent, or where to cease will be more surely determined by the prudence and charity of the Christian captain, given that as a Christian he must demonstrate his faith in word and deed, and rather than worry about his own discomfort or offense he should aim to win for God the *precious commodities of souls*" (453). In the place of wanton greed and vengeance, Acosta advocates the rational calculation of benefits, which is materialized in the metaphor of the soul as valuable good, or precious metal. Vitoria's wise man is here replaced with a prudent "Christian captain" who must guide his conduct toward obtaining the "precious commodities of souls." Ultimately, the extent to which violence will be applied is to be decided by employing rational calculation based on the pursuit of profit, that is, remedying an absence with an excess, or surplus, guaranteed by circulation. Souls are equivalent to precious metals because money is the measure of value, and value is already surplus value, or *beneficio*. This process of abstraction converts natural and human resources into a standing reserve to be subordinated to the form of merchandise, thus Sepúlveda's bellicose hylomorphic teleology is supplemented by commodity fetishism.

PROVIDENCE, EMPIRICISM, AND THE IRREDUCIBLE CONTRADICTIONS OF IMPERIAL REASON

While *De procuranda* is a work that addresses the justification of Spanish power, the possession of precious metals and the exploitation of Amerindian labor from the perspective of missionaries and colonial administrators, it is only half of a larger project that unites colonial government and scientific knowledge in the context of Spanish imperialism. The *Historia natural y moral de las Indias*, published in Spain in 1590, represents the other half of this project, in which Acosta focuses on the natural history and ethnographic description of the New World. Acosta's *Historia natural* was considered a monument to modern science even by Enlightenment skeptics of Catholic superstition and Spanish despotism. William Robertson, for example, bestowed the title of "philosopher" upon Acosta, and Humboldt, who observed and chronicled nature in the Americas, considered the Jesuit's work to be evidence of the importance of the discovery of the New World for the progress of science. Acosta's contribution to modern science, lauded by the likes of Robertson and Humboldt, was nevertheless conditioned by the justification of dispossession presented in *De procuranda* and invoked the same principles of the Jesuit's imperial metaphysics.

The development of Acosta's work on colonial administration and natural history in the context of Spanish America would seem to bolster Heidegger's assertion that Aristotle's *Physics* was the foundational work of Western metaphysics.[10] The principles of Aristotle's physics are constantly cited in Aquinas's *Secunda secundae* in order to explain the relations between the divine, natural, and human laws. In his explanation of the ultimate foundations of sovereignty, Vitoria likewise appealed to Aristotelian physics and hylomorphism. The Aristotelian-Thomist tradition establishes an isomorphic relationship between the world of politics and the realm of nature by grounding human laws and natural regularities on an origin-end structure according to which God is a divine artisan who stands in front of the world in the same way an artisan stands in front of matter. Central to the argument of this book is the claim that Scholasticism provided a frame for what Adorno calls the "polemics of possession." Such a frame consisted of an instrumentalist metaphysics that transposes the sphere of technique to ontology, and then nature and politics. In order to hide this fetishistic transposition, metaphysics must repudiate its own instrumentalism in order provide natural subordination with a universal and necessary character. While natural mastery always presupposes artificial mastery, artifacts and manual labor appear as inert and neutral tools with no capacity to move themselves. For Aristotle and Aquinas, a technical artifact does not have the inner capacity to change. In other words, the ultimate and irreducible contradiction within Scholastic metaphysical instrumentalism is that, while technique is used to illustrate natural law, natural law proceeds to reduce both nature and technical objects to the result of imposing a form upon matter by a preexisting efficient cause that directs it to a proper end. An analysis of Acosta's tropology reveals the materialization of an imperial instrumentalism that attributes a "natural order" to the real. Imperial reason dictates its own foundations and finality, yet it must likewise disavow the arbitrary and violent act of mandating the law with an appeal to a transcendental ground capable of maintaining the totality.

In the *Historia natural*, Acosta uses Aristotelian physics and causality in order to explain the natural order of the New World, but these physical principles, as they apply to precious metals, are the same used by Sepúlveda to understand the place of the Amerindian in the hierarchy of being.[11] Precious metals are, therefore, understood as imperfect, amorphous, and plastic matter that must be corrected (*beneficiada*), in order to refine them into a more perfect and perennial form—that is, precious metal. This underlying hylomorphism contains the essential aporia of imperial reason: its objective, or telos, which determines both the success and failure of the universal Catholic monarchy, is predicated upon the extraction of precious metals and the transformation of Amerindians into forced laborers. As gold is transmuted into money and becomes the object of commodity fetishism, it likewise reveals an absence or fissure in the depths of the mines of Potosí, which are described as "foreign to human nature" and remain as reminder of the violent origins of imperialism. Acosta's encounter with the mines of Potosí represents the impossible task of assimilating the disruptive and violent force of colonialism to the ideality of imperial metaphysics, which nonetheless depends upon this violence as its organizing principle.

The concept of reason in the *Historia natural* includes both the observation and explanation of natural phenomena, as Acosta considered himself to be the first to undertake the project of analyzing the empirical data of the New World with reference to causes. In the "Proemio al lector," Acosta clearly outlines the aims of his project:

> Many authors have written sundry books and reports in which they disclose the new and strange things that have been discovered in the New World and the West Indies and the deeds and adventures of the Spaniards who conquered and settled those lands. But hitherto I have seen no author who deals with the causes and reasons for those new things and natural wonders, nor has any made a discourse and investigation of these matters; nor have I encountered any book whose matter consists of the deeds and history of those same ancient Indians and natural inhabitants of the New World. Indeed, both of these things are of no small difficulty. (*History*, 8)

While Acosta's work offers a plethora of empirical observations, these are ultimately dependent on Scholastic metaphysical instrumentalism and the division between determining form and indeterminate matter. Just as in Jerónimo de Chávez's *Tratado de la esfera* (1545), Acosta understands the world as an intricate machine designed to coordinate its various parts toward predetermined ends. The subordination of the parts is logical, and not merely imaginable, given the presumption of the Supreme Artificer believed to "uphold and enclose this immense machine of the earth" (*History*, 21).[12]

In *The Tropics of Empire*, Nicolás Wey Gómez explains how both geography and politics were intimately connected through a cosmological tradition that imagined the orderly working of the geocentric universe as a "machine of the world" (61). Wey Gómez describes how Scholasticism "connected the concepts of *place*—a subject of philosophical commentary among the ancient and medieval heirs of Plato's and Aristotle's physics—to what eventually became *colonialism*—a set of beliefs and practices associated with Christian Europe's great territorial expansion in the latter half of the fifteenth century" (61). This metaphor calls to mind the all-encompassing "hierarchical causality" that provided the cosmos with a "working coherence," linking spheres of knowledge corresponding to individual aspects of the created whole (Wey Gómez, 93). It was the somewhat permanent assumptions of this epistemic system that legitimized connecting concepts of place to the political ideas that framed colonial practices in the second half of the fifteenth century (93). Wey Gómez explains how Scholasticism inherits a sense of "machine" that maintained its connection to the Doric term "*machana*, which had stood for a medium of sorts, in the physical sense, an apparatus" (94). During the twelfth and thirteenth centuries, the concept of world-as-God's-machine was associated with the vision of nature as a "second cause," "in the context of a technological revolution that sought to mechanize a wide range of human activities" (94). Despite all the variations across the centuries, in the fifteenth and sixteenth centuries, the term *machine* was constantly employed to describe a Christian cosmos as a "working artifact" where God regulated motion and change in the elements and their compound by means of its intermediaries, the celestial bodies" (95). It is possible to draw many important lessons from Wey Gómez's reconstruction of the intricate relations between geography, colonialism, and the philosophical notion of *machina mundi*. First, Wey Gómez's monumental work shows that empirical knowledge that was inseparable from a vast philosophical tradition and geographical exploration did not displace the philosophical assumptions of the imperial agents. On the contrary, the practical culture of the explorers and merchants was backed up by the culture of theologians and natural philosophers well into the mid-sixteenth century. Second, it confirms Heidegger's history of productionist metaphysics as the increasing supersession of the artificial over the natural in the sphere of ontology. Among Heidegger's most important contributions to the genealogy of the relation between philosophy and technology is his analysis of the idea of imperial Latin Roman reason, *imperium*. According to Heidegger, medieval thought inherited Roman imperial reason as a self-regulated search for territorial domination:

> The Greek *aletheuein*, to disclose the unconcealed, which in Aristotle still permeates the essence of *techne*, is transformed into the calculating self-directing of the *ratio*. *This determines for the future, as a consequence of a new transformation of the essence of truth, the technological character of modern, i.e., machine, technology.* And that has its origin in the originating realm out of which the imperial emerges. *The imperial springs forth from the essence of truth as correctness in the sense of directive self-adjusting guarantee of the security of*

domination. The "talking as true" of *ratio*, of *reor*, becomes a far-reaching and anticipatory security. *Ratio* becomes counting, calculating, calculus. Ratio is a self-adjustment to what is correct. (Heidegger, *Parmenides*, 50; my emphasis)

By linking the Roman notion of imperial reason and the Scholastic notion of *ratio* as parts of the genesis of the security of domination and technological calculation, Heidegger also provides an insightful genesis of the discursive metaphysical presuppositions of Western technology in a colonial context. Finally, combining Wey Gómez's genealogy of notion of world as God's machine with the central claim of this chapter, which is about the transposition of technique to nature and politics, makes it possible to understand how Scholasticism is not only a philosophy of nature but also a philosophy of technology. On the one hand, by appealing to an efficient cause that imposes form on matter in order to direct it to its end, Scholasticism transposes technique to nature and politics. On the other hand, technique is a prosthesis—in the Doric sense of medium (*machana*), neutral device, or instrument—that is always controlled by an efficient cause, where God stands before the world in the same way that a technician stands before a machine or an efficient cause stands before prime matter, imposing forms in order to direct it to its proper end. Once again, imperial ideology is grounded in a metaphysical instrumentalism because it borrows the force of persuasion from technique in order to ground natural law and then disavows technique, reducing it to a neutral medium, subordinating the technique and the know-how of the artisan himself. To conclude, for Acosta, the Indies are part of the world machine, an all-encompassing harmonious totality, where everything has its precise place assigned by a divine artisan, who rules the world in a rational way.

Acosta was not only an empiricist who collected useful data but also a natural philosopher guided by the principle of natural subordination of imperfect matter to perfect form. The notion of rational machine, an all-encompassing harmonious unity of elements and compounds dependent on the Divine Artificer, an efficient cause that imposes forms on matter directing them to their proper end, is not an object of observation but, rather, an ontological frame used to organize empirical reality and orient it to the production of profit. The primacy of a Scholastic instrumental reason over the empirical in Acosta can be seen when imagination, the faculty that organizes empirical data, has to be firmly guided by reason in order to avoid error:

> It is surely a marvelous thing to consider that on the one hand it is not possible for human understanding to perceive and achieve the truth without making use of imagination, and on the other hand neither is it possible to avoid error if we follow only our imaginations. We cannot understand that the heavens are round, as they are, and the earth in the midst of them, except by imagining it. But if reason does not correct and reform this same imagination, but instead lets understanding be carried away by it, we must necessarily be mistaken and commit error. (*History*, 30)

In *Religion in the Andes*, Sabine MacCormack argues that Acosta elevates the imagination above the realm of illusion and demonic influence in order to vindicate the Aristotelian concept of speculative thought, which unites imagination with experience and reason (277–78). This, according to MacCormack, would bring Acosta within the sphere of Francisco Suárez and the Jesuits of Coimbra, who argued for the imagination as a faculty of the intellect (277). While Burgaleta questions a direct link between Acosta and the Second School of Salamanca, whose principal exponent was Francisco Suárez, they nevertheless share the same metaphysical presuppositions, particularly as these correspond to the context of Spanish imperialism in the Americas. As a matter of fact, Acosta's rationalization of imagination is a circular and tautological procedure: while the intellect perceives nothing that is not received first from the senses, this empirical matter is nevertheless regulated and perfected through the regulation of imagination by reason.

The first and second books of the *Historia natural* concern the material being of the *Ecumene* (the inhabitable portion of the earth) as the common substrate of both the New and Old Worlds. In the third book, Acosta explains that the New World is composed of Empedocles's four simple elements (earth, air, water, and fire) whose combination determines the geography and climate of the Indies. The fourth book deals with the *compuestos*, that is, mineral, vegetable, and animal matter. While this fourth book consists of one of the first treatises on the theory and practice of metallurgy and mining in Spanish America, it also reveals the relationship between imperial metaphysics and the understanding of the natural world, which the *Historia natural* reorders according to the ontological, theological, and political principles of the "imperial school" of Spanish theologians and jurists. Acosta subordinates his otherwise dispersed empirical observations on minerals and mining to the unifying principle of subordination.

Imperial reason, which orders in both the political and epistemological sense, is based on the Scholastic principles that ground not only Acosta's approach to colonial government but also his understanding of natural history and its relation to Providence in the economic order of Spanish America. This is especially true of Acosta's dissertation on mining in which the providential order of the natural world is directly related to the evangelical enterprise and the administration of a secular colonial government. Acosta interprets both simple and compound elements of the natural world always in relation to ends that are both theological and political. The fact that the Jesuit expounds upon the artificial practice of mining in a chapter dedicated to the natural order—as opposed to the artificial or moral order—of human societies and customs is a clear sign of the indiscernible distinction between technique and the metaphysical ground of imperial reason.

Acosta explains the Aristotelian theory of precious metals, which are understood by an analogy to animated matter that grows within earth by the dispensation of Providence:

> And though there are many other kinds, we will reduce this subject to three, namely, metals, plants, and animals. Metals are like plants hidden in the bowels of the earth and have some resemblance to plants in the manner in which they are produced because the places from which their roots arise, and their trunk as it were, can be perceived; these are the large and small veins that are very well interlaced and organized among themselves. And in a way it seems that minerals grow in the manner as plants, not because they have real vegetable and inner life, for this is true only of real plants, but because they are produced in the bowels of the earth in such a way, by the virtue and efficacy of the sun and the other planets, that over a long period of time they gradually grow old and almost, one might say, propagate. (*History*, 161)

Acosta thus asserts a relationship of similitude between metals and vegetable matter (metals are "like" plants), which establishes identity through the analogy of participation.[13] In fact, Acosta's theory of metals hinges on this simile. In the Aristotelian-Thomist tradition, every difference, determination, or state of perfection can be expressed analogically through the subordination of any part of the whole through a model that links this difference to the totality. Metals, therefore, can be *like* plants without being plants. Both metals and plants are at least partly hidden (*encubiertos*) from plain sight, a property that likewise suggests an anomaly (either moral or epistemological) that remains invisible until it is properly explained or justified. This anomaly is further hidden in the "bowels of the earth," a metaphor that suggests a matter buried within the core or depths of the earth as within an animal organism. In *The Order of Things*, Michel Foucault explains that reversibility and polyvalence endow "analogy with a universal field of application. Through it, all the figures in the whole universe can be drawn together" (22).

This metaphor of the earth's bowels suggests a further connection with animate beings, yet, more importantly, Acosta compares the production of metals to the growth of plants. Nevertheless, metals do not possess "truly vegetative and interior life" but, rather, what may be called the peculiar life of imperfect matter, for true animate and interior being is that which matter receives from form. Aristotelian hylomorphism thus maintains a metaphysical dichotomy between inanimate and animate entities, yet the peculiar life of metals, as Acosta describes them, is neither one nor the other. The life of metals is not a positive or observable phenomenon but, rather, an unobservable supplement whose being is strictly relational within the artificial structure of the natural order. This peculiarity is a result of the ultimate indeterminacy of analogical determination: an entity *like* that of a being-bestowing form but nevertheless deprived of this determining form in the last instance.

Acosta explains the "efficient cause" of metals as the "virtue and efficacy" of the sun and the planets, which influence the simple elements of the sublunary world. The matter of metals is not determined as a purely passive potentiality that awaits an active form to receive the interiority of being but, rather, already possesses a minimum of life (its peculiarity) that corresponds neither to a perfecting form nor to the "truly vegetative and interior life." This peculiarity is the obverse of the "quasi-propagation" of metals, which is a chaotic and undetermined process inseparable from a certain surplus that escapes analogical determination. This fissure within Acosta's metaphysics further widens when compared to his description of Amerindian idolatry in the fifth book of the *Historia natural*.

In Chapter 5, titled "On the idolatry of particular things by the Indians," Acosta explains that the Amerindian worships "whatever natural thing seems remarkable and different from everything else" adding that "they worship it as if recognizing in it a particular deity" (*History*, 308). The Jesuit cites a pair of examples that bring this practice uncannily close to his own analogical determination of the life of metals: "in this regard, anything that was different from other things like it seemed to them to be divine, and they even regarded small stones and metals and even roots and fruits of the earth this way—as with the roots they call potatoes, there are some oddities to which they give the name *llallahuas*, and they kiss and worship them" (308). Acosta says that for the Indians anything strange or different from its own species is something divine. This singularity that seems to be the object of worship is also associated with the convergence of the mineral and living world in the Andean cosmology. Cristóbal de Albornoz, an ecclesiastic inspector general commissioned in the 1560s to conduct an anti-idolatry campaign in the Andes, likewise explains the analogical relationship between metals and plants in the Andean reverence of *guacas* (indigenous objects of devotion that could also be natural locations):

> There is another type of *guaca* that they revere and serve with great care, which are the first fruits reaped from any field that has not been sown. They choose the finest fruit and safeguard it, and then make likenesses of the corn cob or potato out of stones or gold and silver, which they call "mamazara" or "mamapapa"; and the same is done with other fruits and the gold, silver, and quicksilver that was discovered long ago. They have chosen the most beautiful stones and metals and they keep them and chisel them, calling them the mothers of the mines. And the days they work at the mines, before heading to work they chisel off and imbibe some of the stone, calling it the breast of what they are to mine. I have found many of these objects in Guamanga wherever there are mines. (*Fábulas y mitos*, 18)[14]

For Acosta, however, this analogical relationship is predicated on the false analogy between idolatry and the Christian sacraments. The peculiarity of the metal-vegetable analogy is to be explained according to final causes, not as a pretext for idolatry.

Acosta further extends the metal-plant analogy to animal life: "And, just as metals resemble the earth's hidden plants, we may also say that plants are like animals that remain in one place, whose life is ruled by the nourishment nature supplies them at their birth" (*History*, 161). As "stationary animals," plants form part of the analogical economy of imperial metaphysics in which all parts are administrated in view of a finality just as the republic is governed by subjecting intransigent matter to ordering form. While maintaining the mineral-plant-animal analogy, Acosta also establishes a hierarchy of being that is based on perfecting forms: "But animals are greater than plants, for because they have a more perfect nature they also need more perfect nourishment; and Nature gave them movement to seek it and senses to find and recognize it" (161).[15] The nutritive faculty, which is essentially the power to transform nature in order to sustain life, is present in both animals and plants, but it is virtually absent in the case of metals, which are a purely passive potentiality. Each rung on this ladder of being represents an increasing ability to artificially shape the inferior rungs as well as superior mobility: "Hence, harsh and barren land serves as substance and nourishment for metals, and fertile and more amenable land is substance and nourishment for plants. Plants themselves are the food of animals, and both plants and animals are food for men, with the lower order of nature always serving the higher and the less perfect subordinated to the more perfect" (161–62).

This is strikingly similar to Sepúlveda's explanation of the hylomorphic principle in his defense of just war against the Amerindians. When discussing the different meanings of the dominion, Sepúlveda insists that they all derive from one and the same principle, that of the subordination of imperfect matter to perfecting form. The Indians were supposed to obey the Spaniards, just as matter is supposed to obey form. He considered their disobedience as a deviation from reason that could be punished. As Adorno persuasively argues in her monumental *The Polemics of Possession*, "this is the basic presupposition of Sepúlveda's thinking, and it is similar to Vitoria's reasoning on the same subject. For this reason, Sepulveda places great confidence in Vitoria's implicit support for his position" (118). The subjection of the indigenous peoples is grounded on natural law, which results from the fetishistic transposition of artisanal technique to natural causality and to political organization. While in *De procuranda* Acosta rejects Sepúlveda's arguments, which simply lead to unrestrained "military license," he nevertheless uses them to ground his metaphysical interpretation of New World nature and justify the extraction of precious metals. One could say that, in Acosta's hylomorphic and instrumental treatise, the nature of the New World finally becomes part of an imperial machine in which imperfect matter is subjected to the perfect form. Metals are at the base of this hierarchy of subordination, through which nature is not merely observed but, more importantly, interpreted and constructed as a resource by imperial metaphysics. Metals therefore become a pure standing reserve for the sustenance of the *machina mundi*, an all-encompassing harmonious unity, a medium or apparatus mastered by an efficient cause that organizes matter by imposing vertical hierarchies of perfecting forms and directing them to their proper end. While returning hylomorphism to the consideration of nature, Acosta nonetheless converts it into an administrative and instrumental philosophy that was implicit in both Vitoria and Sepúlveda yet not entirely developed in either one.

Although Burgaleta simply rejects MacCormack's association of José de Acosta to Suárez's metaphysics, due to the differences between the two Schools of Salamanca, it is instructive to examine Suárez in order to compare his own solution to the problem of prime matter to Acosta's. As Dennis Des Chene explains, while Aquinas denied any existence or power to matter, Suárez thought that prime matter could exist without a form (*Physiologia*, 91). Descartes deprived matter of any active power but attributed to it an independent existence by identifying it with extension. When examining the powers of the composite of matter and form, such powers may be found either in the form or within matter itself. Des Chene explains that "the parallel with Cartesian dualism need hardly be emphasized" (*Physiologia*, 151). Suárez's proto-Cartesian modernism is therefore found in his reduction of "*dipositiones* to the modes of quantity, and the elevation of all powers to the soul above the capacities of matter" (151). Heidegger also sees the importance of Suárez and the School of Salamanca as clear antecedents to the modern metaphysics of the *subjectum* of Descartes and later Kant. The consequences of Suárez's thinking are evident in a Thomism opposed to any such rupture between matter and form by which matter might obtain autonomy: "The dangers were not hard to discern. Matter, having been given little to the common properties of bodies, might now be free to declare its independence not only from form, but from God" (151). This is not the place to examine the intricate relations between Suárez and modernity, but it is important to mention that Suárez opened the door to Descartes's dualism—the separation of the realm of mind and matter—which ends in a subjectivist turn, where humanity attributes to itself the role of God, the grounding ground, the Master and producer of everything. While Suárez opens the door to subjectivism and Cartesian dualism, Acosta opens the door to considering metals as a "standing reserve" that both serve and sustain the machine of the world. The ultimate ideological and ontological question remains, however, of how a purely passive matter can sustain the chain of being.

The greatest threat to the legitimacy of imperial metaphysics is the wealth that sustains the machinery derived directly from Amerindian labor and the precious metals of the New World, as the source of this wealth reveals the arbitrary and violent origins of Spanish political power. Acosta refers to the confusion of the proper hierarchy of being occasioned by the desire for precious metals: "This makes us understand how far are gold and silver, which greedy men in their covetousness hold so dear, from being a worthy object for man, for they are many degrees lower than man" (*History*, 162). Acosta emphasizes that gold and silver are mere means subordinated to much greater ends, "and man is subject and subordinate only to the Creator and universal Maker of all things as his proper end repose, and all other things are worthy only insofar as they guide him and help him to attain this goal" (162). These paragraphs exhibit the classical Thomist ontology according to which God's machine of the world is an appendix, a means, subjected to an efficient cause or origin that orders it to its end.

The observation of nature itself becomes an act of devotion that reaffirms the unity of being: "The man who contemplates created things from this viewpoint, and ponders them, can gain advantage from knowledge and consideration of them, using them to know and glorify the Author of them all" (162). The very undertaking of natural history is regulated by the ultimate ends of the providential order, which reveals its "marvelous effects" to the imperial metaphysician. To observe is not to look on passively, but more importantly to imagine the finality of the natural order. Acosta insists that this is not merely an instrumental knowledge with a view to profit: "The man who goes no further than to understand [the] properties and uses [of created things], or who is merely curious as to knowledge of them or covetous in acquiring them, will find that in the end these creatures will be as the sage has said 'to the feet of the unwise and foolish they are a snare and a net into which they fall and are entangled'" (162). The reduction of nature to "properties and utilities" is the result of mere curiosity, greed, and folly, an action that confuses the ontological distance between grounding being and grounded entities. This condemnation nevertheless reveals Acosta's anxiety with regard to his own technical, instrumental philosophy, which unwittingly subverts the metaphysical binaries that provide its own ground. His imperial metaphysics, therefore, assumes the task of establishing a further division of intellectual labor that would distinguish between the mere practical observation and gathering of material and a theoretical knowledge that would link this activity to a providential design.

MONEY AND MINING IN THE SPIRITUAL ECONOMY OF THE NEW WORLD

In Chapter 1 of the *Historia natural*, Acosta ranks metals below other entities in the hylomorphic hierarchy of being. Metals are thus deprived of determination and available for the ends of human beings. Notwithstanding their inferiority in the hierarchy of being, metals nevertheless sustain all other more perfect beings. This minimal fissure between subordination and sustaining is aggravated as the inferior, subordinated entity gains the power to represent and support the totality of superior being as a generalized equivalent. Therefore, it is crucial for Acosta to explain the finality of metals, which God created "for medicine and for defense and for adornment and as instruments of man's activities" (162). Acosta explains that, of these four uses of metals, the fabrication of tools is the most important:

> We could easily give examples of all four of these uses, but the chief object of metals is the last of them; for human life requires not mere survival, like the life of beasts, but must also be lived according to the capacity and reason bestowed on it by the Creator. And thus, just as man's intelligence extends to different arts and faculties, the Author of all things also provided it with materials of different kinds for man's investigation and for security and ornament and a great number of activities. (162–63)

Within the finality dictated by the Supreme Artificer through the natural order, metals are an instrument to be used by human beings in the achievement of their higher purpose. The artificial use of metals becomes, therefore, natural and providential. Unlike the animal, human life cannot sustain itself without the use of reason and instruments. While the capacity to create such instruments does not exhaust human possibility, human beings cannot attain their potential or ends without tools. The use of metals as tools is the most important end, because these are the instruments through which human beings exercise the faculty of reason bestowed upon them by the Creator. Human beings realize this faculty (*ingenio*) through their capacity to manipulate means toward ends. The polysemy of ingenio is important, as it not only signifies this creative capacity to invent and wield instruments but also refers to the specific methods of amalgamating metals in the mines of Potosí.[16]

According to Acosta, God buried metals within the earth so that they would be discovered through human ingenuity to serve the uses of men. Metals are therefore an imperial standing reserve, a raw and passive matter to be perfected by form. Within the various instrumental forms that metals assume there are, paradoxically, some that are "simple and natural," while others, such as money, are of principal importance for the order of nature: "But above all these uses, which are simple and natural, communication among men resulted in the use of money, which (as the philosopher said) *is the measure of all things*; and, although by its nature it is but one thing, actually it is all things, for money represents food and clothing and shelter, mounts to ride on, and everything of which men have need" (*History*, 163; my emphasis). Acosta is quoting Chapter 5, Book 5, *Nicomachean Ethics* 1133a: "For it is not two doctors that associate exchange, but a doctor and farmer, or in general people who are different and unequal; but these must be equated. This is why all things that are exchanged must be somehow comparable. It is for this end that money has been introduced, and it becomes in a sense an intermediate; for it measures all things and therefore the excess and the defect—how many shoes are equal to a house or to a given amount of food" (Aristotle, *Basic Works*, 1010, 1011). Commenting on this passage, Aquinas says:

> He says first, in order that the products of the different workmen be equated and thus become possible to exchange, it is necessary that all things capable of exchange should be comparable in some way with one another so that it can be known which of them has greater value and which less. It was for this purpose that money or currency was invented, to measure the price of such things. In this way currency becomes a medium inasmuch

as it measures everything, both excess and defect, to the extent that one thing exceeds another. It is a mean of justice—as if someone should call it a measure of excess and defect. (*Commentary on the* Nicomachean Ethics, 312)

In his book *Economy and Nature in the Fourteenth Century*, Joel Kaye argues that Aquinas regards money as a medium because it serves the purpose of measuring everything by potentially equalizing them (66). Given the importance of calculation and measurement in Aquinas's intellectual context, Kaye argues that Aquinas took the capacity of money to the measure of everything "very seriously" (66). Kaye adds that heterogeneous commodities can be represented by a "commonly possessed quality" (66). Money, then, is regarded as an "invented, artificial measure" (66). Money enables the comparison of commodities by serving as a common standard (66). Therefore, both Aquinas and Aristotle conclude "that money acts as a *medium* and a measure by quantifying the qualities of use, utility and need attached to all goods in exchange" (68). This capacity of money is also an object of attention in *New World Gold* by Elvira Vilches, who notes that sixteenth-century Iberian "economic writing recognized the instrumentality of money" (51). Combining Nicolás Wey Gómez's explanation of the notion of *machine mundi* in terms of *medium* or machine with Aristotle and Aquinas's philosophy of technique as inert and neutral *instrument*, and Aquinas's statement about money being a *medium* of justice, makes it possible to better comprehend Acosta's justification of the mita. Recall that the mita and the encomienda are also *means* that need to be subordinated to the common good. Now, money is also an artificial means of exchange that provides a common measure. The problem is that money seems to be not a merely neutral medium but, rather, a medium capable of equating everything. Karl Marx also paid particular attention to Aristotle's notion of money in *Capital*. In the chapter titled "The Commodity," Marx comments on the same passage, arguing that Aristotle discovered the relation of equality in the "value-form" of the commodity, precisely because he acknowledged how commodities lack a real identity, which means that they also lack the commensurability or equality needed for exchange (*Capital*, 1:151). According to Marx, for Aristotle equation was "something foreign to the true nature of things" (151).[17]

Acosta explains the process by which money came to encompass the natural order: "For this invention, that of making one thing serve for all things, men—guided by natural instinct—chose the most durable and negotiable thing of all, which is metal. And among the metals they decided that those whose nature was most durable and incorruptible, namely, silver and gold, must have primacy in this invention of money" (*History*, 163). In the passage cited above, *tratable* (pliable) conveys the additional sense of "negotiable" or "exchangeable"; that is, precious metals can be used as a general equivalent because they can be transported and exchanged without affecting their identity or determination. In other words, despite its being an artificial thing, that is, a product of human *ingenio*, Acosta grounds the capacity of money to exchange value in material traits such as being durable and negotiable. Although metals are the least determined entities in the hierarchy of being, they become the sustaining force and measure of this order: "And thus, everything obeys money, as the Sage says" (*History*, 211).

Acosta's imperial reason, based on a compulsion to order, becomes explicitly mercantilist. Acosta codifies the expansionist impulse of capital that was already announced in Columbus's declaration that gold opens the doors of Paradise.[18] In the first place, Acosta describes money as an invention or a novelty elaborated by human ingenuity and therefore only indirectly part of the natural order. The development of money, as technique or means of exchange, is an act of shaping indeterminate matter toward the ends of determination, identification, and productivity. Nevertheless, the peculiar being of money, as it derives from metals, depends on both its indeterminacy and its incorruptibility. These characteristics, according to Acosta, permit the measurement of the value of all other entities. That is, indeterminate entities have the power to impose form through the circulation of money. Metallic money is also like a plant without being a plant: although it does not possess any of the qualities of life, it is capable of persisting through time. Metals are alive because money possesses a quasi-living, transcendent quality, which is the protean capacity to measure and *order* everything else.

Why is imperial metaphysical instrumentalism intrinsically fetishistic? It is well-known that, for Marx, being under the spell of ideology equals not knowing what one is doing. Metaphysical instrumentalism is ideological in the sense popularized by Marx, which points to a certain divorce between knowledge and practice: "They do this without being aware of it" (Marx, *Capital*, 1:166–67).[19] When under the spell of ideology, people seem to understand that money is simply an inert object, while in practice they endow it with a life of its own. The ultimate example of the procedure of ideological mystification is commodity fetishism, in which social relations are supplanted by relations between things. This means that the subject of ideology confuses a structural effect of a social network, a set of material practices such as exchange, with the natural property of one of the elements of the network, namely, money as the embodiment of wealth, or value in-itself.

In "The Fetishism of the Commodity and Its Secret" in the first volume of *Capital*, Marx shows how, in order to understand how ideology shapes reality itself, it is necessary to address the metaphysical and theological character of the commodity. Marx states that "a commodity appears at first sight an extremely obvious, trivial thing. But its analysis brings out that it is a very strange thing, abounding in metaphysical subtleties and theological niceties" (163). What appears to be obvious and self-evident is really full of what are, from a dialectical perspective, radical inconsistencies. First, it is necessary to point out that the secret of commodity fetishism, the form of value, is not use value (163). As soon as the product of labor emerges as a commodity sold on the market, "it stands on its head, and evolves out of its wooden brain grotesque ideas (163). This means that the "mystical character" of the commodity is not inherent in its use value or instrumental character. When Marx asks after the origin of the "enigmatic character" of the commodity, he unequivocally answers, "Clearly, it arises from this form itself" (164). He is saying that the equality of commodities is embodied in their equal objectivity once they are measured by money. The value of a commodity is expressed through the value of another commodity, and ultimately, through money as the value of value. Marx argues that "the mysterious character of the commodity form consists therefore simply in the fact that the commodity reflects the social characteristics of men's own labor as "objective characteristics of the products of labor themselves, as the socio-natural properties of these things" (165). The sensuous products of labor become "supersensible" (165).

The analysis of the "metaphysical and theological" character of money as instrument of exchange is the prologue to the notion of fetishism: as the conflation between social relation between people and the fantastic form of the relation between things, while endowing the artificial products of hands with a life of their own. Despite being a mere artificial product of the human brain, money appears to be endowed with intrinsic natural qualities and a life of its own. This fetishistic transposition gives origin to the perverse desire to accumulate wealth, which then proceeds to dominate and regulate the satisfaction of human needs. The limitless capitalist drive of capitalist self-valorization synthesized in the logic M-C-M′ is born out of the transformation of the means (money as instrument of exchange) into an end in itself (pursuing profit for the sake of profit) (*Capital*, 1:253). A particular subordinated thing becomes a new totalizing principle that proceeds to subordinate everything else, disrupting a preordained hierarchical natural balance that can no longer prevent the imbalance introduced into reality by the power of money. Since money represents everything by being the standard of all value against which all values are dissolved, nothing can hold in check the excessive assymetrical power of money. Those who possess money can buy things, while those who posses things do not necessarily possess money. Paraphrasing Acosta, money can buy "clothing and shelter, mounts to ride on," but "clothing and shelter, mounts to ride on" do not constitute money. No matter how useful these commodities are, they lose their value if they are not measured against money as their embodied value. Since money is endowed with an agency of its own, its circulation dismembers reality by reorganizing society according to a new metaphysical and theological logic that promises to produce profit while creating a new commodity economy. As a result, the promise of a theological value that is always to come is nothing but the deferral or postponement of use value itself. Infinite debt, financial speculation, the dematerialization of money, documented in detail by Elvira Vilches,[20] as well as the creation of the networks of commerce and credit described by Kamen, are inseparable from the split between instrumental value and the transcendent metaphysical supersensible character of the commodity. Although Wallerstein locates the rise of the modern world system, which is intrinsically capitalist, in the "long" sixteenth century (c. 1450–1640) as the result of the contingent crisis of feudalism (c. 1290–1450), it is more accurate to locate it in metaphysical, practical, and capitalist subordination of the imperfect matter of the New World to the perfect form of value. Sixteenth-century global expansion was inseparable from the gradual transference of the belief in transcendent (which is to say, more-than-material) forms and ends, which gave sense to a structured and hierarchical universe, into an unacknowledged belief in the mystical aura of commodities as self-subsistent entities that command everything else. Moreover, such transference would have been impossible without the Spaniards' stubborn attachment to money as possessed of intrinsic value. It is instructive to bring Vilches's elegant and succinct explanation of how Iberian Scholastics' theories of economy "embraced the spirit of advancing capitalism, while remaining, paradoxically, attached to gold as substance of intrinsic value" (3). I interpret this paradox as the result of a *practical* fetishistic illusion. While, in theory, social agents know that metal is a subordinated medium, in the real social practice that takes place in the expanding networks of the market, these same social agents act as if money was the natural embodiment of universal value.

Returning to Acosta after this brief detour through Marx, the question that arises is: Where, then, is the fetishistic inversion in Acosta's metaphysical instrumentalism? It resides in the contradiction between the subordinated nature of metals as imperfect matter and the subordinating power of money. Whereas in his first chapter, Acosta regards metals as a passive imperfect principle subordinated to more perfect forms, in the second chapter he shifts from an instrumentalist explanation of metals to focus on their role as instrument of exchange. On the one hand, following a strict Thomist logic, metals are imperfect matter, a subordinated medium, an inert passive instrument that fulfills needs by being deployed by an efficient cause that directs them to their proper end. In this sense, they have only a use value, the result of a goal-oriented activity within a providentalist frame. On the other hand, precious metals are an abstract universal medium of exchange that results from separating them from their concrete use value, attributing their subordinating power to one specific feature, their being incorruptible and durable, endowing them with intrinsic qualities that ground their universal exchange value. When transformed in money, precious metals acquire a commanding power of their own. A particular medium or instrument becomes the measure of everything else, an end in itself. In other words, in Acosta's explanation of how money went from being just one particular thing among others to the universal commodity as stand-in for all other commodities, there is a *salto mortale* by means of which the lowest entity, a subordinated, plastic, raw matter is elevated to the condition of a new totalizing principle and subordinates everything else, even labor and other material techniques, to its own power. While in the first chapter he justifies extractivism in line with Divine Providence, in the second chapter he states that precious metals are in themselves money.

Combining insights from Heidegger and dialectical materialism, it is possible to see how this shift of use value (metals serving an instrumental purpose) to universal exchange value (being a virtual construct that represents everything else) is the result of processes of abstraction in which money retroactively configures the metaphysical presuppositions used to justify the real aim of the empire. In other words, the ultimate contradiction between the merely instrumental character of money and its commanding capacity derives from the abstractions that take place within metaphysical instrumentalism. The ultimate irreducible contradiction within metaphysical instrumentalism is the disjunction between the subordinated use-value and the subordinating exchange-value, or the metaphysical instrumentalist conception of both nature and instruments as passive neutral prostheses and an incipient capitalism in which money commands and measures the value of everything else.

In sum, while in the first chapter of the fourth book of the *Historia* Acosta uses a logic of instrumental domination based on the principle of natural subordination, where the imperfect is subordinated to the perfect, in the second chapter he uses a logic of mercantile capitalism where that which was imperfect now subdues the superior beings, including humans, to its own power. To paraphrase Marx, we can say Acosta knows very well that, in Aquinas's theory, metals are imperfect matter, pure plastic potency, dead inert tools, whose only purpose is to serve the superior form, *but in practice,* everybody treats metals as if they were the living source of value, the measure of value of everything else, even of the evangelical mission. In theory, metals are deprived of consistency, determination, and perfection, being mere instrumental means that serve the goals/ends of the commanding efficient cause, while in practice, they are the measure of all value, the quasi-living source of power that sustains the material networks of production, circulation, and credit. For this reason, Acosta's imperial ideology—with his appeal to the monetarization of work, his enthusiastic support for the introduction of a revolutionary technology like amalgamation, and his defense of the introduction of the mita as forced yet paid labor in order to satisfy the labor supply of the newly implemented refining method—is inseparable from the theological question of value, the ultimate presupposition, the fiction that sustained the local and global system, a transcendent ground fissured by an ultimate lack of guarantee that presented what everybody strove for as already accomplished. The principle of subordination traverses the theological, political, and economic sphere of the colonial enterprise.

Returning to the *History*, it is important to remark that Acosta was not entirely ignorant, though, of the consequences of transforming precious metals into ends in themselves. The Jesuit writes, "But the truth is that their greed was not as immoderate as ours, nor did they idolize gold and silver as much—even though they were idolaters—as some bad Christians, who have committed so many excesses for the sake of gold and silver" (*History*, 163). Acosta reveals his anxiety that precious metals and the development of technologies for extracting them become the measure of all things, including Christian values. The use of the word *idolatría* (idolatry) indicates his concern for safeguarding the faith and reining in the corruption of the onto-theological principles that undergirded Spain's technical and scientific expansion.

Acosta argues that Divine Providence decided "to enrich the most remote parts of the world inhabited by the most uncivilized people, and has placed there the greatest number of mines that ever existed" so that God could "invite men to seek out and possess those lands and coincidentally to communicate their religion and the worship of the true God to men who do not know it" (164). Thus, Spanish imperial expansion had come to fulfill Isaiah's prophecy that the church would expand "on to the right hand and to the left" (164). Saint Augustine likewise had declared that the Gospel would be spread not only through charity but also "by those who proclaim it through temporal aims and means" (164). Thus, precious metals abound among those peoples without *"policía"*—that is, without civil institutions in conformity with natural law—due to a providential will to bring these peoples within the fold of Christianity by means of Spanish imperial expansion. The apparent contradiction in the coexistence of an absence of civilization and an excess of precious metals is resolved through the metaphysical teleology of prophetic narrative. Within this teleocratic and prophetic framework, the history of humanity has a spiritual end, the expansion of the Catholic Church, which employs temporal means. The unnatural, alien, or unformed life is regulated, corrected, and converted to its place within universal history, but only by means that threaten to undermine the very order they supposedly sustain. The Jesuit doctrine of accommodation of temporal means to the ends of evangelization is inseparable from the accommodation of these temporal means to a teleological view of history as always written from the end.[21] Acosta's teleology imposes order upon the apparent preexisting chaos or division that separates regions in which metals proliferate from those in which money is scarce.

Thus, ambition and vice—which ignore the distance between human dignity and perfection and the imperfect life of metals—become the means employed by Providence in order to save the souls of the Amerindians. In Acosta's words, "Hence we see that the lands in the Indies that are richest in mines and wealth have been those most advanced in the Christian religion in our time; and thus the Lord takes advantage of our desires to serve his sovereign ends" (164). The blind forces of history, avarice, and desire for that which threatens to destabilize the moral order are reordered as means employed by the invisible hand of Divine Providence. Acosta compares the Andes to a homely daughter who needs a generous dowry in order to attract suitors: "In this regard a wise man once said that what a man does to marry off an ugly daughter is give her a large dowry; this is what God has done with that rugged land, endowing it with great wealth in mines so that whoever wished could find it by this means" (164).

Acosta marvels that precious metals are so varied and abundant in Peru and that "new mines are discovered every day" (164). The sterile and unproductive nature of the land, which according to the Jesuit is propitious for the generation of metals, is a clear sign that many more mines have yet to be discovered, "and it even appears that the whole land there is sown, as it were, with these metals, more than any other land known at present in the world or in any written about in the past" (165). Once again, Acosta not only describes but also explains, searching for rational causes of phenomena: "The reason why there is so much mineral wealth in the Indies (especially in the West Indies of Peru), is, as I have said, the will of the Creator, who distributed his gifts as he pleased" (165). The availability of raw material is evidence of onto-theological predetermination: "Today this great treasure comes to Spain from the Indies, for Divine Providence has ordained that some realms must serve others and render their wealth and share its use for the good of all men if they use properly the assets they possess" (169). Let us recall that, for Aquinas, the rational pattern of Divine Providence is eternal law, which, again, is exemplified as an act of technical subordination. In Article 1, *Summa Theologica* IaIIae.93, titled "Whether the eternal law is supreme reason existing in God," Aquinas responds to all objections of those who don't think that eternal law is the supreme reason existing in God appealing to the instrumentalist example of the principle of subordination according to which God stands in relation to nature in the same way the craftsman stands in relation to the product of his art. Also, when arguing that every law derives from eternal law, Vitoria appeals to this same line of argumentation. As a result, Acosta's providential design is enframed in Aquinas's metaphysical instrumentalism, according to which the preexisting rational pattern moves things to a proper end, in the same way that a craftsman moves the products of his art, providing an ideological justification for the transatlantic networks of circulation of credit, commerce, and the production of silver.

In *De procuranda* Acosta had defended the Spanish presence in the New World by appealing to Vitoria's right to circulate, communicate, and trade. He defended the encomienda and the mita using Aquinas's metaphysical instrumentalism, by invoking the notions of unity and common good as the ultimate ends of society. In the *Historia natural*, he uses the principle of the natural subordination of imperfect matter to the perfect form in order to explain how metals serve and sustain the *machina mundi*. In other words, Acosta uses the principle of subordination not only as a justification of natural servitude of the Indians but also as a justification of the servitude of nature. Nature itself is conceived as a machine, a standing reserve, a means subordinated to divine and human ends. Metals are an imperfect and subordinated particular thing that subordinates everything else by becoming the measure of common good. In the text there is a transit from the instrumental value to exchange value that threatens to erase the natural hierarchies invoked by natural law. Nature thus becomes a machine-means and money becomes an abstract form of exchange that undermines the verticality of transcendence. While Vitoria disavows the instrumentalism of metaphysics by naturalizing civil power, Acosta disavows instrumentalism by reinscribing teleology in the frame of Divine Providence. Precious metals, which take on the appearance of vital form while remaining imperfect matter, reappear as an excess whose unequal distribution threatens to undermine the natural and moral orders.

INSIDE THE MINES

After describing the discovery of Potosí within an imperial and ontotheological historical framework, Acosta proceeds to focus his analysis on the extraction and refinement of metals. In the chapter titled "On the method of working the mines of Potosí," Acosta explains that each vein is mined by several operators who lend their names to their enterprises. The Jesuit focuses on the abyssal depth of some mines that "are 180 *estados* deep, and even 200" (*History*, 180). These mines reached such depths that the Spaniards began penetrating the veins from below through "*socavones*" or "caves that are excavated from below through one side of the mountain, piercing it in order to reach the lodes" (180). The latter method was more cost effective, yet the mines remained an infernal abyss that disoriented the senses: "And yet they work there in perpetual darkness, with no idea of when it is day or night; and as these are places never visited by the sun, not only is there perpetual darkness but it is also extremely cold, with a very heavy atmosphere unfit for man's nature; and so it happens that those who enter the mine for the first time fell weak and dizzy, as happened to me, experiencing nausea and cramps in the stomach" (180).

The contrast with the optimistic celebration of the role of metals in the rational pattern of the emerging world system is appalling. Following the development of Acosta's discourse, we see how he first explains how metals are subordinated to the good functioning of the machine of the world, then describes the use of metals as a medium of exchange and equivalence, and finally descends into the abysmal conditions of the production of riches that ultimately sustains the empire. The descent into the mines is an inhumane experience, likened to death itself. While the light of Divine Providence generates precious metals within the mountains, this interior is cast in perpetual darkness and produces a corporeal sensation of disgust. Imperial reason finally reaches the limit of the principle of subordination, where vertigo and nausea overwhelm the subject who steps into the abyss. The abyssal experience in the interior of the mines where indigenous laborers extract metals is the exact correlate of the metaphysical instrumentalist justification of the mita. The material conditions of production of mining—the manual labor of the miners inside the mines—are a horrible and immoral thing that has to remain hidden from sight.

At the very bottom of the chain of being, labor sustains the world in an "atmosphere unfit for man's nature." Acosta continues describing the material conditions of Amerindian labor. Laborers extract silver from the mine aided by the tenuous flicker of candles, "for, as I have said, no daylight comes from above" (181). While in the *Historia*, Acosta does not refer directly to the mita, which was justified in *De procuranda*, he nevertheless expresses horror at the conditions in which the Amerindians labor at Potosí as well as the greed that is responsible for their subjection. In describing the depths to which the laborers must descend and then climb up again weighed down with rocks, Acosta remarks, "They climb catching hold with both hands, and in this way ascend the great distances I have described, often more than 150 *estados*, a horrible thing about which it is frightening even to think. Such is the power of money, for the sake of which men do and suffer so much" (*History*, 181). Acosta seems to be conscious of the material genesis of the value of metals as medium of exchange described using the exact same words in Chapter 5, Book 5, *Nicomachean Ethics* 1133a, a process through which the context of greed, nausea, and terror bears precious metals that will sustain the business enterprise of the empire. The depths of the mine and the inhuman conditions of Amerindian laborers are what sustain the imperial business itself. Money bears the trace of its origin in the artificial violence behind metaphysical instrumentalism, an experience of horror and emptiness that reveals the true nature of the subordination of matter to form and means to ends.

Acosta's descent into the mines in the *Historia natural*, which is also a descent into the disavowed consequences of his metaphysical instrumentalism, is directly connected to his observations on the use of Amerindian labor in mining and the necessity of secular administration in Chapter 18 of *De procuranda*. In the latter work, Acosta condemns the forced labor levies as cruel and outrageous and reminds the reader that work in mines was used as punishment for criminals in ancient Rome (*De procuranda*, 489). The Jesuit asserts that such labor is akin to a death sentence and concludes, "To force free men who have done no wrong to perform this work is inhuman and iniquitous" (489). Just as in the *Historia natural*, Acosta refers to the mines as infernal abysses and ruminates that "Not without reason did the ancients feign that wealth was hidden within the realm of Pluto" (*De procuranda*, 489). Acosta gives a dramatic description, punctuated by the use of asyndeton: "Yet all that is smoke and shadows in comparison with what we witness today: eternal horrendous night, thick subterranean air, a perilous descent, arduous struggle against treacherous stone, standing is dangerous, a slip of the foot could be the end, the onerous loads bowing their shoulders, the ascent on inclined and uneven ramps, and other aspects that are frightening to imagine" (489). Acosta admits that the *mitayo* system is a violation of the natural freedom of Amerindian subjects who are uprooted from their communities in order to supply Potosí with a steady source of labor (489). Once delivered over to the mines, the laborers are exposed to pure contingency: "Morever, they are exposed to extreme danger due to the burden of the work, the change of climate and the many hazards from which many die" (489). No free subject of the Crown, Acosta argues, should by exposed to pure chance, accident, or death such as occurs in Potosí.

Nonetheless, Acosta recognizes that this horrifying experience sustains not only mining but also the evangelization of the Indies. The Jesuit argues, "if the mining of metals ceases . . . and the work in the mines is abandoned, then the Indies are finished, and the republic along with the administration of the Indians will perish" (490). Profit from the mines is the motive behind the desire to travel, trade, and survey that is the cornerstone of natural rights and the law of nations. This fundamental desire for personal profit moves secular and ecclesiastic alike: "because this is what Spaniards seek after such a long journey over the ocean: the merchant trades, the judge presides, and often even the priest spreads the Gospel in search of precious metals" (490). Just as in José Luis Capoche's *Relación de la Villa Imperial de Potosí*, Acosta describes colonial society as organized around the fulfillment of a basic want, or absence, of precious metals: "The day that silver and gold disappear, with them will disappear all commerce and abundance, and the gathering of civil adminstrators and priests will cease" (490). Acosta offers the not-so-distant Spanish American colonial past of Hispaniola, Cuba, and Puerto Rico as evidence for the imminent collapse of the viceroyalty should precious metals, or the means to extract them, disappear: in the past they were densely populated while there was silver and god, and now they are almost deserted and savage after the Indians died off and there was no one left to mine the precious metals that abound there" (490).

Thus Acosta confronts the fundamental double bind that underlies the problem he has taken on: "Given this state of affairs, I do not know what is preferable. On the one hand, I should lament the calamity of our times in which charity has grown so cold and, as the Gospel says, faith is rarely to be found on earth. Isn't the salvation of so many thousands of souls to be won for Christ sufficient recompense to awaken and inspire our own souls? Is the desire for gold and silver so great among us that when they are absent the salvation of souls means nothing to us?" (490). On the one hand, Acosta laments that spiritual gains are predicated upon the temporal profit motivated by greed, but on the other hand he marvels at the workings of Providence:

> But, on the other hand, we must stand in awe before the providence and benevolence of our Lord, who employs our human condition to bring the Gospel to such remote and barbarous nations, copiously planting these lands with gold and silver in order to awaken our greed. As charity was not enough to inspire us, then at least lust for gold would be sufficient stimulus. Just as in other times the disbelief of Israel was the cause for the salvation of the gentiles, now the greed of Christians has become the cause of the Indians' salvation. (490)

Here, Acosta makes the decision to abide the inhuman and infernal horrors of Potosí as a temporal and contingent means that, like all earthly events and actions, are stamped with the rational order of Divine Providence. Acosta asks if there is anyone who would not "look up with awe" at how Divine Providence transmutes silver and gold, "plague of mortals," into the spiritual salvation of the Amerindian (491). The encounter with the tenebrous depths of the Potosí mines—that is, with the emptiness, contingency, and inhumanity, or all that is repudiated as cruel and incompatible with Christian morality—is thus incorporated within a providential, teleological discourse that ultimately depends on a constitutive privation that is substantial rather than accidental. The pious and prudent administrator cannot abide the offense to God or the spiritual damage to the Amerindian that might be occasioned by the disappearance of mining.

Moreover, Acosta adds that the matter of Amerindian labor has been "seriously and maturely" deliberated by a council of theologians and jurists: "And there are provincial laws that determine the order and moderation to be observed in extracting metals in order to provide for the health and comfort of the Indians" (491). Before the authority of such provisions, Acosta concludes that it is not his place to "reject the judgment of such illustrious men, and much less propose or petition for new laws that would mitigate the harsh conditions of the natives" (491). Acosta therefore limits himself to insisting upon the necessity of overseeing the work conditions of Amerindian laborers and assuring that they recceive spiritual guidance, sufficient sustenance, proper physical treatment, and a fair wage so that they may "seek their own small profit" (491). Ultimately, punishment for ill treatment of the Amerindian is God's responsibility: "But if they [the Spaniards] do not comply with these guidelines and cruelly treat the Indians like slaves, then they must reckon with God, who is the father of the poor and protector of the orphaned" (491). Acosta finds a solution to the deadlock caused by the practice of mining in the promulgation of laws that would protect the Indians. In sum, Acosta pursues a practice of mining with a human face.

The inhuman material conditions inside the mines, which are cause of scandal even for an apologist of the empire such as Acosta, add a new dimension to the fetishistic character of the metaphysical basis of imperial ideology. As stated before, ideology implies a certain disavowed knowledge, an ignorance of what one is doing. The inhuman conditions in the mines are the material conditions of possibility of the production of money as a commodity. The inhuman abyssal space of the mines is the literal embodiment of the inconsistencies of metaphysical instrumentalism. This means that mining is practiced regardless of the knowledge of its destructive consequences. Acosta knows very well what is going on inside the mines, yet he still states that it is part of a providentalist and teleological frame—metaphysical instrumentalism—that justifies the extraction of precious metals for the sake of a transcendent end. They know very well what they are doing, but they keep doing it.[22] The infernal hell of Potosí is the materialization of a void that condenses the contradictions of metaphysical instrumentalism. The only way to cover this gap will be by reintroducing metaphysical instrumentalism, by naturalizing the subordination of the means to the end. The problem is that this operation does not do away with the contradictions but, rather, exacerbates them, since the material practices now turn around this contradictory void, Potosí.

Acosta confronts the fissure that crosses natural rights theories from Vitoria onward: enjoying the fruits of Amerindian labor is impossible to justify yet nonetheless necessary. While the labor within the mines of Potosí is described as horrid and inhuman, a reminder of the artificial power of money as both medium and measure of riches, it is nevertheless a means created for the realization of transcendent ends. While Vitoria used natural law in order to justify the indirect expropriation of Amerindian property and labor, and Sepúlveda used these same principles in order to unequivocally assert the need to subject the Indians, Acosta concludes that violence is an inevitable and necessary means for the consummation of a providential plan. Acosta's administrative rhetoric is backed up by a metaphysical instrumentalism that subordinates the means to the end. Acosta represents the administrative and practical turn within the history of the principle of natural subordination of matter to form and means to an end. His oscillation between the morality of the means and the supersensible ends is cut by the decision to stick to the means despite their undesirable consequences. The more Acosta descends to the material conditions of the imperial enterprise, the more contradictory becomes the relationship between the material means and the more than material ends. Acosta's last attempt to eliminate this irreducible remainder, the horrifying pain that is inseparable from the mining enterprise, was by describing the functioning of amalgamation in Potosí.

After addressing the issue of Amerindian labor and the infernal conditions of Potosí in the *Historia natural*, Acosta then turns to the amalgamation of silver with mercury, a revolutionary method that Francisco de Toledo implemented in Peru. Before 1570, high-grade ores were smelted in portable ovens called *guairas*, while low-grade ores remained unproductive. This revolutionary process initiated what Bakewell calls "the silver boom of the 1570s."[23] In the terminology used by Acosta, only ores that contained lead could be refined while those that were "dry" were discarded (*History*, [182). With the method of extraction using mercury, these "dry" ores "turned out to be immensely rich" (183). Acosta writes, "and this was a complete success for those mines; for with quicksilver they could extract any amount of silver from the ore that had been wasted, which they called slag. For, as has been said, quicksilver extracts all the silver even though it be poor and of low grade, which is something that smelting by fire does not do" (188). Acosta describes with wonder the "strange and marvelous property" of mercury in seemingly transmuting worthless dross into valuable metal; the *beneficio*, or process of amalgamation, separated the inert and unproductive matter from the incorruptible and lasting form. There is a revealing contrast between Acosta's description of the work conditions of the mines, in which he employs the rhetoric of horror, and his description of this technological advance, in which he uses words like *extraña* and *maravillosa*. While similarly referring to events or objects that are unfamiliar or aberrant, the latter description has a patently positive valence.

In the *Historia* Acosta explains that mercury resembles the properties of both inert beings and animate life: "because it appears to be alive in the way it bubbles and slides quickly from one side to another, has great and wonderful properties among all other metals" (183). Mercury is an exception within its category, for it combines the characteristics of hard metals with those of liquid metals: strangely, it is both heavy and liquid. Acosta explains, "The first is that, although it is a metal, it is not hard, nor does it have a shape and hold together like the other metals but is a liquid and runs, unlike silver and gold, which once melted by fire are liquid and run, and does so out of its own nature; even though it is a liquid it is heavier than any other metal, and so the other metals float on the quicksilver and do not sink because they are lighter" (183). Acosta adds that the technological advance of using mercury in the amalgamation of silver was a secret unknown to the Ancients, that is, a wonder reserved for modern times. Mercury's paradoxically fluid and heavy characteristics inspire awe:

> Among all the wonders of this strange liquid, for me the one most worthy of consideration is that, although it is the heaviest thing in the world, it instantly turns into the lightest thing in the world, which is vapor, and rises straight up. And then that very vapor, which is a thing so light in weight, immediately becomes as heavy as the liquid mercury into which it turns, for when the vapor of that metal strikes some hard body above it, or reaches a cold atmosphere, it immediately takes shape and again falls in the form of liquid quicksilver. (185)

The contradictory essence of mercury, representing opposing extremes within its nature, is what makes it the condition for transforming precious metals into the measure of all things. Speculative and practical reason again confronts the fundamental strangeness of nature: "Surely the cases in nature of such immediate transmutation of so heavy a thing into one so light, and vice-versa, must be very rare" (185). This "rare" thing is analogous to the "refinement" of Amerindian habits, wherein the accidental is corrected through evangelization and administration. Nevertheless, Acosta describes the refinement of metals not as a transformation of their essence but, rather, as the isolation of the incorruptible and eternal from the useless.

In Chapter 12 of the *Historia natural*, titled "On the art of mercury amalgamation and extracting silver from it," Acosta gives a detailed description of the amalgamation method first invented by Bartolomé de Medina in Mexico and later implemented by Pedro de Velasco in Peru. The chapter ends with a direct comparison between this process and ascetic techniques for purifying the soul, which is based on his reading of Psalm 11 where the word of God is described as pure as silver refined seven times over. The process of amalgamation is a kind of penitence to be performed on imperfect matter in order to produce a perfect form:

> Silver undergoes all these torments and sufferings (so to speak) in order to be pure; if we think about it, it is a shaped mass that is ground and sifted, and kneaded and leavened and cooked, and even in addition to this it is washed over and over and cooked and recooked, passing through mallets and sieves, kneading troughs and furnaces, barrels and pans, wringers and kilns, and finally through water and fire.
>
> I say this because, having seen this process in Potosí, I thought of what Scripture says of just men, "colabit eos, et purgabit quasi argentum," he shall purify them, and refine them as silver. And I thought of what it says elsewhere, "sicut argentum purgatum terrae, purgatum septuplum," as silver purged from the earth, refined seven times. To purify silver, and refine it and cleanse it of the earth and clay where it occurs, they purge and purify it seven times, for indeed it is done seven times; that is, many and many times is silver tormented until it is left pure and fine. So it is with the Word of the Lord, and just so will the souls destined to partake of his divine purity be refined. (*History*, 191–92)

The *beneficio*, which is a technology introduced for the purpose of speeding and maximizing the isolation of silver, is explicitly compared to a technique of penitence aimed at purifying the soul and achieving salvation. If, in *De procuranda*, he compared evangelization to commerce by employing the metaphor of the "precious commodity of the soul" (453), the mortification of the soul, gaining the precious commodity of the soul, is illustrated using the metaphor borrowed from a technology employed to increase profit. Acosta resolves the contradiction between the technical means and nontechnical ends by transforming the surplus of pain into a surplus of salvation of the "precious commodity of the soul" (453). In *De procuranda*, Acosta maintains that the moderate use of coercion is necessary in order to reform the bestial habits of the Amerindian;

in the *Historia natural*, this violence reappears in the analogy between amalgamation and mortification as technique for separating out impurities through suffering. Faced with the infernal abysses of Potosí, Acosta naturally turns to the disciplinary, ascetic techniques of controlling the emotions and actions through the application of pain. The horror of the Peruvian mines is therefore justified by appealing to an end as the cause of all being. In Acosta's imperial metaphysics, the transition between political administration, mining technology, and ascetic discipline contaminates any possible distinction between the natural and the artificial upon which imperial reason could rest. Salvation itself becomes both a business and a technology. Mining is good not only for business but also for evangelization.

IDEOLOGY AND REALITY

To conclude, I claim here that a dialectical materialist mutual constitution between actual material networks of production and the virtual metaphysical ideals of unity remains hidden and distorted by both empiricist historiography and the imperial school of thought. Historians such as Henry Kamen demonstrate how labor and international networks of technology sustained an empire that escaped the control of the Spanish Crown.[24] Official theorists of the empire, according to Kamen, gave a Castilian vision of it that hides the truth about the struggle of the Spanish Crown to dominate and control a network as uncontrollable as the production of commodities and the financial debt-creditor system. Kamen clearly sees how the material production (what he calls the "business of world power") sustains the ideal in the same way that Acosta's metals sustained the all-encompassing machinery of the world business. Nevertheless Kamen overlooks the way technique and the know-how of the laborer is immediately subordinated to a higher unity. Let us recall that Acosta declares that the inferior nature sustains the superior one, *by being subordinated to it*, according to the principle of natural subordination of matter to form and means to ends. In other words, while Kamen thinks that all ideas and representations of the imperial school of thought were merely imaginary and symbolic, as opposed to the real material networks of production and extraction of profit, he overlooks the way that ideology does not simply mask reality but produces it.

When dealing with colonial writings, it is important to keep in mind

Adorno's principles, the first one being that colonial writings do not merely reflect social and political practices because they are "constitutive" of them (4). As Adorno emphatically claims, "These works do not describe events; they *are* events, and they transcend self-reference to refer to the world outside themselves" (4). Therefore, it is completely unproductive to oppose metaphysical unity to material multiplicities (a very Platonic opposition indeed) once one sees that this unity is the result of a transposition of the artisanal technique to the realm of metaphysics that is instantly disavowed and subordinated, in the same way the master dominates and subordinates the labor of the servant. The question of the material networks that sustain the empire is the correlate of the question of the ideal unity that attempts to subordinate these material networks. In other words, the transcendent (i.e., more-than-material), ideal unity creates a fetishistic illusion, a certain misrecognition that has an effect of its own. Natural law results from the disavowal of a fetishistic attribution of a magic inner life to the products of human hands, hiding the mutual constitution between the realm of material technical production and the realm of virtual more-than-material ideals. Such transposition/disavowal depends on a certain misrecognition that attributes an immediate and natural property of one of its elements to what is really an effect of the network of relations between these elements. The machinery of the network of social relations is kept together by a performative fiction, a metaphysical instrumentalism that takes over its material means, subordinating them to its own transcendent more-than-material ends. As a result, the network produces an ideological mystification that considers a property of one of its elements as a natural property when, as a matter of fact, that property depends on a mutual co-implication, the structural effects of the relations of the network itself. Once we reinscribe the productionist presuppositions of imperial metaphysics, we are able to see the inner antagonisms and conflicts and irresolvable contradictions that are endemic to the Iberian Empire.

The question that arises is: what is the metaphysical fiction that takes over the material network of production? It is the fiction of the value of money as inherent in precious metals that serves and sustains the empire. I have tried to trace these contradictions associated with the decline and fall of the Spanish Empire back to Acosta's contradictory attempt to consider metals as both passive raw material destined to serve the empire and as money, the embodiment of value that sustains the network of the first global empire. Briefly put, metals, the lowest material means, become the highest measure of everything else.

Aquinas's metaphysical instrumentalism provides a sense of unity and coherence that bridges what Cañizares-Esguerra calls the pragmatic and utilitarian character of Acosta's science with his political, administrative, and ethnographic writings. The deep presuppositions of metaphysical instrumentalism constitute a kind of methodological a priori that schematizes the chaotic, fragmented, and changing realities of the New World. Acosta's geopolitical and historical context, the discovery of the mines of Potosí, and the changes introduced by Francisco de Toledo are all facts that contributed to a radicalization of Vitoria's and Sepúlveda's imperial uses of Aquinas's metaphysical instrumentalism. While Vitoria and Sepúlveda turned to the basic principles of the natural subordination of matter to form to answer the question of the subjection of the natives, Acosta represents a sort of administrative turn, according to which the question about *how* things function replaces and answers the former questions about the *why*, the reasons of the conquest. Acosta knew that the ultimate *why* of the conquest was unjustifiable so he proceeded to seek for a retroactive justification in the *how*. If the valid origin of the conquest was absent, sense and finality conferred meaning retroactively, after the fact, by appealing to unconditional ends. Yet, once again, unconditional ends can only borrow their power of persuasion from the examples of technique. Technique worked as an ultimate reason behind the reasons provided by political theology, yet it was necessary to hide the artificiality of technique in order to give natural law the appearance of universality and necessity. There is more concreteness and determination in the example of technique than in the exemplified abstractions, principles, and causes of metaphysics.

The consequences of Acosta's hylomorphism and administrative use of the principle of natural subordination are multiple. One of these consequences concerns the nature of the New World, and another one concerns the nature of its inhabitants. For Vitoria, precious metals were already an available resource. Acosta takes another step in this direction, stating that metals are an imperfect matter that serves the higher purposes of the perfect being, using the same language employed by Sepúlveda when discussing Aristotle's hylomorphism. Besides, indigenous laborers also become indispensable means, as Acosta's teleological defense of the mita and the encomienda makes evident. The centrality of money, the glue that kept the empire together, also bears witness to Acosta's modernity. Acosta knew that mining for money was the fuel that kept business and evangelization running. In the *Historia*, money is a means that risks being transmuted into an end, yet without its being transformed into an end it cannot remain an effective means. Although he pretends to subordinate technical means to spiritual ends, he knows that Spaniards and merchants travel to Peru in order to obtain profit. The elevation of the instrumental use of metals to money as medium of exchange exhibits the inner contradiction of the principle of natural subordination. In Acosta, it becomes evident that natural mastery presupposes artificial mastery: the ideological role of metaphysical instrumentalism is justifying the means. For all practical purposes, the machine of the world becomes independent and autonomous. Yet metaphysics provides a language that helps to naturalize hierarchies and divisions presupposed by technique. Finally, the contradiction between the subordinated material means and the subordinating power of money is materialized in the abyssal hell of Potosí.

Acosta was conscious that the abyss of Potosí is the center around which all metaphysical instrumentalist ideology is organized. The centrality of money is a constant reminder of a fissure within the imperial system. Imperial reason is organized around this absence of a metaphysical ground, which is replaced with the metaphors of artisanal technique and, in the last instance, ascetic torment. In the next chapter, I will examine the ideological foundations of the reforms introduced by Francisco de Toledo. These reforms represent a radicalization of the metaphysical instrumentalism of the authors considered here as well as an administrative turn within the attempts to ground Spanish dominion and the entire enterprise of mining. By examining the ideological foundations of the Toledan reforms it is possible to show a radicalization of the instrumentalist core of metaphysical instrumentalism because the aim of the empire was no longer simply to justify the conquest and the expansion of empire but to conserve the means (that is, the mineral wealth of the Indies) and the extraction of tribute.

4

FROM IMPERIAL REASON TO INSTRUMENTAL REASON

The Ideology of the Circle of Toledo

Francisco de Vitoria laid the metaphysical foundation of the Spanish Empire by appealing to Aquinas's principle of natural subordination of matter to form and means to an end. The inherent inconsistencies of metaphysical instrumentalism produced an impasse that became visible in the debate in Valladolid between Ginés de Sepúlveda and Bartolomé de Las Casas. José de Acosta responded to the impasses of metaphysical instrumentalism by creating a pragmatic and rational program for controlling nature and the resources of the Spanish Empire. He defended the exploitation of metals and the use of forced labor introduced by Francisco de Toledo by employing a teleological frame in order to justify the disposition of metals and the centrality of mining. The ideological framework of the Toledan reforms was introduced between 1569 and 1572. Francisco de Toledo's (1515–1584) fame as a "supreme organizer of Peru" is due to a series of radical measures, including the introduction of the *mita* (a regime of tributary work, simultaneously obligatory and nominally remunerated), a revolutionary method of amalgamation called *beneficio*, and the establishment of *reducciones* (congregations of Amerindians brought together for the purpose of introducing them to Christianity and to a Western way of life). By perfecting the management of an ideal structure of imperial administration and labor regimes, these reforms ironically resulted in a greater dependence of the empire on the material and technical means of production. The more the administration took control of the means of production that had formerly been in the hands of indigenous people, the more it became dependent on a system intended to run by itself.

In this chapter we will analyze three works that represent the ideological basis of the Toledan reforms. The first is the *Anónimo de Yucay* (1571), a detailed exposition of Toledo's views on Inca tyranny and the centrality of mining for the Spanish Empire. The second is Juan de Matienzo's *Gobierno del Perú* (1567), where the author sustains that the Andean people are servile and proposes a system of rotational work in order to compel them to work. The third is José Luis Capoche's *Relación general de la villa imperial de Potosí* (1585), which is a retrospective defense of Toledo's reforms that frames the transition from Andean methods of refining to the system of amalgamation introduced by Toledo in terms of technical necessity. Examining these texts will show how the ideology of Toledo's intellectual circle represents a radicalization of the instrumentalism found in Aquinas, Vitoria, and Sepúlveda's ontological frame. Finally, contemporary academic work that pays close attention to the Toledan reforms (such as that of Jeremy Mumford and Nicholas A. Robins) tends to use two different logics in order to explain the colonial origins of modernity. One is a logic of instrumental domination, and the other is a logic of the search for surplus value. I argue that both logics appear in the primary sources examined here and this can be seen in the way they manifest the inherent impasses and contradictions of metaphysical instrumentalism.

According to Henry Kamen, the story of the rise and fall of an all-powerful Spanish Empire driven by a Catholic will to power and in firm control of the globe is simply false. Viceroy Mendoza's suspension of the New Laws, Philip II's capitulation to the demands of the *encomienda*, and the huge transoceanic gap between the Old World and the New World illustrate the incapacity of the Crown to exercise direct control over the colonies through either threat or coercion. And yet the vast imperial network of credit, commerce, production, and circulation managed to survive in spite of the Crown's attempts to control it. Kamen has an insightful explanation for this paradox: "The world's greatest empire of the sixteenth century, consequently, owed its survival to virtual absence of direct control" (142). Kamen's analysis shows that the virtual absence of direct control was the dysfunctional kernel around which the whole network of the empire was organized. As a matter of fact, this virtual absence of direct control was supplemented by the outsourcing of authoritarianism to viceroys such as Francisco de Toledo, who ruled Peru from 1569 to 1581. What is shocking is that Kamen's image of a decentered, almost rhizomatic empire finds its best example in figures such as Toledo, a champion of vertical subordination of the indigenous peoples to the economic needs of the empire. His impact in Peru is recognized by all scholars in the field of Andean studies; he has been called the "Supreme Organizer of Peru" by Roberto Levillier and a Machiavellian ruler that "succeeded in stamping and impress on Peru that was in varying degree to endure until the close of the colonial period" by D. A. Brading (*The First America*, 146). As a matter of fact, he was credited by his contemporaries and through centuries of scholarship with introducing the resettlements that gathered the indigenous peoples, separated Andean communities from their lands, and subjected them to Spanish rule. Toledo is also associated with the introduction of the infamous mita, a system of compulsory labor that obliged the indigenous peoples to work in the mines in exchange for a wage, as well as the revolutionary technique of amalgamation, two measures that dramatically increased the Crown's revenues and impacted the very ecological balance of the region.

First, a brief detour is needed in order to clarify the indissoluble synthesis between Toledo's reforms and Philip II's new international politics. From the beginning of his reign, Philip II sought to affirm the authority of the Crown and secure adequate finances and mobile capital when dealing with the New World. Philip no longer appealed to the papal donation in order to secure his authority. It is possible to consider this a triumph for Fransico de Vitoria, whose aim was to put limits on papal authority in order to make room for the autonomous machinery of civil power. As Kamen explains, the days of the *requerimiento* were gone, and the last independent Andean state of Vilcabamba had submitted to the authority of the Crown alone (192).

An example of how Toledan deeds and reforms in Peru materialized Philip II's new imperial politics is the capture and execution of Tupac Amaru, the last Inca son of Manco Inca, an act that "encapsulated both the philosophy and the practice of Spanish imperialism" (Kamen, 193). In 1571 Titu Cusi, the son of Manco Inca and founder of the rebel Inca state in Vilcabamba, died after the Spanish missionaries failed to cure him. The Indians accused the priest of poisoning Titu Cusi and killing him, which gave Toledo the excuse to intervene and repel force with force, as Francisco de Vitoria would put it. Toledo sent a small army and captured the last Inca, Tupac Amaru, Titu Cusi's brother, and beheaded him. The act was condemned by writers of different generations such as Guaman Poma, Inca Garcilaso, and Antonio de la Calancha, who even declared that Toledo's actions were guided by the poisonous reason of state and not by Christian piety. Toledo's execution of Tupac Amaru was an example of how there were no limits to the Crown's authority. The novelty and impact of Toledo's deeds won him a certain proximity to Machiavellism and the idea of "reason of state." [1]

The best way to understand the historical construction of the historical figure of Toledo is through the adverse reactions to him from writers such as Guaman Poma, de la Vega, and Calancha, who portrays the execution of Tupac Amaru as a cold act of reason of the state in his *Crónica Moralizada* (1638):

> The Viceroy sentenced him to death, hailed the Republic, and did not consider his plea; the Inca asked to be exiled to Spain so he could serve as a page to our King; all were moved by his plea, but the Viceroy was not persuaded. Our holy Bishop, Fray Augustine, considering such circumstances of piety, as well as the laws of justice, came to the Viceroy on his knees, eyes full of tears, and begged him to be less severe and reverse the sentence or to allow [Tupac Amaru] to be exiled in Spain. If the Viceroy was moved, he was not overcome, declaring that this was how he would best serve the King. The holy man appealed to laws of conscience and defenses of justice, not wanting to rise in order to achieve by kneeling what he could not negotiate through reason. The Viceroy refused it all, saying he would not be worn down, that this was the ultimate resolution. *Oh, how many noblemen's hearts have been hardened by the pestilent reason of the state, the venom of monarchies, with convenient deceptions, and the blade of knowledge in the hands of conservation!* Looking at the Viceroy, the determined holy Bishop said to him: I rise with a heavy heart, but supposing that this sentence is based in the reason of the state, Your Excellency will see that our King will not thank you for doing this to please him because he is a Catholic King and he fears God, and you will see some punishment for this death. Everything happened as the Bishop predicted, as Don Francisco's fall from the King's grace and the sadness with which he died were the results of the Inca's death sentence. This Gentleman tarnished a thousand commendable ac-

tions by obliging Peru to carry out such a severe sentence when he could have achieved peace with less criminal measures. (Calancha, *Corónica Moralizada*, 3:1586; my emphasis)

According to Kamen, while governance in the first half of the century relied heavily on the authority of the pope and on discussions around the Spanish dominion, during the second half decisions were made by the Crown and the Crown alone (193).[2] According to this new policy, there was only one emperor and owner of the lands and riches of the New World, whose obligation was to defend and conserve the native Indians through just laws.[3]

In 1568 the king formed a committee to discuss the government of America; Juan de Ovando and Francisco de Toledo were among the members of this council. Among the results of this committee was a new imperial policy that was clearly stated in the Ordinance on Discovery and Population issued on July 13, 1573.[4] The ordinance used a Lascasian language and "reflected in some measure the aspirations of Las Casas, who had died seven years before but whose writings we used in framing its text" (Kamen, 255).[5] The main results of the law were the consolidation and control of the settled areas, the conversion and protection of the Indians, the prohibition of further conquests, the redefinition of the Patronato Real (the power to appoint church officials), and the enforcement of the Council of Trent between 1573 and 1574.[6] Missionaries backed up by military protection were the only people authorized to explore the frontier. With this ordinance, Philip II applied the same defense politics as were employed in the Mediterranean, whose frontier consisted of a "network of small defensive garrisons" (255).

The "Junta Magna," convened in 1568 and composed of members of the Council of Indies and the Council of Castile, was also in charge of finding solutions to the empire's economic problems. The Junta Magna had a "distinctly modern feeling," as Jeremy Mumford explains, "the air of a government task force or blue ribbon panel today," where a group of "experienced officials" discussed relevant policy issues "gathering the most reliable information and recommendations from the field" (79). The viceroys Martín Enríquez (Mexico) and Toledo (Peru) were instructed by the Junta Magna (1568) to carry out its reforms and to restore order, becoming the "perfect embodiment of Philip's sober and dedicated system of government" (Truxillo, 81). The new target of the reforms was to increase the Crown's revenues through the combined results of improved mining and the congregation of Indians in settlements in order to protect and Christianize them. Geoffrey Parker, in his *Grand Strategy of Philip II*, explains that the Junta Magna was not merely part of an expansionist program but, rather, a component of a "global strategic vision" (9).[7] Carlos Sempat Assadourian, explains in a seminal article that the Junta resulted in a redefinition of the nature of the colonial system to be imposed in the Indies.[8] According to Assadourian, the core of the problem was that the Crown considered itself entitled to a royal fifth, the 20 percent tax levied by Spain on colonial mining, that required the colonial settlers to implement efficient changes in order to preserve an imperial enterprise dependent on the mines, which were the source of the riches of the American kingdoms. As González Casasnovas explains, the new guidelines were based on a preexisting program that emerged from the need to regulate the problem of the encomienda (32).[9] In his article, Assadourian analyzes the instructions given to Toledo and concludes that there was a shift in the Crown's political economy, which materialized in Toledo's reforms. Assadourian explains the empire's metapolitical shift in terms similar to those used by Kamen. While the first half of the century was characterized by military invasion and discussions around the Catholic dominion aroused by the abuses of the *encomenderos*, the imperial policy of the second half of the century was "grounded in the economic utility of the Crown, identifying the interests of the Crown with the interests of the Indians" (6). The mita, implemented in Peru by Toledo, represented the "archetypical element of the policy of economic utility" (8). The production of silver was the "principal and most substantial business for the greatest good and benefit of those kingdoms" (9). Like Kamen, Assadourian stresses that the economic motivations and goals of Philip II's politics directly influenced the Toledan reforms. Both scholars see Toledo as an embodiment of Philip's new policies.

Although this image of monolithic synchronization is very appealing, it is necessary to make some adjustments to it. For instance, Jeremy Mumford also observes that Philip's "global" strategic vision and Toledo's "social engineering" in Peru only apparently look like the "two sides of the equation that seem to fit together" (79). He questions this picture that presupposes certain divisions between the conscious planning and organization at the imperial center and the execution of power over the subjects at the colonial periphery (79). Mumford notes that Toledo's three most polemical policies, "personally leading a General Inspection, regularizing the forced labor draft, and carrying out universal resettlement," were not endorsed by the Junta Magna (79). Mumford categorically negates the assertion that Toledo's project was a "straightforward implementation of decisions made by the king and his advisors" (79). On the contrary, "Toledo was in fact operating in a relative vacuum of central direction" (79). In sum, while the Junta Magna inspired the goals of the Catholic monarchy, it did not directly mandate the Toledan reforms. Rather, it was Toledo himself who made suggestions to the Junta Magna, including the General Inspection through the central and south central Andes, keeping well in mind the campaign for universal resettlement (79).

The Junta Magna had to tackle "the subject of government and justice" because Pope Pius V (1566–1572) was paying close attention to Spain's treatment of Indians (Mumford, 80). The pope's position was in line with the arguments of Bartolomé de Las Casas, who condemned the encomenderos' abuses and challenged the Spaniards' presence in the Indies in the name of natural law. Nevertheless, while Charles V could benefit from Las Casas's assault on the encomenderos because it expanded his own power, Las Casas's radicalization in the 1560s—as well as his assertions about the need to restitute stolen property, and authority, to the locals—did not exactly benefit Philip II's politics. After Las Casas's death in 1566, his ideology remained influential among Dominicans in Peru.[10]

In the 1570s, from the Dominicans' perspective, the need for accountability was just as pressing as it was at the beginning of the discovery and conquest of the New World.[11] In order to block the pope's challenges and the remnant of radical Lascasian claims, the Junta imitated the structure of the juntas summoned under Charles V in order to give their arguments a level of authority and unimpeachable quality. The purpose of the Cardinal Diego de Espinosa, the president of the Council of Castile who also presided the Junta, was to erase any ethical or theological doubts about the legality and morality of the Spanish enterprise based on Las Casas's critique (Mumford, 81).[12] The Christian ideology of the Spanish Empire was a double-edged sword because it served to question the Spanish dominion while also being the Spanish Empire's ideological basis (81).[13]

As Adorno explains in her *Polemics of Possession*, when Toledo arrived at Lima on November 30, 1569, he found that the Lascasian arguments in favor of the restitution of power and lands to the Incas were widely popular among the friars who defended the Indians. As a matter of fact, in 1550s Las Casas's ideas were generally prevalent among both theologians and bishops in Spain and among friars in the colonies. An important obstacle for Toledo and his goal of strengthening the political and economic power of the Crown was the specter of Las Casas, who advocated the restitution of the riches and sovereignty to the Incas in *Los Tesoros del Perú* (1654). Toledo's strategy consisted of not only justifying Spanish dominion but also delegitimizing Incan sovereignty in an active and systematic way (Adorno, 51).[14] Toledo recollected testimonies beween 1570 and 1572 trying to prove, against the Lascasian position, that the Incas were tyrants and not the natural lords of the region (52). Toledo ordered the suppression of all of Las Casas's pamphlets written between 1552 and 1553 (52). Philip was willing to excommunicate anyone who read Las Casas, but Toledo considered any punishment insufficient when trying to defend the Spanish sovereignty against Las Casas (59). In Toledo's words:

> The books of the Bishop of Chiapas and other works printed without the permission of the Council of the Indies will be collected. The Bishop's books have caused much damage, and His Majesty would do well to be advised that he should address and raise a complaint about the relation between them at the council to be held in Lima and that he send some sort of reprimand or order of compliance against those monks to show what they have because, even though the priests have been censured, it is not enough. (Levillier, *Gobernantes*, 4:462)

Despite the triumph represented by the execution of Inca Tupac, Toledo could not control the circulation of ideas "whose goal was the return of Andean sovereignty over Peru" (Adorno, 59). As Adorno concisely puts it, "the strenuous battle Toledo waged against the ideas of the potential Inca restoration is a testimony to the vitality of those ideas, not only as presented by Las Casas but as advocated by his colleagues and followers in the 1570s in Peru" (59). Toledo and his supporters attempted to eradicate Las Casas's doctrine of restitution, which, let us recall, appealed to the tradition of natural law advocated by Vitoria and Aquinas.

Toledo's war against the Lacasian doctrine also took place on ideological grounds. Toledo surrounded himself with a "circle of trusted lieutenants," including his cousin the Dominican priest García de Toledo and

Pedro Sarmiento de Gamboa, author of *Historia Índica* (1572) (Brading, 129). In this work, Sarmiento de Gamboa argued that Inca rule was illegitimate, and he considered the Incas to be tyrants both in origin and in practice. He depicted a relatively recent Inca empire that expanded during the reign of Pachakuti and was welcomed only with rebellion since it treated its subjects poorly, subjecting them to continuous resettlements and depriving them of their property and freedom. The *Historia Índica* is an example of the way Toledo "sought to rewrite history, or best to say, extend his command over the Peruvian past" (Brading, 143). MacCormack points out that, by rewriting the indigenous past, these intellectuals also rewrote Andean religion, motivated by a secular agenda that tried to consolidate Spanish sovereignty in the region (186). Two authors that influenced Toledo in his war against the Lascasian doctrine of restitution were Juan Polo de Ondegardo (a magistrate in Cuzco in 1559), whose work was published with other documents in the Third Council of Lima under the title "On the errors and superstitions of the Indians" and Juan de Matienzo, a judge in Charcas who provided Toledo with a blueprint of the most important solutions to the problems of Peru in his *Gobierno del Perú* (1567).

According to Mumford, Toledo had already mastered Andean policy debates before traveling to the New World (79). The Junta Magna had a library that included most reports on colonial affairs, which Ovando had solicited from experts in both Spain and America. Toledo had access to many of the same documents historians do, including the writings of Ondegardo and Matienzo, through the Junta Magna's library (Mumford, 79). As a result, the Junta Magna library provided Toledo and his companions with the ideological background of colonial policy debates of the time. According to Brading, "the circle of lawyers, priests and soldiers who surrounded the viceroy subscribed to the imperial, humanist ideology of Sepúlveda" (Brading, 138). Moreover, it is possible to read Toledo's battle against Las Casas in Peru as a continuation of the debate between Sepúlveda and Las Casas in Valladolid (which Adorno calls "polemics of possession") by means of a revival of Sepulveda's imperial ideology. Sepúlveda's radical imperial politics and his denigration of the Indians was grounded on a metaphysical instrumentalism (also shared by Las Casas) that derived artificial subordination from natural subordination. The ultimate paradox, also noted by Brading and masterfully demonstrated by Mumford, is that when trying to get rid of the specter of Las Casas and the Inca insurgency, Toledo introduced his most important reforms in Peru appealing to "key institutions modeled on native practice" (Brading, 138). The principles that served as a base for the discussion around the problem of dominion in the 1540s returned under an increased pragmatic and utilitarian form in the 1570s. In other words, the theoretical background of Sepúlveda-Vitoria-Aquinas served as a framework for the practical experience of colonial administration under Toledo.

Analyzing three works that represent an ideological defense of the Spanish Empire and Toledan policy (the *Gobierno del Perú* by Matienzo, the *Anónimo de Yucay* edited by Pérez Fernández, and *Relación general de la villa imperial de Potosí*, by José Luis Capoche [1585]) will demonstrate that the accountability of the concrete reforms introduced by Toledo (the General Resettlement, the mita, and the introduction of amalgamation) depends on a metaphysical instrumentalism put into the service of the imperial school of thought and resulting in the subjection of the Andean peoples. The ideology of Toledo's circle represents a radicalization of the instrumentalism found in Aquinas, Vitoria, and Sepúlveda's ontological frame.

Such radicalization consists of a shift of emphasis analogous to that of the imperial metapolitics described by Kamen. It is my understanding that, in this new context, if the Crown could ignore the requerimiento and the papal donations it is because the ontological ground of imperial reason took a "technical turn" that is implicit yet disavowed in Vitoria and Sepúlveda's use of Aquina's metaphysical instrumentalism. That is to say, there was a shift of emphasis in imperial ideology that started to privilege administration over the earlier attempts to justify dominion, making visible the impasses of instrumentalism. In Aquinas's metaphysical instrumentalism, technical manipulation was implictly presupposed by the very structure of the law and the state. The statesman stands in relation to the population as the craftsman does to the products of his art (Aquinas, *Political Writings*, 102). In the apologists of the Toledan reforms there is a slight shift of perspective, according to which the need to justify the extraction of revenue from the mines replaced the need to justify the incorporation into the Spanish Empire.

This shift of perspective brings to light the excess of artificiality over the natural, creating a sort of circularity between the means and the ends. What was presupposed by the teleological and hylomorphic principle of natural subordination was the technical manipulation that now will be invoked as necessary by itself. The explicit aim of invoking transcendent ends, the civilization and evangelization of the natives as well as the common good of both the Spaniards and the Indians, was to justify the material and technical means, the reforms introduced by Toledo such as the mita and the amalgamation. The invocation of common good and the well-being of the natives, central to Vitoria, Sepúlveda, and Las Casas, became entangled with the attempts to justify the production and the extraction of profit. The cost of such contradiction that dwells at the heart of imperial reason is that imperial unity will become dependent on a series of practices that were both destructive and incompatible with the invoked ends.

Ultimately, Toledo's reforms represent not only the materialization of the imperial consequences of the principle of the natural subordination of matter to form and means to an end but also a moment when the Spanish Empire became consciously dependent on the means in such a way as to create a circular contradiction, where the means are elevated to the status of ends in themselves. The machinery of the empire started to become an independent entity beyond the control of the Crown at the same moment that the Crown decided to exercise absolute control over its lands (Kamen, 142). This paradox gives way to the first visible aporia of the principle of natural subordination itself: with Toledo, natural mastery becomes explicit technical mastery.

THE *ANÓNIMO DE YUCAY* *FRENTE A BARTOLOMÉ DE LAS CASAS*

Among the key texts for understanding the ideological foundations of the Toledan reforms is the *Anónimo de Yucay frente a Bartolomé de Las Casas* (1571), an anonymous text attributed partly to Francisco's cousin the Dominican García de Toledo and partly to Juan Polo de Ondegardo. Written in March 1571 in the Yucay valley, the three manuscripts consist of one primary text and two additional sections. Monique Mustapha attributed the text to García de Toledo, Francisco de Toledo's cousin. Isacio Pérez Fernández challenged this attribution in a 1995 edition titled *El anónimo de Yucay frente a Bartolomé de Las Casas*, and he attributes it instead to multiple authors. Pérez Fernández affirms that the text is not the original signed by García de Toledo, since all the multiple additions were introduced subsequently by different authors, at different dates, forming a "heterogeneous document" (*El anónimo*, 81). Pérez Fernández offers a reconstruction of the drafting process of *El anónimo*, based on the additions made to an original text signed in the valley of Yucay on March 16, 1571. Between this date and March 25 of the same year, the text was transcribed in three copies, one of them for Cardinal Espinosa and another for Sarmiento de Gamboa. A copy of the copy that remained in Peru included an "argument against" in the part that claims that the Incas were tyrants. According to Pérez Fernández, this addition was made by Juan Polo de Ondegardo, who was also responsible for adding a section on the mines. According to Pérez Fernández, this copy was used by Pedro Gutierrez Flórez, who added by 1571 an appendix on the treasures and *guacas*, or indigenous objects of devotion (which could also be natural locations). A number of different additions were made to this copy between 1578 and 1623.

The text tries to demolish the arguments presented in Las Casas's *Tesoros del Perú* (1564), where he denounced Spanish tyranny, defending indigenous restitution and the right of the Indians to repel force with force. The anonymous author attacks both the figure and the work of Las Casas, who is characterized as a good religious man who is nevertheless blinded by his stubborn desire to defend the Indians. In this text, the reader learns how the Lascasian doctrine harmed the reign of Charles V by attacking the conquistadors and twisting the words of the theologian jurists like Francisco de Vitoria. Las Casas's biggest mistake, according to the text, was to ignore how the Incas were tyrants who subjugated their neighbors, treating them like slaves for nearly eighty years. The government of the *kurakas* lacked legitimacy because they were not natural rulers and and represented an obstacle for conversion and civilization.

The text begins by explaining that the belief that the Incas were the "natural rulers" of these lands is where the error about the truth and legitimate lordship of Peru originated (*El anónimo*, 116). After acquitting the king, the theologians, and the jurists, all of them blinded by a false doctrine, the text ultimately blames Las Casas and the royal counselors for causing irremediable damage to the Crown of Castile, the Gospel, Christians, and the Spanish vassals of the Indies. First, it grounds the true and legitimate dominion of the universal monarchy in the fact that the Indians were recent tyrants who constructed an empire based on mere caprice. Before the conquest of the Incas there were no universal or particular lords, all the Indians lived in a state of "behetría"—confusion. The author appeals to the papal bull issued by Alexander VI, which gave the Spanish Crown the right to evangelize the kingdoms of Peru.

Pérez Fernández claims that García de Toledo—being a Dominican, and knowledgeable of Montesinos and the *Sublimus Deus*—would never say that the Indians were bestial. He infers that such attribution of bestiality, an ideological device used by the Spaniards to justify their transformation into civilized laborers, was a later addition and had not been written by García de Toledo. He calls these arguments, apparently added by Polo de Ondegardo, "cheap theology," not worthy of a Dominican trained in Aquinas's theology. Rather than establishing who the author of the text really is, I am more interested in showing how its arguments, whether originally made by García de Toledo or added later by Polo de Ondegardo, are perfectly compatible with Aquinas's metaphysical instrumentalism. In other words, both metals and indigenous peoples are understood in this text as being imperfect matter, available and utilizable, deprived of determination, and in need of being ordered to a perfect form or end.

In the section titled "Exculpation of the king, the theologians, the lawyers, and others for having followed the opinion of Father Las Casas," we read how lawyers and theologians are exempted from knowing the truth or falsity of Lascasian doctrine (*El anónimo*, 121).[15] The author points to the example of Vitoria who mistakenly grounded his statements in wrong facts, which necessarily led to wrong conclusions (121). The theologians of the School of Salamanca tried to place limits on both the conquest and the power of the pope, and opened the door to condemnation of the Spanish presence in the New World. According to the text, the ultimate and indemonstrable principles upon which such statements were grounded are not false. The problem was that the theologians inferred wrong conclusions based on the wrong facts they had received, which is a very common mistake among those who categorically state something only in order to immediately change their minds, condemning their former position because they find out that these conclusions were not supported by facts (121). Briefly said, theologians make mistakes when the general principles do not have the power to capture the true facts.

The text goes on to assert that the "prince of darkness" manipulated the facts in such a way that Las Casas caused the emperor as well as the theologians and jurists to doubt Spanish imperial legitimacy to the point that they all advocated for power to remain in the hands of the Incas (122). Apparently, the emperor, moved by these doubts caused by the Lascasian error, wanted to leave the kingdom to the Incas, but Vitoria dissuaded him from making this decision. Pérez Fernández reminds us that Vitoria doubted the legitimacy of the Spanish presence in the Indies, but he adds that in the Memoria de Remedios of 1542 (CX, 120a), Las Casas states that it is necessary to conserve the police in the Indies. It is not clear that Francisco de Vitoria had any input on the formulation of the New Laws, and he lacked sufficient reasons for recommending that the emperor abandon the Indies. Clearly, the author of the *Anónimo* displaces Las Casas's authority with Vitoria's authority, which is closer to Sepúlveda and therefore diametrically opposed to Las Casas. The defense of the Incas is presented as a defense of inhumane practices: "Look, what a subtle ruse in order to turn to the defense of the ignorance of faithlessness and idolatry and human sacrifice and eating human flesh and living like beasts!" (*El anónimo*, 122). The arguments in favor of Inca rule are transformed into arguments in favor of the inhumane practices that Sepúlveda considered sufficient reasons for war. In the section titled "Damage to the royal Crown of Castille," we read that the Incas were, in some way, "perfect lords and rulers," a statement that somehow limits imperial power by affirming that the encomienda should be counterbalanced by the consent of the natural native rulers. Yet the undesirable consequence of this position was that "it raised doubts about the mines and about many other matters of the land and ways of living, because it was founded on the legitimate damage thereof, being tyrannical [the Inca] and the largest in the world" (124–25). Briefly put, Las Casas harmed the Crown because he obstructed the mining business.

Besides harming the Crown of Castile by creating false doubts, Las Casas also harmed the natives of Peru: "Who can but doubt that, if it is advantageous for establishing the kingdoms, for their security, for the eternal preservation of the Catholic faith, to make Spaniards lords, but that he could do it because he is a true lord?" (129). The text makes use of the classic Sepúlvedian-Aristotelian argument of subjecting the Indians for their own benefit since it would be convenient for the natives "to have eternal Spanish lords and the kingdom as their foundation" (129). The text says that the Spaniards would treat the Indians well, and that they would grow fond of these lands and forget about Spain (129). In this way they would prosper without ever exhausting the gold and silver provided by this land. The gold and silver would act as a "perpetual force" that would attract the Spaniards, helping God's plan, which is to evangelize the natives (129).

The document also examines the arguments against indigenous tributary labor, among which appears the contrast between the services demanded by the Incas and the "excessive" tributes demanded by the Crown. Those who wanted to restore sovereignty to the natives argue that taxes imposed by them were spent inside the kingdom and "for the profit of their vassals" (154). On the contrary, the colonial state is "inverted" since it charges the natives with excessive tributes that are spent outside the kingdom and not "for the profit of the natives" (154). Against this position, the text argues, the Crown of Castile asks only for "just and saintly" tributes in order to "sustain" the justice of a government that is not tyrannical but Catholic and also puts those tributes to use "in other allied kingdoms and elsewhere in defense of our holy Catholic faith and in support of the Christian religion" (156). Tributes are a way of sustaining the universal Catholic government, being as central to the common good as mineral amalgamation (156). The discussion takes place within the frame of metaphysical instrumentalism, where the benefit of the local part is subordinated to the benefit of the global whole. On the one hand, the local government is subordinated to the common good. On the other hand, their tribute is used to benefit the Christian commonwealth.

In Appendix 5, apparently written by Polo de Ondegardo, the author claims to be moved by feelings of admiration and devotion when he sees the way God's providence takes care of the Crown by giving it the Indies "and their riches as reward for the work and expenditures the kingdoms of Spain made in conquering them" (157). The justification for the exploitation of the mines appears after the conquest, in the use that the Crown gives to the silver of Potosí:

> Because, although one may look for one, from the time these kingdoms of Peru were acquired until today, almost forty years later, no justification for the labor in the gold, silver, and mercury mines (which are incredible if one hasn't seen them) has been seen nor can one be achieved, even to this day, when the king determined with the divine spirit and extraordinary movement of God, together with that of our Holy Father, so filled with the Holy Spirit that his own works declare it—this Holy Alliance against the enemies of our Catholic faith, while not considering the material riches that

his Majesty can spend on the apportioning of expenses (which were three million every year, in the amount of two hundred fifty thousand *ducados* every month) as the great riches of the glory of God and the souls which they made submit to the kingdom of Christ our Lord; for these reasons I believe and can confirm that the labor of these mines and treasures is justified. (157)

In the above paragraph we read that forty years had passed since the conquest of Peru, and the author had written this appendix by September 1571. The mines of Huancavélica were discovered in 1566, and the viceroy decided to employ mercury in amalgamation in 1570. The Christian league against the enemies of the Catholic faith, promoted by Pius V and Philip II, triumphed at the Battle of Lepanto on October 7, 1571. The coincidence between the discovery of mercury and the triumphs of the Christian alliance are retroactively interpreted within the structure of a providential view of history. In the *Anónimo*, these events attained their meaning, sense, and finality after they occurred, and the creation of the imperial order retroactively confers signification to a preceding chaos, making things look as if the final end was inscribed in a divine origin. The providentialist rhetoric of imperial ideology is retroactively grounded in metaphysical instrumentalism. At this point, we can see how the principle of the natural subordination of material means to transcendent, more-than-material ends is an ideological illusion that consists of retroactively organizing the past by assigning it a meaning after the fact. The violence of the conquest is an event that cannot be cleanly integrated into the official ideology, therefore it demands a retroactive justification that assigns meaning and purpose to it by giving an air of necessity to a contigent combination of events.

The coincidence between the endeavors of the pope and Philip II on one side of the Atlantic and Toledo's efforts to revitalize mining on the other are part of a "divine plot." The viceroy summoned the most important ecclesiastical and secular authorities to a junta in order to discuss the problems in Peru. Among the secular were "Your excellency and *licenciado* Castro from your Majesty's Council, all the Audiences of judges and mayors of the court and Inquisition" (158), and among the ecclesiastical were "the archbishop of the kings, with all the providentials of the religions and other very learned men who went with them, and everyone to the last man, after having aired the matter, took to work in the mines, and so it was" (158). This junta, celebrated in September 1570, was attended by Viceroy Toledo, Lope García de Castro (ex-governor of Peru), Juan de Matienzo (author of the *Gobierno del Perú*), and Fray García de Toledo. This junta decided to mobilize the indigenous peoples by making them work in the mines in exchange for a wage. According to Pérez Fernández, the minutes of the junta omit any decision to force the indigenous peoples to work, probably because they were being paid a wage. These decisions were implemented in February 1574, when the *Ordenanzas* on indigenous labor in the mines were written.[16]

The author of the *Anonymous* continues, writing that the Crown extracts so much gold and silver, "due to its very Catholic and liberal determination, which is sufficient not only to conquer the Turk, but also to establish large land grants in those kingdoms" (158). The author shows particular interest in conveying the sense of awe produced by the combination of the immense quantity of metals and Las Casas's political blindness: "And it is a marvelous thing that, in beginning to work the mines, a frightening and admirable amount of wealth is discovered. And to me, it is much easier to see the blindness of this priest and bishop from Chiapas in his condemnation of extracting these riches from the mines, now that they have established order and means" (158). He continues attacking Las Casas, saying: "because, properly considered, what does it mean to have brought God into the miserable and defenseless souls of these Indians, who are so incapable and so bestial and whose large kingdoms and criminal valleys and lands are so full of gold, silver, and other mineral wealth? And not just here and there but all the mountains are full of it, and all the land in which there are houses or fields and everywhere else is earth mixed with gold dust" (158). Like Sepúlveda and José de Acosta, the author implies that the combination of excessive riches and the Indians' lack of understanding are two sides of the same coin. The presence of excessive quantities of precious metals in these lands inhabited by people deprived of perfection seems to be a sufficient reason for inferring the transcendent end of these material means. Together they are the proof that God placed the metals in the New World for the Spaniards. He continues asking: "And what does this mean other than that God was with these miserable gentiles and with us, like a father who has two daughters, one very fair, discreet, and full of grace and finesse and the other very ugly, filthy, stupid, and beastly?" (159). Like Acosta, he compares the excess of precious metals in lands inhabited by people who lack understanding to a dowry: "For the ugly, clumsy, stupid, and disgraceful daughter, a large dowry is necessary: many jewels, fine clothes, a luxurious house—and all this along with God's help" (159). The author continues, expanding on the interrelation between God's teleocratic design and the attraction of silver: "God did the same with them and with us. We were all unbelievers: Europe and Asia" (159). Nevertheless, unlike the Indians, the European possessed "great beauty, science, discretion" (159). There was no need for more motivation "for the apostles and apostolic men to unite these souls with Jesus Christ through the faith of baptism" (159). Unlike the Europeans, the Indians, "were ugly, crude, stupid, incapable, and filthy, and so a large dowry was necessary" (159). God compensated for the inferiority and imperfection of the Indians with "the mountains of gold and silver, fertile and delightful lands because this attracted the peoples God wanted to go there to convert and baptize them and make them spouses of Jesus Christ" (159). There is a definitive Sepúlvedian quality to this kind of argumentation, in which the Indians are assimilated to an imperfect society that has been perfected by God through the placement of mines in the New World. The excess of metals and the Indians' lack of perfection are two sides of the same ideological coin. Both are the effect of reading the situation from the point of view of the end, the accumulation of precious metals for the sake of the common benefit of the whole globe.

The text continues, explaining how teleology can be known not only through supernatural signs but also through material ones: "in the order of predestination, not only the assets of grace, such as grace and charity and virtue, are means of predestination and the salvation of men, but some material goods are also the means of predestination and salvation, and inversely, their absence is the means of damnation. For some there is a chance that their riches will save them, and for others the lack of wealth will condemn them" (159). The overabundance of riches is a material means for a predestined end: "Thus, I say that one of the means of these Indians' predestination and salvation were these mines, treasures, and riches, because we clearly see that wherever these latter are, there goes the Gospel with great haste and vigor" (160). Experience indicates that evangelization requires the lure of gold and silver, since where there are not rich mines, "the Gospel never goes, as one sees through great experience, to lands where this dowry of gold and silver does not exist; neither soldier nor captain who wants to go, not even a minister of the Gospel" (160). For this reason, "God gave the mines to these barbarians so that they could be brought faith and Christianity and conservation in it for their salvation. And they always used them to work the mines because it has always been that way" (160). Like Acosta, the author associated evangelization with civil administration, the conversion of souls with mining and commerce. Also, like Acosta, the author argues that mines are an explicit technical means for a preexisting end—and with no means, there is no end. The subordinated means sustain the civilizing and evangelizing end.

The *Anónimo* summarizes Las Casas's doctrine in one sentence: "had there been no mines, it would be the same as it is today: demonic" (160). The demon that moved Las Casas also moves the Indians to hide the mines and treasures, telling them that, in the absence of riches, the Spaniards will leave the lands, and they will be able to return to their idolatry (160). The Indians prefer to kill themselves before revealing the mines, because the demon knows that the mines are an "efficient means" to evangelize these lands, and that the Indians are also saved through their riches (160). The demon, moved by the desire to harm the universal monarchy employed a "religious man" as an "instrument" in order to hide the mines, by threatening them with pains of hell. Las Casas was tempted by the devil like so many other "holy men" (160). As a result, "in one stroke, these mines are justified with the sound methods that your Excellency has used, which are so necessary, morally speaking, in those kingdoms if no king or God existed or will exist [in them]" (160). The authorities have the responsibility to preach the Gospel and "maintain what they have received" (161). Nobody would like to face a task without interest or "some utility, because, absent the royal fifths and the storehouses, there would be no willing king" (161). The mines are clearly material means for the end, which is salvation. In this sense, the *Anónimo* text is a classical example of metaphysical instrumentalisnm, where the material means are supposed to be subordinated to the ends as the imperfect matter to the perfect form. All these arguments have a striking resemblance with the ones advanced by Acosta in *De procuranda Indorum* and the *Historia natural y moral de las Indias*, where he appealed to the principle of the natural subordination of matter to form and means to an end in order to justify the appropriation of precious metals. Like Acosta, the author of the *Anónimo* sees a strict correlation between the mining enterprise and evangelization, making the latter dependent on the former. In both authors, the principle of subordination becomes administrative and technical. Yet, in the *Anónimo*, there is an explicit verbalization of the consequences of this administrative turn: "were there no God this would be understood much better because, in these kingdoms more than others, the spiritual depends on the secular" (160).[17]

Evangelization depends on civil power "and so the preaching of the Gospel and conservation—by which I mean God—cannot be achieved without a Catholic king" (160). If there is no Catholic king, "what could have sustained the justice that so many ministers, the garrisons and forces of the kingdom, the doctrine of so many clerics and religious men, provided for the security of these kingdoms, by sea and by land, from pirates that surely will continue to be attracted, ever more well-equipped, by the great interest that they are offered, as we see every day?" (160). While in Aquinas and Vitoria, instrumentalism remains an implicit presupposition of metaphysics, the *Anónimo* presents a clear-cut decision that emphasizes the necessity of the technical for the spiritual, where the spiritual ends up dependent on the technical and material dimensions of power. The means are no longer contingent but necessary. Since the purpose of the text is to justify the mines, profit is not just a by-product of evangelization, of the end, but something to aim for in order to be able to evangelize. The ends are supposed to rule the means, but by depending on the means, the means become necessary. The end becomes the means of the means.

Moreover, evangelization depends on a Catholic civil power, and civil power depends on mining and the extraction of riches: "Thus, without the king, it is clear that the Catholic faith would come to an end in these kingdoms because the kings in these realms were instituted as the necessary means for those who depend on Christianity" (161). The sanctity of the end grounds the morality of the means, making necessary the supposedly contingent means: "then the mines, morally, are as necessary as the king because without them, indeed, without his Majesty, the Gospel will not be conserved" (160). The mines are inherently good, and to deny this is "the blindness of men" and "the malice of the devil" (160). What I propose here is to interpret this text as a symptom of the intrinsic paradox of Scholastic metaphysical instrumentalism: although the preexisting unity (the origin-end structure that provides accountability and normative binding) grounds the morality of the mines (the temporal and merely material means), the latter become retroactively necessary for achieving the end. In other words, a materialist reading of metaphysical instrumentalism shows how the metaphysical presuppositions are posited through the artificial means that retroactively constitute the principles that would otherwise remain abstract and indemonstrable. Presenting the colonial situation as a world upside-down is a way of hiding the aporetic and contradictory character of the universal principle of natural subordination by displacing these inconsistencies onto an external obstacle, namely, the periphery where exceptions rule. The reason for this apparent inversion is that the New World was conceived in advance as deprived of perfection, as imperfect matter in need of correction. Since the natives lack the capacity to rule and administer themselves, it is necessary to civilize them before evangelizing them, and this requires money to sustain both temporal and spiritual power. In order to remedy the natives' inability to rule themselves, it was necessary to introduce a series of reforms, some of them proposed by Matienzo (1520–1579), a judge of the high court of Charcas. Matienzo's *Gobierno del Perú* (1567), another source of inspiration for the Toledan reforms, draws on Sepúlveda's metaphysical instrumentalism in order to defend the introduction of a rotational system of compulsory labor called the mita.

JUAN DE MATIENZO ON NATURAL SERVITUDE

Crisis and a decline in profits hit Potosí in the 1560s when the *yanaconas*, the natives that extracted metal and refined them in their *guayras*, exhausted the rich ores. The introduction of amalgamation in 1572 was the "technological solution to the crisis." Pedro Fernández de Velasco adapted the *beneficio de cajones* invented by Bartolomé de Medina in New Spain. Amalgamation required not only mercury but milling equipment, and milling equipment "created a demand for a large pool of less skilled workers to mine and handle the ore" (Brown, 49). In the absence of abundant rich ores, the yanaconas did not want to work in the mines, yet "rather than admitting that the crisis and new refining technique had removed most of the Indians' economic incentive for going to Potosí, Spaniards mounted a propaganda campaign to justify a greater coercion of labor" (49). By the late 1560s, while the Lascasian front was fiercely opposing compulsory labor, the representatives of colonial power such as Licenciado Castro continued their battle for the legalization of this practice in order to increase economic profitability (González Casasnovas, 28). The Lascasian position became weaker as an effect of pressure from mine owners and the publication of *Gobierno del Perú* in 1567, which led to a clash that culminated in the defeat of the Lascasian front by the end of the decade (28). This text represented the most solid and articulate defense of the policies of the colonial state, laying the foundations for the new economic mercantile structure (28). Before Toledo's reforms, Matienzo thought that the intervention of the state was crucial in order to counteract the hegemony of the kurakas and to increase profit and taxes for the Crown. Among the proposals contained in the *Gobierno* were the transformation of the Indians into a labor force by compulsory means, the monetarization of the economy, individual taxation, and the central role of the extraction and circulation of Potosí silver (28). According to González Casasnovas, the centrality of Matienzo's practical and philosophical measures for the Toledan reforms cannot be emphasized enough (28). As a matter of fact, the *Gobierno del Perú* put the expansion of the labor force and the activation of Potosí mining at the center of the discussion of mercantile development. Undoubtedly, Toledo had in mind all the principles and practical recommendations of the *Gobierno del Perú*, such as the organization of labor in four-month shifts (34).

While Toledo used historical arguments from Gamboa's *Historia índica* and the document of Yucay against the Incas, he also appealed to the well-known philosophical arguments about the natural servitude of the Indians that were present in Matienzo's *Gobierno del Perú*, in which Matienzo argued that fear was the basis of the Incas' devilish tyranny, presenting the Spanish conquest as an act of Christian liberation. In the very beginning of the text, Inca tyranny is depicted as an obstacle to private property: "For having in all more with respect to their benefit than to the public good because the Indians still have something, and all the gold and silver that they dug up was for the Inca, whose intention was to enrich himself and that his Indians were poor and had nothing of their own, making them labor continuously without any reward on large-scale works" (Matienzo, 12). The Incas were tyrants who governed their subjects with a concern only for the interests of the rulers. Tributary labor was the means for keeping their subjects poor and without any kind of private property. And war was the best way to keep them poor and busy: "They were not allowed to be lazy, and to make sure that they were poor and occupied, they were always at war. Stores of corn and weapons along the royal roads at the Indians' expense. They paid tributes without being allowed any excuse and took them to Cuzco at great expense. And those who did not have to pay tribute brought lice in their small containers. They were not masters of their women or children because the Inca did what he wanted with whomever. These Incas were cruel" (12).

All public work was directed toward the accumulation of surplus and not toward the creation of necessary goods. The Incas asked their subjects to pay with valueless and insignificant things. Matienzo not only advanced the thesis that the Incas were tyrants, but he also offers an explanation for their tyranny:

> These Incas made it so the Indians feared them, were timid, and lacked power. They weren't allowed to use their free will even in eating and in choosing women. They did not let them keep their own children, they gave them away beforehand to whoever wanted them, and to better subjugate them they made them go from one place to another, and they called these people *Mitamaes*, and to this day they are away from their lands. To maintain their tyranny they accompanied their relatives and servants, and the governors they installed and sent to visit the land were the Incas' own family members, and it was the same with the captains as they did not trust anyone else. They did not govern by laws but, rather, by their appetite and will. Although these Incas were natural kings of Peru, they were tyrants for their wickedness. (12)

Inca tyranny is associated with the pusillanimous nature of the Indians, who are depicted as lacking power, self-determination, and freedom of choice. Part of the Inca strategy for dominating the Indians was the forced mobilization called the mita, and the mobilized Indians were called *mitimaes*. According to Matienzo, this was a way of keeping the Indians docile and under control. In other words, for Matienzo, the Indians' servility was the exact correlate of the Incas' tyrannical government.

Matienzo revisits the problem of the Spanish dominion over the Indies in Chapter 2: "The Indies were justly won. Because of the concession of the pope or because those desert kingdoms were discovered by the Spanish. Or because the Indians did not want to receive the faith. Or because of their abominable sins against nature. Or because of their infidelity, and even though this is sufficient, the aforementioned tyranny of the Incas would by itself be sufficient to establish that the kingdom of Peru was justly won and that his majesty has a very just title to it" (13). He provides a quick list of the valid titles for Spanish rule, echoing the arguments invoked by Sepúlveda. Unlike Vitoria, Matienzo attributes full validity to the papal donation. Yet like Vitoria, he also defends the right to appropriate things that do not belong to anybody. Like Sepúlveda, he thinks that deviations against natural and divine law must be punished. He remarks that, although one of these reasons would be a sufficient reason, all of them combined are even stronger. Matienzo's aim is to demonstrate that Pizarro not only conquered Peru for the Catholic monarchy but also liberated the Indians from Inca tyranny. Like Vitoria, protecting the innocents and liberating them of unnatural oppression is a universal and necessary divine mandate: "For whomsoever God sent to deliver them from continued oppression and force is sufficient for the Indians to change to a new prince, and tyranny is sufficient cause to make war, being the mandate and authority of the king that recognizes no higher authority, as it is his majesty and the kings of Spain, it cannot be said that the Indians were content, and they did not ask for help from the Spanish because they were oppressed and could not declare their will, so that to have freedom they did it" (13). Matienzo thinks that the conquest was an act of liberation from an oppressive system that was against both private property and self-determination. He confronts Las Casas's claim that only the Indians' acquiescence constituted a sufficient reason for the Spanish rule. Matienzo astutely says that, in order to accept Spanish rule, the Indians had to be free. Since they were not free (because they were servile and subjected to the tyranny of the Incas) they could not transfer their civil power to the Spanish Crown.

Finally, he handles the problem of the violent excesses of the conquerors:

> Nor can it be said that the Spanish committed many excesses, deaths, and thefts or that they had no intention to help them but instead rob them. The response is that the kings gave very just and holy instructions to those they sent and that the crimes of the cacique do not harm the Lord, and perchance it was God's will that that kingdom was conquered since those tyrants were punished or so that that barbarous people would not remain perpetually forgotten, signs of what little knowledge their ancient sages had to instill in that famous Genovese's heart, the miracles that occurred among the people. (13)

Matienzo provides several reasons for sustaining that the private acts of violence of the conquistadors are not an obstacle to Spanish dominion. First, the kings gave "saintly and just" instructions to the people they sent. Second, the superior ruler cannot be blamed for the sins of the subordinated agent. Third, divine law is served by the punishment of tyrants.

Not content with declaring that Spanish dominion was well grounded in Inca tyranny and the servile character of the Indians, in Chapter 3, Matienzo argues that the Indians' subjection is in their best interest, making a list of the advantages of their new situation:

> In that kingdom the Indians are greatly content to be subjects of his majesty and to live in conformity with his laws, all are in favor of him and liberty, and because of the tyranny from which they were liberated and the benefits they have received. Very just laws have been restituted in their lands, their kings and lords are caciques and political chiefs, free from the Devil's and the Incas' servitude. They have now begun to live politically. They have their own possessions and do with them and their children what they want. No one dares to cause offense or to make another work without pay. And now almost everyone has received the waters of baptism and are instructed in the faith. (14)

The Indians are in a better situation since they now enjoy freedom and benefit from the just laws of the Spanish. In what seems to be a perverse appropriation of Las Casas's language, Matienzo employs the word "restitution." That is to say, not only was the conquest not an act of dispossession, it was also an act of restitution of what naturally belongs to the Indian subjects: freedom from tyranny and the freedom to work in exchange for a wage. In other words, the Spaniards did nothing but return to the Indians what was theirs in the first place. Evoking Aristotelian language, Matienzo states that now the Indians live a political life, which appears to be immediately associated with the possession of private property. Briefly put, for Matienzo, the line that separates Inca tyranny from Spanish rule is the introduction of private property and money as a medium of exchange or "common good." Recall that for Aquinas, money was a means for justice, for giving each its own. While tyranny is associated with governing for the benefit of the ruler, justice is associated with freedom to possess private property and receive a just remuneration in exchange for work.

Matienzo is particularly preoccupied with arguing that justice does not benefit only the Spaniards but also the Indians: "Administer justice to them equally as with Spaniards, as the Indians who hire themselves out for money in the city numbered three hundred, and in Potosí and Porco there will be one thousand five hundred, with a regular annual wage of one hundred and sixty thousand pesos not including the money that they earn from farming. They have learned to try and farm using mechanical techniques and farmworkers, which is no less just a title than that declared by his Majesty" (14). Justice appears to be a neutral and universal value that benefits both Spaniards and Indians because now the latter receive a fair wage for their work. Among the advantages made possible by Spanish rule, Matienzo counts the education of Indians in the mechanical arts. For Matienzo, this by itself would be a valid title, echoing Vitoria's eighth possible principle for subjecting the Indians, according to which the Indians' lack of mechanical arts is a clear sign of their incapacity to rule themselves. Teaching the Indians crafts and techniques is another valid title for subjecting the Indians to the Spanish rule.

In Chapter 4, Matienzo deepens his inquiry into the problem of the nature of the Indians, which appears to be intimately associated with Inca tyranny: "The Indians of so many nations that have been discovered are timid and fearful, which comes from their melancholy state, naturally they have less than they could have, they do not think that they deserve good or honor, and therefore they do not have it or seek it, even though it is essential, they are not hurt when they are whipped, nor when their wives, daughters, sisters, or female relatives are taken, they are filthy, and they eat the lice that fall from others' heads" (14). Matienzo is not content to declare that the Indians are melancholic, timid, and pusillanimous; he also says that they consider themselves inferior to the Spaniards because of the abuses they were used to under Inca rule. Again, their servile nature seems to be inseparable from the tyranny of the Inca. Matienzo continues his denigration of indigenous peoples, stating that:

> They are very naive, simple and easily led, fond of novelties, slow, and are hurried by nothing. They move well if you give them coca, they never go anywhere without a load in which they carry their food, they drink the most salubrious water in the inlets they find, from the time they are children they are taught to carry loads and each one carries his *quipu*, which gives us to understand that they were naturally born and raised to serve

and that this is better for them. They have stouter bodies than the Spaniards and suffer more than them. They always serenely sleep in the open and no harm comes to them, and their bodily strength is as much as their understanding is little. They participate in reason in order to feel it and not to have it or to follow it. For them there is no tomorrow. (15)

They are credulous, changeable, slow, and used to bearing heavy weights. All these qualities are offered as proof that they were born and "raised" to serve and that they would benefit from being subjected to Spanish rule. Matienzo is clearly working within the Aristotelian frame, since he appeals to the classical dichotomy between body and soul, matter and form. Like Sepúlveda, he transposes this division onto the division between Indians and Spaniards. The former participate only in the sensuous aspect of reason, without being in command of their actions. They live in the present without time and without future. Moreover, an inability to live in the future is immediately associated with their lack of drive to accumulate riches: "They content themsleves with what they need for a week, they do not work more than is necessary in order to eat and drink that week, they are enemies of work, fond of idleness and drinking and drunkenness and idolatry, and when drunk they commit serious crimes—they are prone to vice with women" (15). Their servility is strictly associated with their inability to form abstract concepts in their minds and to undertand a time that is not immediate, an inability that condemns them to live in a survivalist mode. This survivalist mode causes them to be idle, enemies of work, and prone to vices. Here, Matienzo activates the central axiom of Spanish imperial ideology: everything must be subordinated to the common good, whose measure is money. Lacking inclination to work, being incapable of abstracting and imagining a different future, and ignoring the value of money as a universal medium of exchange and justice are all traits that remain intimately related, all of them depending on the principle of the natural subordination of matter to form that is inherent in metaphyisical instrumentalism.

Matienzo goes on to register an apparent contradiction between the Indians' lack of inclination to work and their obedient character:

> They obey their superiors well and thus it is imperative that they are commanded, ruled, and governed so that they will work and not be idle and avoid excesses, they have little goodwill toward their fellows, they do not

help each other, they do not cure the sick nor the elderly, even if they are their parents.... They are skilled in mechanical tasks of all kinds, such that when they are commanded they make very good farmworkers. Given these conditions and customs they are better off being subjects of the Spanish and governed by them rather than by the Incas. (15)

Despite being enemies of work, the Indians are obedient. Like any servile subject, the Indian is capable of learning mechanical arts. On the one hand, Matienzo's use of the Aristotelian notion of natural servitude presupposes metaphysical instrumentalism. The Indians' place in the economy presupposes the subordination of matter to form and means to an end proper to the division between subordinating and subordinated crafts. They need a master capable of compelling them to work and of taming their excesses. On the other hand, this ontological presupposition is intrinsically ideological, not only for the obvious fact that it is used to justify and naturalize the use of an Indian labor force but because it attributes the central antagonism of colonial society to one particular element. In this case, the violence of colonial antagonism is transposed onto one figure of the indolent yet excessive and passive-aggressive Indian. It condenses all the violence intrinsic to the civilization (grounded on colonial antagonism) into what appears to be some inhuman excess external to society. The source of all dysfunctionality (the inherent failure of mercantile politics to lure the Indians into becoming workers) is attributed to one particular element: the Indian. As a result, the Indian condenses all the contradictions of civil society: Indians are both obedient and enemies of work, melancholic and excessive, potentially good Christians and potential apostates. Recall how Scholasticism invoked a corporatist notion of civil society where each part of the social body is subordinated to a superior one, both fulfilling a function and making possible a harmonious whole. The figure of the Indian accounts for an unacknowledged yet factual division, that colonial society is grounded on a preexisting crisis, an antagonistic split. By exchanging all evils for one, racist imperialism invents an external cause of all evil, the dysfunctional Indian who is paradoxically both obedient and an enemy of work. As the author of the *Anónimo* would say, the Indian is the reason that, in these kingdoms, the spiritual ends depend on the temporal means. In short, the Indian is the fetish that condenses all opposite features and accounts for all evils.

In Matienzo's view, the Indians are a humble and obedient matter impressionable to any doctrine: "They are patient, humble, and obedient, any doctrine or teaching can be *imposed* on them without taking away from what they can understand" (15; my emphasis). Like Sepúlveda, Matienzo conceives the Indians in terms of an imperfect prime matter that needs to have the perfection of form imposed on it. The ultimate presupposition of Matienzo's ideological justification of compulsory labor is Sepúlveda's dogma, the subordination of imperfect matter to the perfection of form. The encounter between natural right (imperial metaphysical instrumentalism) and capitalist mercantilism (made materially possible due to the mines of Potosí) adds another twist to the principle of natural subordination. Indians can become obedient and docile bodies subordinated to a common good under the form of money, a just payment, in a just medium of exchange. Matienzo thinks that by simply paying Indians a wage the Spaniards are liberating them from the oppression and tyranny of the Incas. Money is not only a medium of justice, it is justice itself, the line that demarcates Inca tyranny from the just rule of the Spanish.

The consequence of this doctrine appears in Chapter 5 of the *Gobierno*, where Indians have to be compelled to work: "Based on the Indians' idleness and their condtion and the harm that ensues from it, it should be understood that it is good to force and compel them to work, and thus the Council's provisions that forbid the Indians from being idle and mandate that they hire themselves out to work in the fields and the justices and officials compel them to work at their tasks and make them pay for work and their wages themselves" (15). Given the Indians' condition and their lack of inclination to work, it is necessary to compel them to work in exchange for a wage. Matienzo also explains that the cause of their indolent nature comes from their lack of private property:

> Also, because owing to their idleness the Indians have not had up to this point anything of their own but, rather, everything in common, put things in order and give each one his own lands and his own money in pay for his work with which he can buy sheep and cattle from Spain and other things so that they will grow fond of work and to live in a political society, and there are many things that the Indians should occupy themselves with, understanding that especially in the district of that audience there are many kinds of Indians, caciques, political chiefs, or princes, others are yanaconas, others *atunrunas*, others *tindarunos*, others *mitayos*, others *vros* or fishermen and each one of the diverse occupations. (15–16)

The reason they have little desire to enter the market economy and become laborers is the absence of individual property, a consequence of their collective subjection to Inca tyranny in which the product of their labor was often appropriated for communal, not individual, purposes. Here, Matienzo adds a decisive twist. By providing the Indians with the incentive of private property, the Spaniards can teach them to be free to buy and sell things in the market. They have to be taught to own things, to work, to earn wages, and "to live in a political society." The dual solution to the problem of the Indians' "condition" is to compel them to sell their labor on the market in exchange for a wage. Briefly put, the solution consists of compelling them to be free.

Matienzo knew that it was not possible to rule the Indians without the help of the caciques. In Chapter 6 he states that the *curacas* are the "natural lords" of the Indians and that they are effective despite being tyrannical: "Caciques, political chiefs, and other chiefs are the Indians' natural princes and govern them with great harmony, although also with great tyranny" (16). Caciques are natural, and somehow efficient, rulers. Unlike Sarmiento de Gamboa, who prefers to delegitimize the Incas, saying that they were not natural rulers, Matienzo prefers to admit they are natural because they are somehow efficient rulers. As Mumford explains, "Matienzo despised the caciques and worried that they were learning too much from the Spanish. . . . In spite of his contempt for Andean culture and tradition, or in another sense, precisely because of it—he appreciated the Incas and believed Spain should learn from them" (70). Matienzo points out that the Spaniards can learn from the way the Incas imposed their language and culture: "As they established the general Inca language and learned the doctrine better, even though it is their language they lack the vocabulary to understand what they intend to teach and undertake and with the provisions given for it. Other than this it was necessary for them to occupy themselves with handicrafts, such as being painters or silversmiths, or other similar jobs. It is also important to try not to steal or rob from the Indians or take their daughters or have many wives or concubines" (Matienzo, 17). According to Mumford, Toledo's General Resettlement was partially inspired by the Incas, since "the continuing necessity of the archipielago networks implied the continuous necessity of the caciques, and a degree of indirect rule" (181). In this passage we see how the Incas knew that the Indians had to be taught to remain busy and work with their hands. Despite acknowledging the need to use the caciques as intermediaries between the Spanish officials and the Indians, Matienzo sees them as constituting an obstacle to the Indians' freedom:

> The tyranny of the caciques with their Indians is notorious because after the Incas were liberated from oppression, learning from them that it has made each one another Huayna Capac, and even though the audiences reprimand them for some things but not for others, they want to hide their wickedness. Those who are indoctrinated and their encomenderos, who try to treat them well, in their intentions and in the contracts that these caciques make with their Indians they hinder them and prevent them from having liberty, property, power, or the understanding to be able to complain about them. And the Spanish do not want to look among themselves for this or go to Potosí or to the audiences where they are free and earn money and deal with Spaniards who teach them, and if some relent, they try to stay for a short time and send others to follow because they do not learn how to stay in one place, nor do they become rich and they remain subjects forever. (17)

The caciques oppress the Indians, depriving them of their freedom, property, and capacity for understanding, and most importantly, they prevent them from going to Potosí and the *audiencias*, which were courts that administered justice with civil and criminal jurisdiction and were empowered to hear complaints against officers. Potosí and the *audiencias* are places where Indians can enjoy freedom, make money, and deal with Spaniards who provide them with education. In other words, the caciques do not want their subjects to be free and rich. Caciques, the enemies of Indian freedom, want them to spend less time in the mines and in Potosí in order to prevent them from becoming autonomous economic agents.

A good example of the prosperity of the Indians who worked in the mines are the yanaconas of Potosí and Porco, as explained in Chapter 8:

> The third type of yanaconas are those who are in the mines of Potosí and Porco. What they do nowadays is work in their masters' mines and mill and extract silver so that everything that comes out of the earth in the box together with the metal that they call pulverized ore and debris, which is ten times more than what their masters take from the box and and the seam, although in Porco they do not give them this pulverized ore and instead they buy it and they sell them giving as much as they do for the other, and

they are now so rich and well-paid that they lend with no deposit four and five thousand pesos, and their masters pay them no wages other than these benefits, which is a vast amount of wealth, nor do they give to their masters more than what is extracted, which they don't even extract themselves but with hired Indian labor. (20)

In this passage, Matienzo describes the system of production of silver before 1570, when it was in the hands of the yanaconas who refined metal with the guayras. They were like free agents who did not receive a wage and could easily become rich. Moreover, they even subcontracted Indians to help them to extract metals from the mines:

> The yanaconas themselves work all the time so there is no metal, and Indians they hire for pay help them because the ore they extract is said to be for them and not for their masters and they work of their own free will in the mines as so much yield comes to them, they make profit for their masters in digging until they reach fine metal with no cost to them, and they use hired Indians to reach it. These yanaconas do not expel them from the mountain, although they do kill them, but there are other newcomers who do not have the same advantage and who they easily remove and take to small farms, which is severely harmful and should not be permitted. (20)

The extracted metal belongs to the yanaconas, and not to the Spaniards. They work voluntarily, according to their own interests, and do not want to leave the *cerro* under any circumstances. They produce a surplus for the mine owners by extracting even the smallest amount of metal without losing money. Matienzo paints a rosy picture where Indians enjoy better life conditions in the mines than on their lands:

> And in the mining centers they are better off than in their own lands. They eat well, they are well-dressed and well-educated, they have all the same freedoms the Spaniards have, they are rich and while most of them do not work for themselves, they used to bring in hired Indian laborers to excavate and amalgamate loads of ore and when they reached the metal they took it at their masters' expense. They are very healthy there and without illness and when they are ill they are given the best treatment. They have many children and so no one can say that it would be good to remove them from the place where they earn so much profit and send them elsewhere where

they will not be treated as well. Those who govern these mining centers must pay close attention to them as well as to the yanaconas who are now there as neighbors and those who are born and raised there because more harm would come from relocating them. (21)

This passage is an apologia for mining, which appears to benefit not only the Crown but also the yanaconas. Among the benefits enjoyed by the yanaconas is that their essential needs are taken care of, they receive instruction in Christian doctrine, and finally, they enjoy the same amount of freedom as the Spaniards. Of course, the silver of Potosí benefits both the Indians and the whole kingdom, which depends on the power of silver to function:

> The great wealth of Potosí and Porco is notorious, and if these mountains were insufficient and working them came to nothing, his Majesty would lose much. Thus they are the key to the kingdom, these centers can be easily governed with much prudence and by persons of high quality and trustworthiness, and the truest and most advantageous rule for their preservation is the same for all who will govern well and will not make changes even if they would not be harmful, and this is to increase the number of Indians, and this had to do with what is governed. He who will govern these centers must be a well-known person and someone of whom his Majesty has knowledge and who can be trusted, someone who has experience with mines and can be given wide jurisdiction and can distribute orders throughout the entire province of Charcas. (85)

Like the *Anónimo*, the *Gobierno* sees the mines of Potosí as the key to the kingdom. Compared to the silver of Potosí, the rest of the riches amount to nothing. Basically, everything depends on the mines, which sustain the kingdom with their power. The government of Potosí is so important that it needs a prudent and loyal ruler who is not friend of *novedades* and knows about mining. One of the main tasks of such a ruler is to increase the number of indigenous workers. Finally, such a ruler should enjoy an important amount of power and jurisdiction. Matienzo's discourse opened the empty place that later came to be filled by Toledo.

A section in Chapter 44 where Matienzo discusses the problem of the use of coca leaves among the miners is illustrative of the centrality of mining business for the imperial network:

If the coca were taken away now they would say that Bolivia is a bad mita and the tyranny of the Incas, and if it were taken away they would not go to Potosí nor work nor mine silver, and that which they would take would be buried in their *huacas* and nothing could be done to take away their power. And to say that food or clothes would suffice is nonsense since all of that would amount to three hundred or four hundred thousand pesos while coca brings in a million or more and being without it is clearly to be without everything and the land would be abandoned.... [A]side from this is the [matter of the] Indians' currency with which they conduct business among themselves. Ultimately, to want there to be no coca is to want there to be no Peru and that the land is abandoned and the Indians return to their heathen ways. And as for the claim that this is superstition, it can't be known as such before one sees the profit that the Indians make as is said. And regarding their offering of it to the devil, they also offer everything they eat. (90)

Coca is necessary to keep the Indians working in the mines. Without coca, there will be no silver and no Potosí because otherwise the Indians would bury their *guacas* and treasures, returning to their idolatries. By stimulating mining, coca consumption prevents the region from becoming depopulated. For Matienzo, this is not a real problem because they offer not only coca to the demon but also everything they eat. Coca seems to be the fuel that maintains the smooth functioning of Potosí. The *Gobierno* provides an image of the empire where everything is interconnected and money has the instrumental role of keeping the networks together. The issue of coca consumption in fact bears witness to the importance of Indian labor for Potosí and therefore of Potosí for the Spanish Empire. As in José Luis Capoche's *Relación* (1585), things that the Spaniards found morally questionable become necessary instruments for the end, which is the common good. The increasing instrumentalization of metaphysical instrumentalism ends up elevating the means to the character of an end in itself.

Brading sees the circle of public figures trusted by Toledo as part of the "imperial school of thought," which is the tradition of denigrating the Indians in order to deny their dominion—the best example of which is, of course, Sepúlveda. Brading detects the trace of Sepúlveda behind Matienzo's emphasis on "cash economy and individual property as the basis of civil liberty, combined with the same concept for the servile condition of the native peasantry" (*The First America*, 143). Although Brading sees a connection between Matienzo and Sepúlveda's humanism, it is necessary to explain the philosophical origin of the illusion that fuels this ideological misrecognition. The source of this illusion is the ontological presupposition behind servility, the division between commanding, perfect forms and commanded, imperfect matter. Only metaphysical instrumentalism can explain how their refusal to work is a self-evident sign of their need to be commanded, obliged to work in the mines in exchange for a fair wage. Therefore, Matienzo's arguments in favor of compulsory labor must be situated within the history of the imperial uses of the principle of the natural subordination of matter to form and means to an end.

Going one step further, in Matienzo, we witness the encounter between two propositions that are two sides of the same coin: Inca tyranny is inseparable from the servile character of the Indians. Since, despite identifying them as tyrannical, he considered the Incas natural rulers, the line that separates Inca tyranny from Spanish rule is the circulation and private accumulation of commodities made possible by money. Within the context of Thomist political philosophy, not every act of subordination is tyrannical, only those carried out for the benefit of the ruler. As Mumford explains, the Spaniards did not consider that there was a contradiction in appealing to Inca surveillance and declaring them tyrants because surveillance could be used for the benefit of the ruled and not just in the private interest of the ruler. Yet I understand that the specificity of this kind of imperial subordination is the paradoxical encounter between the principle of the natural subordination of matter to form and the emergence of capitalist economy. The transformation of precious metals into a necessary medium of justice, the measure of all value, as Acosta would say, is the condition of possibility of retroactively declaring the Incas to be tyrants.

Let us recall that, in Acosta, a particular element (prime matter, imperfect amorphous plastic material, the lowest particular things) becomes the measure of all things, overpowering everything else by becoming the commanding principle of everything else. An analogous ideological mutation takes place in Matienzo. First, his racist argument about the servility of the Indians presupposes the natural subordination of matter to form and means to an end, or as Sepúlveda would say, the dogma of dominion of the perfect over the imperfect. Second, the ideological component of natural subordination appears both in the attempt to hide the real causes of colonial antagonism and in the indigenous lack of economic motivation for working in the mines. Matienzo condenses within the Indians a number of contradictory characteristics: they are both obedient and enemies of work, both potentially good Christians and ready to return to their *guacas*. The indigenous worker thus becomes the embodiment of all the inconsistencies of colonial society, where all fears and problems are exchanged for only one fear: the reluctance of Indians to work in the mines. The ideological solution to this problem becomes to compel them to be free, to liberate them from their former oppression to Inca tyranny and their own servile nature, by civilizing them and teaching them the use of money while transforming them into a pool of deskilled labor.

While in Acosta's *Natural History*, the principle of the natural subordination of the material means to transcendent ends paradoxically raised plastic amorphous material to the status of universal and necessary measure of all value, in Matienzo's *Gobierno del Perú*, the same principle of natural subordination treats Indians as an imperfect matter that embodies both excess and lack. Their servility is the consequence of a metaphysical instrumentalism that results from the fetishistic transposition of the contradictory qualities of docile obedience and laziness and inclination to vice onto the Indians in order to justify not only compelling them to work but also subsuming them into a wage system that measures their labor time by means of money. By analyzing both Acosta's *Historia* and Matienzo's *Gobierno*, it is possible to reconstruct another step in the constitution of the object and subject of imperial ideology: money and the indigenous worker. Both precious metals and Indians become imperfect matter available to be mobilized through subordination to other, higher, ends.

Until now I have presented Matienzo's arguments in favor of implementing the method of compulsory labor that would later be known as the Toledan mita. I have shown how these arguments were from the very beginning aligned with Sepúlveda's metaphysical instrumentalism. Now I will examine the impact of these arguments on Toledo's epistolary account of the removal of the Incas that was sent to Philip II in 1572 and that implicitly relies on Gamboa's and Matienzo's arguments.[18] Toledo cites both Inca tyranny and the servile nature of the Indians as justifications for his severe reforms. Toledo affirms that the Andean Indians lacked any kind of masters or government before their recent conquest by Pachakuti Inca Yupanqui: "Until Topa Inca Yupanqui, who possessed and tyrannically subjected these kingdoms, the aforementioned natives had neither Lord nor chief to command them or govern them in times of peace, nor did anyone subject them and they were as perfectly equal communities having among themselves no form of government other than that each enjoyed what was his and lived as he wanted" (Montesinos and Toledo, *Memorias antiguas*, 186). The Andeans lived in a state of continuous chaos, war, and anarchy until the creation of the Inca Empire by Tupa Inca. He ruled tyrannically, arbitrarily appointing kurakas and displacing entire communities (187). Toledo emphasizes that the Incas used these harsh measures and tyrannical methods because they knew about the inner indolence of the natives: "and also regarding the customs that these kingdoms had before the Spaniards entered them and what method of governing the Incas had, compelling them to work so that they would not be idle" (191). The rigors of building and carrying out public works kept the Indians occupied.

Toledo concludes that it is possible to infer a series of "authentic principles" based on the "proven reasons and facts" (201). The first is that Philip II is the legitimate ruler of these kingdoms and not the "tyrannical and, as such, intrusive Incas and chiefs" (201). The second is that the Crown can rule the Andeans "temporarily or perpetually," with or without jurisdiction, "without respect for bonds; and this would be one of the most important things for the spiritual and secular government of these Indians because their caciques and political chiefs will always be what they were, both in virtue and in vice" (201). The third conclusion is that, presupposing the true dominion of the Crown over these kingdoms, it is possible to distribute these lands among the Spaniards without the scruples of those who affirmed that the Incas were legitimate rulers, "all this being false, as the evidence shows" (201).

The fourth conclusion is that, as it has true dominion over these kingdoms, the mines, minerals, treasures, lands, and goods that do not have particular owners therefore belong to the Crown (202). The fifth conclusion is that, as their legitimate king, Philip II has the obligation to defend the natives of this kingdom, and because of their "weakness of reason and poor understanding, Your Majesty can order laws for their proper conservation and make them comply, even though they may disobey and seem to contradict their own freedom, as it would keep them from idleness and occupied with things that are good for them and for the republic, and govern them with some fear because otherwise they will do nothing, as can be seen and as the facts of the investigation show" (202).

Toledo provides both proven facts (*hechos probados*) and philosophical ground (*fundamentos*) in order to conclude that Philip II has true dominion over Peru and the mineral riches that do not belong to any particular individuals. Matienzo had provided the ideological foundations for Toledo by associating Inca tyranny with the servile nature of the Indians. The tyranny of the Incas and the servility of the natives are in this view two sides of the same coin. As a result, Toledo sees the combination of the Indians' inability to rule themselves and the tyrannical rule of the Incas as arguments in favor of the dominion of Philip II. In sum, the solution of the crisis of 1560s produced a change in the history of the ways Spanish dominion was justified.

The grounding principle of dominion, the natural subordination of imperfect matter to the more perfect form, is the ultimate presupposition behind the Toledan reforms. In Francisco de Vitoria, this principle grounded civil power and the capacity of a perfect community to defend commerce, evangelization, and the appropriation of precious metals that do not belong to anybody. All the arguments in favor of just war resulted from the impasses of instrumentalism in Sepúlveda's and Las Casas's interpretation of the hylomorphic doctrine. Acosta gave this principle a technical dimension by providing a retroactive justification for Spanish dominion and a causal and rational explanation of the role of metals and workers in the network of production and circulation of commodities. Finally, in Matienzo and Toledo, this principle is used by the state in order to impose capitalist mercantilism in Peru and to produce profit and taxes for the Crown. I will now analyze a letter written by an expert miner who makes a passionate defense of both the Toledan introduction of the mita and the method of amalgamation by mercury.

JOSÉ LUIS CAPOCHE'S *RELACIÓN GENERAL DE LA VILLA IMPERIAL DE POTOSÍ* (1585)

One of Toledo's greatest triumphs was the introduction of the method of amalgamation, which, combined with the mita, resulted in what Bakewell calls a revival of Potosí and an increase in the royal revenue. The introduction of amalgamation (which was both effective and practical) and the problematic and infamous mita were attempts to find a solution to the crisis of the 1560s.[19] According to Peter Bakewell, one of the most intriguing questions in the history of colonial mining is the delay between the introduction of large-scale amalgamation in New Spain by Bartolomé de Medina in the 1550s and the introduction of this process in Potosí in 1570s (18). While the Portuguese Enrique Garcés tried to introduce the method in 1559, the easy money made possible by the overabundance of metal and the effectiveness of portable ovens delayed the introduction of a method that required a greater investment, as well as greater technical knowledge (Bakewell, 18; Bargalló, *Amalgamación*, 172). It might be added that Viceroy Toledo was the only viceroy who had firsthand knowledge of the specific problems of mining in Potosí, which resulted from a trip through the heart of colonial Peru (1570–1575) (Bakewell, 61). In February 1572, Toledo ordered the delivery of a sum of money to Pedro Hernández de Velasco naming him "chief master of mercury" (Bakewell, 19). The amalgamation process was a true technological revolution that allowed one to refine the base metals that had been discarded as "debris" (19). The efficacy of amalgamation derives from its capacity to transform useless ore into value, making the unproductive productive. Amalgamation employed massive mills called *ingenios* to grind the ore that were driven either by water or by animal or human power (19–20). This machinery was already in use in Europe, as can be seen in "book 8" of *De Re Metallica* by Georgius Agricola, in which an illustration of a "double-headed" mill appears (Bakewell, 13). Amalgamation also had a big impact on the environment since it required the creation of artificial lakes and aqueducts that became the "aorta of Potosí's industrial organism" (13). As a result, Huancavélica emerged as the main source of mercury to meet the needs of the mills of Potosí (25). The effect of the introduction of amalgamation by Toledo was that of an economic miracle, the result of "the combination of cheap ore, cheap labor, and new and efficient technology" that made the period from 1573 to 1582 "a remarkable decade for Potosí" (26). Bakewell emphasizes that "such an explosion of production and prosperity has never been seen there since" (26). In sum, the introduction of amalgamation was a veritable technological event that divided the history of Potosí in two.

Such an explosion of "production and prosperity" did not depend solely on the technical aspects of amalgamation but also on the massive mobilization of human resources. In fact, the mita had been a subject of scandal, giving rise to a local "black legend" that was attacked by Guaman Poma de Ayala and regarded ambivalently by Solórzano Pereira.[20] Toledo appealed to an argument that proved crucial, claiming that the natives already provided personal service in the time of the Inca Empire (Bakewell, 64–68). Bakewell explains that when Toledo went to Potosí, there were clear antecedents of the mita that facilitated its implementation (59). Trustees had used forced Indian labor in Potosí since 1540 (59). Bakewell explains that, in its organizational aspect, the mita was a continuation of service to the Incas "and was so perceived by the Indians" (59). And around 1570, the mayor of Potosí had the habit of assigning "indigenous labor conscripts to the extraction of precious metals." Toledo only "sought to standardize much that had doubtless been variable before: length to stay in Potosí, wages, conditions of work" (59). For this reason, Bakewell concludes that it is the "formal culmination of many preexisting practices and notions, not [a] new creation" (59). By demystifying the novelty of Toledo's measures, Bakewell shows that they were less the effect of a mythical iron will than the crystallization or formalization of an earlier and larger process. The presuppositions that resulted from the retroactive reconstructions of these "preexisting practices and notions" made possible both the exploitation of "natural resources" and the mobilization of indigenous labor. The *Historia general de la villa imperial de Potosí* (1585) by José Luis Capoche, a mine owner, made a radical defense of Toledo's policies. Capoche was one of many Spanish mine and mill owners who wrote reports for the purpose of influencing the decisions of the authorities (Hanke, "Prólogo," 43). Capoche completed the manuscript by August 1585 and, at that time, already "possessed some mills and was a man of some substance" (45). In this document, Capoche narrates the social and economic history of the Cerro de Potosí. The *Relación* is dedicated to Hernando de Torres y Portugal, Count of Villar (43).[21]

Capoche warns that his intention was to provide the authorities with a firsthand understanding of the many difficulties that plagued the "imperial villa":

> And as this is the most important and most burdensome villa Your Excellency possesses, and as the preservation and growth of everything else depends on it, it struck me to begin serving Your Excellency by establishing this relation between this mining center and mountain, the state in which your mines and all the mines of the province are in, and the law of metals and other matters having to do with governing it, with respect to some things that have happened so that the difficulty of this business in these new lands can be better understood—this has been my principal intention;

although Your Excellency is absolute and it would be possible to represent them in many false and inaccurate ways, and it is a place where for some time it was dangerous to seek and know the truth (which is what is common and consummate within them), they are so different for their singularity and subject to matters they cannot comprehend if it is not from experience in the practices that usually take place in new governments. (72)

This passage repeats the commonplace that Potosí mining is the most "important" business for the "preservation" of the kingdom. The authorities must be convinced that Potosí is not just a city among others, because the entire region depends on it, and for this reason, the information provided should be considered crucial for future administrative decisions. The technical knowledge of a mine and mill owner will be critical for the smooth functioning of the colonial economy, as well as for the preservation of the Crown's power. In short, his service is to cooperate with the government of Potosí, guiding the decisions of the authorities, helping to prevent, as the viceroy reports, "adulterated and off center." Capoche adds: "But Your most prudent Excellency will overcome the difficulties and traps of this labyrinth, which are many, achieving victory, that in God's eternally great name I beg Your Excellency to receive this small service, paying no mind to how insignificant it is but, rather, to the will with which I offer it, with which I have dedicated to the service of Your Excellency, whose most excellent personage Our Great Lord may protect and place in the greatness you deserve with the growth of a greater nation" (72). Potosí look likes a "labyrinth," a series of almost insoluble problems, and the *Relación* is presented as an Ariadne's thread that holds the key to the solution to these problems.

According to Lewis Hanke, the document is part of the literature known as "claims of Potosí," whose purpose is to influence the decisions of the Crown in favor of the local interests of the mine owners (50). Hanke sustains that the audience was not indifferent to Capoche's claims, because after the completion of the draft, Don Pedro de Cordova Mesia, Count of Villar, who was an agent and inspector appointed by the Crown to "bring order" to the region, took the manuscript to Lima with the purpose of disseminating its ideas (64). Among the problems and solutions proposed by Capoche were the confirmation of the rights of the Indians to the *rescate* (sale of stolen metals), the maintenance of the tunnels, the conservation of *oficio de protector de naturals*, and the reduction of the price of quicksilver. All these requests were considered by the authorities to a greater or lesser degree (Hanke, 64–65). An analysis of Capoche's narration of the transition from the method of smelting ores with portable ovens to amalgamation by mercury will show how the celebration of amalgamation and the defense of the mita were presented in tandem, with the latter being the necessary consequence of the technical aspects of the first one.

HUAYRA, A TYPE OF SMELTING OVEN

Both the Indians and their refining method are the protagonists of the history of the first phase of the exploitation of the Cerro before the crisis of the 1560s:

> All the mines benefited from this refining process in the time when precious metals were found and the Indians possessed all the riches of the kingdom because it was dependent on this business and there was no other relief there other than the silver that the Indians refined with the *huayra*. And not all the miners could avail themselves of this, because the Indians used it only in the rich mines that were known to produce high yields. And those who did not possess rich mines, their wealth now steadily declining, worked at their own expense and peril with Indians forced into labor by ordinary justice and others whom they hired, [and] they settled for what [the metal] was worth with people who had mined it at their own expense. But in the beginning, all the mines had metals refined by the huayra because what they mined were four main seams. (Capoche, 109)

During this early period metals were refined by the yanaconas with their huayras, which were wind-fueled smelting ovens. The extracted metals were very rich, and the huayra was very fast and efficient. Capoche suggests that these indigenous *huayradores* enjoyed an important degree of technical agency before the introduction of the method of amalgamation. Moreover, most of the riches were in the hands of huayradores and the whole kingdom depended on this type of deal, which allowed these *indios ventureros* to remain in control of the business. The kingdom depended on the goodwill of people who, according to the official ideology, were not inclined to work. In other words the period of the huayras, represents the moment when Spaniards had indirect control over wealth depending solely on the erratic will of the Indians. Not all miners benefited from Indian labor, only those who owned mines with veins abundant in rich ores. The mine owners who had no contract labor Indians (*indios varas*) who wanted to work had to use Indians provided by "ordinary justice" or they could just rent their mines to the *indios varas*. Sometimes some the *indios varas* asked permission to mine owners to exploit a vein (109). The Indians paid for the instruments such as the "barreta" (a digging tool) or the candles (109). Capoche explains that this type of extraction and refinement by huayras fell into disuse as the richest veins became scarce (109). Capoche stresses that in the "present" both *indios de cédula* (*mitayos*) and *mingados* (free workers) work for wages (109).

To explain the operation of huayra and compare it with other methods of refinement, Capoche proceeds to narrate its history. Capoche begins by stating categorically that Indians lack inventiveness: "The natives of this kingdom and all the western nations of the New World of the Indies by nature have little ingenuity and lack sufficient imagination to invent the tools that are necessary and useful for the work that they do, and so they lived in great ignorance of what the world had to offer, as if they had not been born there" (109). Like Matienzo or Sepúlveda, Capoche argues that indigenous people lack "ingenuity" and "imaginative" capacity to invent instruments. Let us recall the importance given by Acosta to this capacity. Capoche reproduces the intellectual division of labor present in Matienzo: while the Spaniards have intellect, the Indians provide material work. The attribution of lack of imagination and inventive capacity to the indigenous follows the natural subordination of superior to inferior crafts. But Capoche goes beyond Matienzo and Sepúlveda, who limited themselves to dispossessing the Indians of their self-determination but still thought the Indians possessed some technical skill. For Capoche, Indians provide manual labor because they lack understanding and technical imagination. The Indians are deprived of the tool-inventing aspect intrinsic to human nature.

Note that, when explaining how the smelters operated, Capoche stops at what he sees as shortcomings and errors in using indigenous methods: "And as they did not have use of bellows to make their smelters, the Indians of Peru used copper tubes in three-hand lengths through which they would laboriously blow" (109). Paradoxically, although he explains indigenous techniques in terms of what they lack, he must admit that using portable furnaces requires taking advantage of wind (110). When smelting required more fuel, the Indians employed ovens made of stones in the upper parts of the hill, "stacked one on top of the other without mortar turret-like shafts two hands in height" (110). To facilitate combustion they employed manure and wood, "as they did not have charcoal; and with the wind rushing through the openings in the stones, the metal was smelted" (110). The huayra works when it is necessary to generate more force with less effort, since the wind does all the work.

Regarding the origin of the huayra, Capoche is the only chronicler who attributes the huayras to a Spaniard: "And as time is the master and inventor of the arts, through his industriousness, Juan de Marroquí, native of [blank] taught how to make from clay some form of manufacturing this demonstration, which we call *guayrachina* or *guayra*, which remains in use today wherever metals are easily smelted without the use of bellows, which is very costly and not as advantageously used for the metals here as is the *guayra* in a climate with such strong wind, since it is not like that, it is not profitable to use it" (110). Although the bellows worked in Porco, they failed in Potosí, and this is the reason why they were replaced by huayra, an invention of Juan de Marroquí. Capoche even says that the Spaniards call these ovens *guayrachinas* or guayras.[22]

The following passages discussed the functioning of the huayra (Capoche, 110). The metal goes through several stages. First they "grind" and "wash" it in order to separate it from the *tierra muerta* (110). It is not necessary to wash the "very rich" metal because the cost would be greater than the benefit (110). Also, the richer the metal, the less necessary it is to wash it. For example, for the *tacanas*, which is a very rich metal, it suffices to combine it with the mixture of lead and silver (110).[23] Thus, the rich metal is mixed with the *soroche* in order to soften it and make it easier to melt by virtue of its moisture and softness (110). Finally, silver flows more easily because it is dry, cold, and hard and the *soroche* facilitates the process of incorporation into a mixture that otherwise would just evaporate (110). Capoche explains that the wind speed increases the speed of the refinement, implying that it even surpasses the bellows in efficiency and speed because the force of the wind heats the metal faster than bellows (110). Its efficacy is the result of the combination of small windows and strong wind that is very common in Potosí (110). In a gnomic phrase that ends the description of the huayra, Capoche states that the four elements contribute to the profit of silver: "In this way the four elements are involved in the refining of silver; the earth gives us the metal, fire refines it, water washes and helps it, the wind blows and works the bellows, and it seems that they are serving and helping man, helping him with the silver that comes from there in order that he secure the necessities of life" (110). The cooperation between the primordial elements makes evident that nature is experienced in instrumental terms. Moreover, technique makes possible the convergence of the four elements that are the material basis of the physical word, subordinating them to the production of silver and fulfillment of human needs.

Huayras are placed in the "peaks and slopes of the mountains and hillocks that are in view of and surround the villa" (111). In the darkness of the night you can see "fires throughout the countryside, some placed on the tops and pinnacles of the mountains like signals and others confusedly situated on the hillsides and ravines" (111). The number of huayras is six thousand four hundred and ninety-seven (111). Finally, he points out that the majority of the huayras remain on the Cerro, although they are no longer in use.

When Capoche wrote the *Relación*, the use of huayra as the primary means of refining metals had already been displaced by the process of amalgamation by mercury. The displacement of the huayra meant a decrease of indigenous technological agency. Indeed, in another section of the document, Capoche explains that the mine owners realized they could increase profits by adopting amalgamation "without having to share them with the *indios varas*," and they decided to "put doors and locks on some of the mines" (162):

> The invention of amalgamation by mercury caused them much sorrow because it deprived them of their profits, and they had no other way to support themselves or pay their rates except through wage labor, having possessed all the former wealth and abundance with which they had preserved and predicated the profit and general benefit of the kingdom, and the method of refining with the huayra had caused them to lose all this, as important as it was, for reaping what they had sown by not using mercury amalgamation. (162)

The Indians lost not only control over the means of production but also the material means of paying the tribute. In order to pay these taxes they needed a wage because they did not have control over the circulation of mercury or enough capital to buy it. Mining, which was central to the sustenance of the whole kingdom, depended on the huayras. The introduction of amalgamation meant not only the expropriation of the product (silver) but also the Indians' loss of control over the means of production. In other words, it meant the subsumption of the laborer into the network of the debtor-creditor system. The history of the transition from the huayra to amalgamation as narrated by Capoche cannot be entirely contained within some kind of naive technological determinisim, where the most efficient technologies are destined to replace the less efficient ones. It was the result of the demands of capitalist mercantilism, which uproots and deterritorializes even technology itself. Accumulation for the sake of accumulation implied the Spaniards' increased control over technology itself and the annexation of the body of the Indian to the machine itself.

MITA AND TECHNICAL NECESSITY

Spanish mining was on the verge of chaos when the rich ores of the Cerro were depleted. When Toledo arrived at Potosí, "he found the earth's potential for producing silver had greatly diminished through having depleted the precious metals of that mountain" (115). The lack of precious metals had a decisive impact across the entire region, "and with this lack the entire land and republic changed, blessed as it had been by its riches, heralded throughout the world for the rich spoils that had come from it, and with so many men who had become rich from it" (115). The excess reverted into lack. The decadence was augmented since "this general ruin and harm was felt more every day for the scant silver that remained and the low price and poor rate of commerce that all goods had" (115). Potosí mining and the imperial economy were on the brink of collapse and destruction, a fact that was visible in the inefficient circulation of commodities throughout the empire (115). Since commodities did not circulate, taxes decreased (115). The lack of money meant that the region had to return to the "redemption and exchange of one thing for another, a custom which these peoples had used since before our arrival" (115). The mercantile economy was threatened by the specter of the return of the precolonial barter economy (115). Capoche describes this as a "miserable event" (115). The locals regretted the "present calamity, foretelling a future destruction, declaring that Peru had already depleted its riches" (115). Capoche continues his gloomy portrait of the situation saying that there were no "precious metals for the huayra" (116). The royal fifth was diminished, and the silver that circulated "no longer had more than half of the law" (116). There was a labor shortage due to the lack of economic incentive for the Indians to work in the mines (116). The mine owners could not think of a solution since the "process of refining by mercury amalgamation did not occur to them nor had they any memory of it" (116). There was a rumor that "one could not refine with mercury amalgamation, and they sought no other methods, and every day the necessity grew" (116). In sum, the imperial order was on the eve of a "future destruction" caused by a crisis in the mercantile economy that sustained the whole colonial administration.

The mine owners "exhausted all options," failing to find alternatives for the high cost of amalgamation and the "milling of metals" (117). The miners "had no experience with or knowledge of these mills." The miners were not satisfied with the results and "were skeptical of spending their fortunes on them at such a high cost" (117). Amalgamation mills needed water, and "among a few ravines they found some nearby flatlands where some quantity of water had accumulated as lagoons" (117). The engineers made "repairs in the form of strong bulwarks ... in the narrowest parts of the ravines, retaining and damming the currents" (117). Toledo also made just laws concerning the mining business and the discovery of new mines (117). He regulated "the Indians' pay and other things worthy of Your Excellency, making amends for the poor, making sure they were paid for their sweat. And to believe that it is from their zeal that they came to be sold as they are today, it would make an example of the oppressors" (117). The mine owners "sought alternatives for the cost of the method of amalgamation" (117). According to Capoche's narrative, the miners could not find technical solutions to the problem, as they were accustomed to a quick and easy way to make profits. It seems that the technological revolution represented by the introduction of amalgamation was not exactly favored by mine owners. In Capoche's narrative, the mine owners were incapable of finding a solution to the crisis, and Toledan reforms appear not as the result of continuous bottom-up series of practices but as a top-down, vertical solution provided by a prudent master.

Although there were precious metals in other part of the region, their number remained small compared to the quantities of metals extracted from Potosí, "where universal benefit continues and increases the benefits for Castile and the Continent" (116). The lack of silver and money caused a tremendous economic deflation since "all the commodities would lose the value they once had, and the fleets would cease coming, as would duty tax interests" (116). The decreased commodity prices could not be solved by the local commerce alone "because, as there was no silver in the kingdom, it was not possible to communicate with or have contracts or commerce in Castile, because the presence of two seas in the middle of so much coastline requires meticulous navigation, this land does not reap from its harvest anything that would be necessary in any other" (116). Capoche seems to be conscious of the interconnected character of the whole enterprise, the network of credit, commerce, and production of silver that linked Spain and Peru. The fissures materialized by the lack of metals aggravated the oceanic fissure between the kingdoms of Peru and Castile. From Capoche's point of view, the whole transoceanic imperial network depended on Potosí. The machinery of mercantile capitalism in general depended on the Potosí mining machine.

Moreover, the smooth functioning of the state machinery depended on the royal fifth provided by the New World silver:

> His Majesty would not be well served, and the authorities and *audiencias* that he has established in the kingdom, such as the exercise of the most eminent responsibility of the viceroy to defend his royal conscience, the administration of justice, and the public good could not with any decency be sustained without silver. And thus this would bring about great insubordinations and disturbances that would be the ruin and destruction of these kingdoms, and the preaching of the Gospel will not continue because, without the Catholics who form the backbone of the Holy Gospel in this new and remote land, the Indians would return to their errors and idolatries in which the devil had kept them. And their devil is as powerful today as he was fifty years ago, being that the relief of their salvation remains in sight. (116)

Like the document of Yucay and José de Acosta, Capoche thinks that the achievement of the transcendent ends of the universal monarchy, the salvation of the Indians, depends on temporal means, materialized in the technical colonial administration. Money sustains the colonial institutions, and without it a crisis of sovereignty would cause chaos and anarchy. Without mining, the Indians would return to idolatry. Just like José de Acosta, the author of the *Anónimo*, and Matienzo, Capoche sees a strict correlation between mining, the market, and the evangelizing mission. The whole system was organized around the accumulation of riches, and in the absence of money there was chaos pure and simple. The conservation of the kingdom became an end in itself, and the imperial "machine" (to use Deleuzian terms) became dependent on the mining machinery. If metaphysical instrumentalism aimed at justifying the means, now the implicit instrumentalist presupposition emerges under the form of the necessity of the means.

Capoche continues, explaining that, having been in Lima for only one year, Toledo tried to implement his program "through his royal instructions, as Christian as they are necessary" (116). Included among the most important benefits for the region is the mission to evangelize and educate the Indians (116). Capoche depicts a Toledo who exhausted all of the most effective means to achieve the "preservation and increase and civility of the republic of the Indians" (116). He organized the "natives in unity and human life, taking them from the highlands and wilderness where they lived densely and in dispersion, congregating them in villages so that they can be shown the ways of Our Holy Catholic Faith" (116). This direct reference to the General Resettlement—which was, according to Mumford, Toledo's most relevant reform—appears explicitly associated with the Spanish task of administering and civilizing the life of the Indians. The threat of chaos is inseparable from the metaphysical imperative of imposing unity and sense to what appears to be disjointed, dispersed, or disseminated. The "effective and supreme" means of the imperial order and its technical dispositions have the double purpose of evangelizing and civilizing the Indians. Toledo procured the "conservation and increase" of power through disciplining and correcting the Indians through civility. Toledo imposed order, harmony, and unity to the dispersed Indians, initiating them in Catholic doctrine, and liberating them from the Inca tyranny. Also for Capoche, Toledo is the supreme organizer of Peru, a master who gave the Indians human life. Relocating the Indians in settlements made the priests' evangelizing task easier, thereby assisting in the salvation of the souls of those who died in infinite numbers with no baptism or confession (116). Toledo's decrees protected the Indians, liberating them from the vexation in which they were subjected by the kurakas as well as the encomenderos (116). The implications of Capoche's narrative, given the passage analyzed above, can be easily rendered in retrospective terms. Capoche depicts Toledo as an all powerful agent, a prudent master, who imposes order on chaos and multiplicity, in the same sense that a Divine artisan imposes a preconceived form on an amorphous matter.[24]

The ideological efficacy of metaphysical instrumentalism consists of grounding a preexisting final goal (the common good) to a preexisting origin as a valid source of domination (God). A materialist reading not only takes into consideration the preceding practices that made the erection of the new order possible, it also makes visible the retroactive movement according to which events receive their meaning afterwards and the resulting order gives sense to a preceding chaos. The appearance of natural subordination and the iron necessity of a hierarchical world is an *après-coup* effect, an illusion that arises out of aleatory historical events. This ideological move depends on a tautological gesture that turns chaos into order, creating the illusion that this order did not emerge out of a fluctuating network of notions and practices but was already well grounded in an origin. In Capoche's text, the dysfunctional crisis that hit Potosí prompted one element to assert itself as a dominant ordering force that organized its authority over the rest. Crisis prompts a fetishistic exchange of a multiplicity of problems for one problem: in this case, Indian nature was purported as the source of all problems. Necessity is an ideological illusion, an effect of the imperial discourse that hides its performative and tautological character in the potential catastrophe represented in the lack of precious metals.

Amalgamation appears to be the solution to all the problems. Toledo also "began trying to refine the metals from this mountain with mercury amalgamation" (116). Capoche adds that, "he made them conduct assays of the precious metals and debris that they brought from this villa that provided good results" (116). Toledo "demonstrated care in fixing the village," and he ordered "them to bring mercury and that they begin refining with it" (116). The owner of mines took the amalgamated silver to Toledo who received it "contentedly and every day they continued to know the riches that this refining method promised" (116). Capoche continues, saying, "and taking on this business with the enthusiasm and zeal he gave to other things he was offered, including the good government of this land, he departed from Cuzco, and proceeded to visit thanks to this villa" (117). The locals organized a solemn reception and received his "presence and authority" with "joy and contentment" (117). His authority increased, as did his reputation as a wise and prudent ruler (117).

Toledo put the same effort and energy into amalgamation as he did in the rest of the "business" of the empire. By attributing the introduction of amalgamation to Toledo, Capoche remains within this picture of a powerful master who offered technical solutions to the crisis. Technique itself is subordinated to the preexisting master and relegated to the status of a neutral appendage that depends on the power of an external efficient cause to be set in motion.

Capoche declared that Toledo ordered the mita as a result of the increased demand for laborers required by amalgamation:

> And as he saw the wealth and prosperity that was hoped for with the new refining method, he ordered a great number of Indians to be brought for general compulsory labor for those persons who wanted to make use of it, as will be shown in due time. He climbed the mountain and entered the caverns where he saw the labor the Indians were undertaking, ordering visits to the mines to take stock of them and repair them where necessary; and he erected a church on the mountain where they say Mass as there were more than one hundred sixty houses that populated the seams that contained metals and where the miners lived. (117)

In Capoche's words, when Toledo *saw* the effectiveness of the method, he *ordered* the mobilization of a great quantity of Indians. When Toledo arrived at Potosí in 1572, "there were few people and the center was almost deserted, and the buildings were quite ruined and the residents had little silver" (135). Capoche continues with his exposition of the relation of necessity between the mita and the process of amalgamation: "And not finding a number of Indians sufficient for the need anticipated for the new invention of mercury refining, by finding the Indians in this practice and if they had gone absent for not having the large estates of the past, he ordered and commanded that the same assigned indentured labor be increased to more than the number of Indians that had to serve in this villa" (135). Capoche argues that the sufficient numbers of Indians necessary for the new invention did not exist. In other words, the mita appears as an unavoidable, inevitable, and necessary requirement for the smooth functioning of the amalgamating machine.

Reconstructing Capoche's apology of the Toledan reforms, we can observe the following chain of arguments. The evangelizing mission and the administration of the kingdom depend on commerce and the circulation of money and commodities. The crisis of the network of commerce makes it necessary to extract precious metals. The extraction of riches depends on amalgamation, and amalgamation depends on the forced mobilization of indigenous workers. At numerous points Capoche invokes the need for forced Indian labor in order to keep the ingenios working: "the necessity that the residents of this villa of Indians have for the earning power of their estates, as the number of indentured laborers is not sufficient for the many mines that exist there and the mills that they keep making every day" (173). The mita and amalgamation were so closely linked that the increasing number of mills ended up demanding an even greater number of mitayos, as "every day the necessity will grow" (173). As a result, technical necessity, the disavowed inner core of metaphysical instrumentalism, justifies the forced mobilization of the Indians. The mita is a necessary means for the process of amalgamation, amalgamation is a necessary means for the economy, the economy is a necessary means for evangelization and the "conservation" of the kingdom, and the conservation of the kingdom is necessary for spreading universal salvation and fighting the Reformation. In Capoche, the Toledan reforms were forms of social engineering that were necessary for the entire imperial enterprise. The necessity of the mita-amalgamation complex consists in its capacity to decide the fate of the whole. Ultimately, what confers necessity to technique is not only the unconditional character of the more-than-material means but also the threat of crisis. Therefore, the threat of the catastrophe elevates the technical means to the category of a necessary end in itself. If natural mastery and the subordination of metaphysical instrumentalism presupposed the transposition of technique to nature, now technical necessity becomes the disavowed consequence of artificial mastery itself. The Toledan reforms materialized the paradox of the principle of subordination already present at an abstract level in the metaphysical instrumentalism of the imperial school of thought: imperfect matter is subordinated to perfect form, the *medium of exchange* ultimately *sustains* the superior forms, subordinating everything to its universal value.

Capoche continues, explaining that soon the area was populated by ingenios—amalgamation mills that required water in order to function

(117). In the long list of ingenios provided by Capoche, some are drawn by animals and others are hydraulic. Capoche also provides a list of human-drawn ingenios: "by young men who they said can move them with their feet," "by hand," "by horse," "with stones in the manner of a gypsum mill," with "counter wheels," "with horses with certain wheels that move mallets," "with cranes, which the Indians carry like a spring wheel," and lastly, "by water with an axle and a large wheel like a water mill, building them in the creeks that run through the villa and in the river that runs through Tarapaya, Pilcomayo, and Tauaconuño" (117). Capoche adds that the only remaining mills are those fueled by "blood and water, which is to say horse-drawn and water-powered mills, as the most convenient way to mill" (117). Since water did not flow regularly during winter, the locals decided to build hydraulic ingenios on the riverbank (117). Ultimately, the division between water-drawn and "blood-drawn" ingenios bears witness to the process of the humanization of nature and the denaturalization of the human that is central to metaphysical instrumentalism. Blood and water become resources, passive matter, cogs in the machine.

In this ongoing conquest of nature and transformation of everything, including human life, into a technical means, the Indians went from enjoying certain mastery over the huayra to being an appendage of the mill. The transition from the huayra system to the amalgamation system is also the transition from a system that depended mostly on Indian know-how to a system that subordinated indigenous technical skills to the pre-existing will of the master, turning the body of the Indian into a living tool of the amalgamating machine. In Aquinas's commentary on Aristotle's *Politics*, the nonexistence of self-moving automatic devices required servants who could understand the orders given by the master. Aquinas understood natural servants in terms of living tools. In the 1570s, Potosí, Capoche argued, needed Indians in order to operate mills that obviously could not move themselves. The introduction of amalgamation and its social corollary, the mita, as a means of providing the labor force required by this technique, meant an increase in the ongoing technological and political alienation of the Indians. As we can observe in Capoche's *Relación*, the Andean workers went through a series of dispossessions. First, Indians were expropriated of the product, the precious metals themselves. Then, they were expropriated of their know-how, their technique, the huayra, as well of the means of production. And finally, they were compelled to work in order to fulfill the exigencies of the new method of amalgamation.

From a Marxian perspective, the introduction of machinery in the production of silver is the logical conclusion of the capitalization of production, the completion of a process that started when Indian labor was formally subsumed through the imposition of the wage economy. Formal subsumption of labor into capital (measuring labor time by means of a payment with an abstract universal commodity, money) precedes machinic subsumption: "In handicrafts and manufacture, the workman makes use of a tool; in the factory, the machine makes use of him. There the movements of the instruments of labour proceed from him, here it is the movement of the machine that he must follow. In manufacture the workers are the parts of a living mechanism. In the factory we have a lifeless mechanism which is independent of the workers, who are incorporated into it as its living appendage" (*Capital*, 1:548).

Another Marxian insight useful for understanding the Toledan reforms is that, both historically and conceptually, formal subsumption preceded real subsumption. This means that the introduction of machine technology is a logical consequence of capitalism, because before being an appendage of the machine, the worker was formally subsumed through the imposition of the wage economy, which transforms living labor into dead, frozen labor. While Marx is obviously thinking of the Industrial Revolution, if we examine Capoche's narrative of the transition of the huayra to the process of amalgamation and the justification of the mita in satisfying the needs of the new refining technology, we can see how Indians were reduced to a necessary component of the amalgamation mills, a standing reserve ready to be mobilized in the name of a greater good. This ongoing process of the naturalization of technique and the denaturalization of the human is an effect of the alliance between the Scholastic principle of natural subordination and the exigencies of mercantile capitalism. In other words, metaphysical instrumentalism is also the ideology of the Spanish Empire because exploitation (subordination) derives not only from the explicit effects of metaphysics but also from an implicit core, a silent presupposition, which is anchored in the fetishistic elevation of money to a social link.[25]

In sum, technical necessity is invoked to subsume the know-how of the indigenous miners in a mechanical system of production that reduces them to the status of deskilled workers. Kendal Brown provides us with an instructive synthesis of the situation that illuminates the process of machinic subsumption and the creation of a large mass of deskilled laborers:

> Amalgamation even marginalized the yanaconas, whose technical expertise with the guayras had given them a relatively high status and considerable independence at Potosí. Instead, the new technologies created a demand for a large pool of less-skilled workers to mine and handle the ore. The crisis of the 1560s, however, had stripped Potosí of most of its labor force, and Spaniards doubted the Indians would return. Rather than admitting that the crisis and the new refining technique had removed most of Indian's economic incentive for going to Potosí, Spaniards mounted a propaganda campaign to justify a greater coercion of labor. (49)

THE COSTS OF MINING: MORE BLOOD THAN METAL

Although Capoche's text is a defense of Toledo, he also dwells extensively on the cost in lives, both inside the mines and in the mills. Inside the mines, some workers extract metals in the interior of the mines using the *barreta*, others bring it to the surface by means of "triple branched ladders made from leather twisted like thick ropes and sticks going from one branch to another like a ladder so that one Indian can climb up while another descends" (Capoche, 109). The Indians bring up the metal by tying it to their chests and backs (109). They work in groups of three, and there is one who uses a candle "to be in the dark mines without any light, and the scant light of the candle is often extinguished by the wind, and by using both hands they grip it and help as best they can" (109). Workers go up and down the mine with much difficulty because of "the distance a man can carry a load before he tires, not to mention climbing and descending with so much effort and danger" (109). When they arrive, "sweating and breathless and gripped by heat, and the refreshment they usually find to comfort their fatigue is being told he is a dog and to go back because they brought too little metal or moved too slowly or that all he brought was dirt or that he had stolen" (109). Capoche then tells the story of a miner who wanted to hit a worker who escaped into the mine and then himself fell in and died (109).

When considering the risks faced by the workers inside the mine,

Capoche declares his astonishment before an apparent contradiction: "being naturally cowardly over little things, they have to take great risks where the very brave will not go" (158). They suffer "one thousand deaths and disasters every day," to the point where lawers and theologians, "people of thought and speech," are concerned about "the danger they put up with in working in the mines" (158). Not only are the Indians subjected to these miserable working conditions, they are also uprooted "from their villages and natural homes, leaving their homes, farms, and livestock, the fathers and mothers shedding many tears for their sons, believing that they will not see them again" (158). They are obliged to do things they do not understand because they are not greedy (158). The mitayos suffer "unfortunate events that commonly occur" (158). Their women "grieve without their husbands, and many children are orphans without their parents" (158).

Capoche knows that the mitayos go to work "forced and quite against their will, as it was impossible to remove them from their houses and lands" (158). Some, in order to evade the mita, "often give over fifteen or twenty head of cattle, which is their entire estate, so that other Indians will go in their place" (158). In another story, an Indian was working in a water-powered ingenio on the bank of the river in Tarapaya when "a wall fell and killed four Indians who were standing on the mortar, crushing them" (159). He also recalls when "an Indian went to lift himself from the mortar, the cams caught him by the head and smashed him to pieces" (159). In just a few days, eight Indians died, "and if one were to write about it in great detail, one would use much paper" (159).

Accidents and death occur not only inside the mine but also in the mills that refine metals. Capoche reports the terrible working conditions in the mines, and the radical imbalance between the production of riches and the material conditions of their extraction in order to defend the right of the Indians to "recovery," that is, trading stolen metals in the plaza of Potosí, a practice according to which Indians worked in the mines during the weekends in order to supplement their low wages: "And so that Your Excellency sees and can better understand the rights of the Indians in the production of metal, I will put forth the risk they take in the mines and the blood that it costs them and the danger to which they put their lives in the depths and false steps, where justice is better inferred, and what a rigorous executioner this mountain has been for this nation, as every day it consumes and finishes them and their lives are awash with the fear of death" (158). Capoche confirms that there is a structural imbalance between the risks suffered by the Indians and the profit enjoyed by the Spaniards with a famous lapidary phrase: "and among the painful things that have befallen this people, fresh in their memories, I will make Your Excellency aware of [some things] so it can be understood that the work they suffer through and what the metal costs them can be said to be more blood than metal" (158). The extraction of riches that sustains the colonial and imperial economy depends on a disproportionally unequal exchange between money and living labor.

The Indians "regularly fall over dead and others with broken heads and legs, and every day they are injured in the ingenios" (159). The nocturnal work, "in such cold earth," working the "mortar, which is above all working in the dust that gets into their eyes and mouths and causes significant damage" (159). The hospital is full of Indians, and there are more than fifty deaths every year as a result of "this savage beast that eats them alive" (159). There are "more than seventy criminal deaths of Indians in the native tribunals and the courts of the mines" (159). The endless lists of deaths produced inside and outside the mines show that death is not an accidental exception but a structural effect inseparable from the creation of wealth. Indian deaths are not just an accidental exception. Capoche notes with concern that "were these peoples political and logical it could be intimated that this case can be considered typical and that one should be careful with what is done with them" (159). As a result, in exchange for "such significant work," it is convenient that they "have some benefit, as Your Excellency's intention in ordering them to come to these mines, and Your Majesty's permission for it was so that they may benefit and become rich by their own will, because without this Your Majesty would not remain for even a day in Potosí" (159). He reminds the authorities that, if the Indians were reasonable and political people, they would sue the authorities, multiplying the already enormous amount of litigations.

Capoche repeatedly declares his intentions when reporting the deaths of Indians to the authorities. In the passage quoted above he states that his purpose is to show that extracting silver costs more blood than money. Now, Capoche compares the disproportion between the benefits and the costs in lives and the violence suffered by the Indians with the purpose of persuading the authorities to allow Indians to sell stolen metals in the Plaza del Gato:

> As the natives have faced so much risk and danger in extracting metal from the mines, it would be just and right to make Your Excellency aware that the fruits and profits that issue to other Spaniards who depend on this relief, as a republic that has no other resource than the refining of and commerce in metals, and all because of the Indians and the mercy that His Majesty and Your Excellency bestows upon them, advising the Indians how to sell in the [Plaza del] Gato, as well as about the laws that exist and the amount [of metal] that they will be allowed and the method the Spaniards have for selling and buying it, and because of which much profit flows to the royal estate and the Indians by their own will in this center, which is what His Majesty intends, along with other very useful and no less necessary and advantageous things, as related by the same views that, for their part were much more in favor of it if they had claimed to have populated the center with the Indians and brought them there against their will, and considering the lives that it costs to extract metal, in the things I write [have written] on this, readers can better understand the rights of the Indians and the justification for this relief. (160)

In this passage, Capoche first declares that he explained the risks and perils of working in the mines and mills in order to show how the interests of the Spaniards depend on the Indians provided by the Crown. Second, he explains that "recovery" stimulates the economy and thereby increases royal revenue. Finally, he concludes that, since this activity is voluntary and not coerced (unlike the mita), it is in the best interests of both the Indians and the Royal Hacienda. In other words, Capoche is conscious that the illegal circulation of metals is part of the business and the vast economic network that is the lifeblood of the empire. While it would be too easy to celebrate the practice of "recovery" by saying that was an act of reappropriation or resistance by the Andean people, the truth is that this transgression was co-opted in advance by the colonial power. The enumeration of the abuses, accidents in the mines, health problems caused by the mills, and cost in human lives is not part of a humanitarian agenda. It is part of a calculation of benefits, an axiomatic application of the cold evaluation of the useful means for creating more profit. Since mining costs more in blood than in precious metal, it is necessary to let the Indians participate in the production of profit, which ultimately benefits the Spaniards and the Crown.

Up to this point, I have analyzed the most representative moments of the doctrines that either inspired or defended Toledan reforms like the mita and the introduction of amalgamation. Our examination of the *Anónimo de Yucay* shows how the impasses of instrumentalism present in the dispute between Las Casas and Sepúlveda are resolved by emphasizing the centrality of the mines to the larger mission of evangelization and the general well-being of the Indians. The instrumentalist metaphysics called to mind by both Las Casas and Sepúlveda is invoked directly against Las Casas. In *Anónimo de Yucay*, Peru appears as the world turned upside-down, the place where the supremacy of the spiritual over the temporal world breaks down. Our analysis of Matienzo shows how he revitalized Sepúlveda's arguments about the servile condition of the Indians to transform them into a labor force. Such servility is not only an argument for their transformation into a labor force but also a sign of how their nature is an obstacle to their integration into the market. Since they do not enjoy their freedom and they do not want to accumulate riches or possess private property, they lack incentive for working in the mines. Therefore, the only solution is to force them to be free. Lastly, our analysis of Capoche shows how the transition from the huayra system of smelting to the process of amalgamation by mercury represented not just a substitution of one technique for another but also a dispossession of Indians of their technological agency and relative control over the means of production. In Capoche's narrative, the mita appears as a consequence of the workers' need for amalgamation, which was the technical solution to a looming catastrophe that affected all the levels of political and economic life in the kingdom. The Indians who once had relative control over their tools are now appendages of the ingenios. They become living tools, parts of a machine, which was precisely how Aquinas conceptualized those who were born to obey.

While in metaphysical instrumentalism, technique remains implicit and somehow disavowed, in the defense of mining accomplished by these authors, instrumentalism becomes necessary. Mines are depicted as means that have destructive effects and yet are nevertheless necessary since their absence would equal the breakdown of the colonial system. As in Vitoria, who insisted that commerce and tribute would not cease if the Spaniards failed to meet the valid titles, Capoche insisted on the necessity of employing mitayos in the mines even if mining cost more in blood than in precious metal. Within the limits of Aquinas's metaphysical instrumentalism, the temporal benefits and means are subordinated to the common good and transcendent ends. Nevertheless, the defense of the practice of colonial mining accomplished by Toledo's sources and supporters, as well as appeals to the tenets of imperial metaphysical instrumentalism, subverted the same metaphysical basis that served to justify it. While metaphysical instrumentalism appeals to the means for the sake of the end, in the apologies of colonial mining accomplished by these authors, we witness an excess of the means over the end, since the profit or surplus becomes the problem around which the entire colonial system is structured.

BIOPOLITICS AND THE NOVELTY OF THE TOLEDAN EXPERIMENT

According to Mumford, the General Resettlement, which was the central and most important of all the Toledan reforms, brought a new vertical conception of government, one that was centralized and authoritarian, a form of "social engineering unknown in Europe" (9). Such radical innovation in the history of social engineering was the product of the Spanish view of the Incas as a by-product of the vertical landscape (9).[26] The Toledan reforms were inspired by an ideology that combined "authoritarianism and surveillance" with a "network based form of community economy," specifically the Andean economy (9). Paradoxically, the modernity inherent in the project was a consequence of the Spanish vision of the Inca Empire (9). Far from reducing the Toledan General Resettlement to a case of ethnocide, Mumford's work demonstrates that Toledo and his companions borrowed certain aspects of Andean culture—such as governing by surveillance or compelling the Indians to work—while clearly rejecting others (10). At the same time, Mumford explains, the General Resettlement constituted an act of hubris, constructed upon the belief that it was possible to change Andean culture and language employing rapid and efficient technologies of power (176). It was in Peru that Toledo "embraced a concept of State power emanating irresistibly from a single point" (176). In other words, the centrality of the state was less a Spanish concept than a product of Toledo's improvisations and redeployment of the Inca past, which resulted in a project that ended up being "remarkably modern" (176). Borrowing the concept from James Scott, Mumford characterizes the General Resettlement as "authoritarian high modernism" (176).

Mumford thinks that the notions of verticality and modernity should not make us believe that the agent of the General Resettlement was the state. Instead, he invokes the now fashionable image of a decentralized Spanish monarchy where individuals made decisions and were part of the political institution (178).[27] Nevertheless, Mumford also invokes the idea of the state "as a construct, not a reality," that is still useful for thinking about the General Resettlement. It is possible to argue that, like many other fictions, the fiction of the state portrayed by Thomist philosophy, a perfect community, an autonomous machine composed of an efficient cause that imposes forms upon a commonwealth directing it to its proper end, the common good, is a guiding fiction, an ideal with structuring effects that reverberates through the entirety of the colonial world and the transatlantic imperial network. It is partially misleading to say that the unity of the state as perfect community is a mere construction beneath which there is a multitude of networks as Kamen and, to some extent, Mumford argue. It can be argued that the fiction of an efficient cause that imposes unity on the Andean vertical landscape through technique gives rise to performative and structuring illusions that are necessary presuppositions of a series of real and material changes. Following Foucault's concept of governmentality, Mumford distances himself from state centered explanations in order to make room for a better understanding of what was really a novel experiment never witnessed in Europe.

Mumford follows Foucault's theories about the changes in the functioning of European governments that took place in the sixteenth century with the development of newly powerful monarchies.[28] The convergence of the pastoral model of authority with a notion of a monarch who should administer national resources resulted in an engineering of both the individual subject and the collectivity. As Mumford reminds us, this paradoxical conjunction of totalitarianism and individualism is at the core of what Foucault call "governmentality." Governmentality is a theory of the state in which the role of governor does not consist in passively administering justice but in an active engineering and molding of society by means of the regulation of everyday habits of the individuals called *police*. As a result, Mumford argues that "the early modern Spanish monarchy, which commentators have sometimes portrayed as backward, experimented quite early with new conceptions of governmentality" (179).[29]

The totalitarian flavor of the "authoritarian high modernism" was developed in the periphery by Toledo before it took place in Europe. Toledo and his circle invoked divine law and human civilization in order to impose a state model that was "ahead of its time" (Mumford, 179). Mumford credits Sara Castro-Klarén, in her essay "Historiography on the Ground," for being the only scholar who points out how Toledo's reforms in colonial Peru were part of the laboratory of techniques of power that preceded European totalitarianism (164). This conception of state power pioneered by the General Resettlement was grounded in "divine law" and "human civilization" (179). Despite being ahead of its time, this specific form of governmentality was old, since it consisted of a Spanish notion of restoring the precolonial policies of the Incas (179). What was abnormal in Europe was the form of rule in Peru. Toledo resorted to Inca tyranny because it was a necessary means for ordering the Andean landscape and its inhabitants: "For Toledo and his companions, the reducciones' tyranny was not radical but recuperative—not alien aggression but local tradition" (180). This intrinsically racist view conceived of the Indians as servile barbarians who had to be vertically dominated (180). Mumford emphasizes that the distinctive side of this racist argument was the vertical archipelago, an infrastructure made of a mobile network of exchange and settlement that Spaniards identified with the Incas' *mitimas* (180).

Spaniards had a twofold conception of Andean culture, which was composed of authoritarianism and the notion of an archipelago, Inca power, and networks of labor and resources that covered the gap created by otherwise insurmountable distances. There was a mutual co-constitution between the techniques of surveillance and mobilization of labor and resources and the physical nature of the landscape and its inhabitants. Mumford makes a decisive observation that is central for the purposes of my own interpretation of the process as a correlate of metaphysical instrumentalism: "Early modern Europeans tended to assume that societies' organizing structures did not evolve organically, but were the creation of lawgivers. As a watch implies a watchmaker, so the complex organizations of the Andean community seemed impossible to Spaniards without Inca authority and planning. To have created a social structure so complex and effective, the Incas must have combined a level of state authority and knowledge seldom seen elsewhere—so colonial observers reasoned" (180). Mumford is implicitly alluding here to the theories of natural law, according to which the origin of the law is the commanding principle of the natural subordination of matter to form and means to an end.

It is necessary to supplement Mumford's insights concerning the novelty of Toledo's biopolitical innovation with the instrumentalist presuppositions of Spanish imperial metaphysics. If the watch implies the watchmaker, to use Mumford's metaphor (which is more Cartesian than Thomist), it is because everything that is moved is moved by a superior mover. Authority and planning demand a superior master artisan that directs its subjects to a common good. These kinds of metaphors are the result of a certain transposition of the domain of technique at the heart of Toledo's biopolitical experiment. The governor is positioned against the population in the same way that God is positioned against His creation: they are both craftsmen who impose order on raw material in order to make a statue or any object of art. There was an authoritarian biopolitical project in the margins also because the margins of Europe were interpreted according to the exigencies of metaphysical instrumentalism.

In sum, this notion of governmentality presupposes the principle of the natural subordination of imperfect matter to perfect form, the ideological side of metaphysical instrumentalism. The biopolitical dimension of the Toledan experiment, the capacity to mold the habits of the subjects as different from dispensation of justice, is inseparable from an excess of instrumentality implicit in imperial metaphysics. The question that arises is: Can the logic of domination implicit in biopolitics exhaust the novelty, or even the motivation, of these measures?

The texts analyzed above exceed the biopolitical interpretation, because they contain an economic logic that follows the demands of mercantile capitalism. In other words, these texts can be read as part of the Heideggerian history of technological domination, antecedents of Theodor Adorno's dialectic of the Enlightenment and instrumental rationality, or examples of Foucauldian biopolitics, as Mumford asserts. But they can also be read as part of the disruptive power of the market that reconfigured the region as capitalist deterritorialization, as theorized by Karl Marx. As a matter of fact, Capoche is capable of giving an air of necessity to amalgamation and the mita only by employing a rethoric of catastrophe that emphasizes the possiblity of an econimic and political crisis due to a short circuit in the circulation of money and commodities. The importance of amalgamation and the mita for lubricating the economy is also emphasized by contemporary scholars. In a book titled

Mercury, Mining, and Empire, which shows in amazing detail one of the "largest and longest-lasting ecological disasters ever known," Nicholas A. Robins concludes:

> The silver mines of Latin America, and specifically those of Potosí, were a vital component in the rise of modern global capitalism. The tens of thousands of tons of silver that traveled the globe, and were exchanged in countless markets and salons, reflected the hopes of traders while eclipsing the despair of those who had produced the metal. This flow of silver transformed the world, prying open new trade routes and enabling people to experience the tantalizing flavors of exotic spices and teas, the smooth sensation of silk, and the translucent delicacy of porcelain. As Potosí and other mining centers began to produce literally fabulous amount of silver, the peso of eight quickly became a ubiquitous currency by sheer force of numbers if not by consistent purity and quality. (177)

The silver extracted from Potosí and refined in the amalgamation mills circulated in "Europe, Russia, the Levant, the Middle East, Africa, India, and Indonesia, and China served as the ultimate magnet and final destination of much of the New World silver" (Robins, 178). Silver traveled not only via Europe but also directly from New Spain and Peru, "impelled across the Pacific by wind and the force of desire, where it was eagerly traded for China's mythical merchandise" (178). Just as the "mineshaft hollows out that which is inside" (178), so too did the mercury and silver mining economies eviscerate communities and leave tens of thousands of people to die slow, painful, and anguished deaths. Amalgamation by mercury prompted mining economies that uprooted communities and cost the lives of thousands of peoples. The transition from the huayra to amalgamation "stimulated the regional economy" and also facilitated the circulation of commodities from around the world (178). The revitalized production of silver described by José Luis Capoche, among others, increased Potosí's centrality "as an integrating force in global economy, as the silver extracted from the Cerro Rico again traveled the globe as it was traded for myriad goods and thousands of people" (179). This summary helps us to understand the disruptive and "deterritorializing" effects of the introduction of amalgamation by mercury and its centrality in the creation of an incipient global economy. Nevertheless, it is necessary to go one step further and remember that both the amalgamation and the mita were introduced in order to fulfill the needs not only of the Crown but also of the private agents who could never profit from the mines without cheap technology and cheap labor. And, as Capoche reminds us, the accumulation of riches cost more in blood than in metal. The irony is that by invoking the common good, metaphysical instrumentalism triggered an inversion whereby the elevation of the means to an end ended up destroying that end.

Comparing the academic works of Mumford and Robins, it is possible to see two distinct logics at work when examining the colonial origins of modernity. On the one hand, Mumford sees the colonies as a biopolitical experiment that anticipated the authoritarian kernel of modernity. On the other hand, Robins sees the colonies as the scene of a technological experiment that disrupted the local by fueling global capitalism. Their differences cannot be reduced to their different objects of study. Biopolitical domination and capitalist deterritorialization are two aspects of modernity, two irreconcilable narratives, and each one exists by abstracting the other. There is a superior synthesis, a neutral metalanguage, capable of accounting for both aspects of modernity.[30] On the one hand, Heidegger emphasizes will to power and the logic of technological domination, which influenced Foucault's notion of biopolitics as the regulation and administration of life as well as Adorno's search for answers in the logic of instrumental rationality. On the other hand, Marx emphasizes how the excessive movement of capital dissolves all bonds and disrupts all values by measuring them against the value of all values. The logic of surplus power (the way that indigenous people are instrumentalized and manipulated by apparatuses of subjection) and the logic of surplus value (the logic of production and appropriation of an excess that is ultimately self-disruptive) are not identical and cannot be translated into one another. In order to highlight one, it is necessary to bracket the other. Nevertheless, these two logics are two sides of the same coin, the two irreducible aspects of imperial expansion. My thesis is that both aspects of modernity arise both historically and dialectically from the inherent contradictions within metaphysical instrumentalism. In other words, it is necessary to redouble the gap between the logic of domination and the logic of capitalist deterritorialization within metaphysical instrumentalism.

The Heideggerian critique of Western technological domination can be historically articulated with a Marxian critique of commodity fetishism by means of a critique of the inner impasses of Aquinas's metaphysical instrumentalism. The inner impasses of instrumentalism generated a monster that instrumentalism could not control, an instrument that subordinates all instruments to itself and therefore undermines its own metaphysical presuppositions. This instrument that escaped imperial control was nothing other than money itself, and the indigenous worker was its objective correlate. In our analysis of José de Acosta we observed how precious metals become bearers of a value-in-itself at the very moment they cease to be instruments subordinated to a higher end, thereby becoming a commanding principle that subordinates all other instruments to itself. A close reading of Matienzo's *Gobierno del Perú* also shows how the line separating Inca tyranny from natural and legal compulsory labor is the payment of a wage. In both texts, a particular thing, the silver of Potosí, is elevated to a medium of justice, a measure of everything else, including indigenous labor. Mining as the production of money is the surplus of the means over the end that subjects even transcendent ends to itself. The logic of technological domination—the Toledan mita, the forced mobilization of the servile Indians as the imperfect matter of the perfect empire—is inseparable from the logic of disruptive mercantile capitalism, the endless accumulation of a material instrument, money, capable of measuring everything, even labor itself.

As Marx explains in *Capital*, when we witness a reversal of the cycle C-M-C, where merchants buy cheap in one place in order to sell in another place, into M-C-M´, where the telos of the process is producing more money, one particular subordinated thing, precious metal, is transformed into an overaching genus that directly embodied the one subspecies. Marx explains this reversal in terms of transforming the means (money) into an end (capital, circulation of money that engenders more money): "the circulation of money as capital is, on the contrary, an end in itself, for the expansion of value takes place only within this constantly renewed movement. The circulation of capital has no limits" (*Capital*, 1:254). The means of the empire is turned into an end in itself, the endless circulation around a literal hole, a void, called Potosí. As a result, capitalism deprived metaphysical instrumentalism of its further end, making both possible and necessary the emergence of ways to justify the empire based on technical necessity. The limit of metaphysical instrumentalism, its dysfunctional inner core, is a result of instrumentalism itself, which ends up being deprived of its transcendent ends. Conserving this autonomous network of circulation, avoiding the crisis and the catastrophe, become sufficient reasons for implementing necessary, yet problematic, techniques such as amalgamation and the mita.

5

THE EXHAUSTION OF NATURAL SUBORDINATION

Solórzano Pereira and the Demise of Metaphysical Instrumentalism

The inner contradictions of metaphysical instrumentalism present in Vitoria's attempt to find a solid and rational ground for the empire in natural law haunted the future of the discussions around the justification of mining. Vitoria's project consists of grounding the subjection of the Indians in natural law as opposed to appealing to the authority of the pope or the papal bulls of donation. Nevertheless, the instrumentalism inherent in the principle of natural subordination made it evident that the aim of invoking transcendent ends was intended to justify control over the means. The instrumentalist presuppositions of metaphysics become more obvious and evident in the work of José de Acosta and in the reforms of the viceroy Francisco de Toledo. Toledo's apologists performed a displacement of the principle of natural subordination in which the technical means become more important than the ends. In this chapter we will examine the works of the jurist Juan de Solórzano Pereira (1575–1655), author of the most influential treatise on Spanish imperial law. I argue that his writings exemplify the parallels between the metaphysical impasse around mining and imperial wealth and the political and economic crises of the Spanish Empire in the first half of the seventeenth century. Solórzano's *Política indiana* (1648) is the last great defense of the Spanish government and represents not only the exhaustion of Vitoria's instrumentalism but also its completion. Solórzano makes explicit the internally antagonistic and contradictory nature of the imperial project by examining the arguments in favor of and against the practices of the *mita* and amalgamation. We begin by contextualizing Solórzano's work within the crisis of the Spanish Empire and go on to examine the arguments in favor of and arguments against the rotational system of draft labor called the mita, which was introduced by the viceroy Francisco de Toledo in the 1570s. The arguments in favor of the mita rely on the principle of the natural subordination of matter to form and means to an end, while the arguments against show how the mita fails to achieve the invoked ends. Potosí mining becomes a paradoxical object produced by the principle of natural subordination: it is both the necessary means for and the reminder of the inconsistency of the metaphysical grounds of the mystic body of society. It is an obstacle engendered by instrumentalist metaphysics as the core of imperial ideology. We then analyze Solórzano's attempt to find a way out of this deadlock by continuing to practice the mita while searching for more mines. Such a solution is deeply connected to a belief shared by both Spaniards and Indians about the living power of metals inside the mines. The belief in the capacity of metals to reproduce themselves inside the mines is interpreted by Solórzano as proof that the Crown is the owner of the subsoil. Therefore, I argue that the belief in the self-reproductive capacity of metals is not only compatible with but also inseparable from their imperial appropriation.

Finally, we examine the confluence between Andean and Spanish beliefs in metal's capacity to reproduce itself. Some scholars interpret these beliefs in terms of a pure difference that resists Eurocentric appropriation. Since this position runs the risk of remaining uncritical of actual technological and capitalist forces that mediate this belief, I examine contemporary academic interpretations of Andean vitalism in order to show how the belief in the life of metals becomes retroactively readable through metaphysical instrumentalism. When attempting to describe the specificity of Andean beliefs, scholars cannot circumvent the hylomorphic vocabulary that was essential to the imperial enterprise. At the same time, indigenous ritualistic uses of metals are interpreted within an instrumentalist view of money, while also describing the subsoil as a place where money produces more money. Therefore, while trying to avoid the reduction of Andean beliefs to European and capitalist interpretations, the attempts to theorize indigenous thinking inevitably color it with ideological and even fetishistic overtones. As a result, Andean vitalism appears to be an ambivalent phenomenon, since it is both the condition of possibility of integrations of indigenous beliefs to modernity and the point of inflection of a critique of modernity that refuses to reduce nature to a mere means by conceiving it as an organic whole and an end in itself.

The introduction of amalgamation by mercury had an important impact on the Spanish economy, dramatically increasing the quantity of silver received by the Crown from the New World (Elliott, *Imperial Spain*, 269). After the bankruptcies of the 1557 and 1575, the 1580s and the 1590s were witness to a bonanza. The silver from the New World made possible Philip's attempts to recover the Netherlands, launch his army against England, and intervene in the civil wars in France (270). In addition to the American silver, the annexation of Portugal in the 1580s would prove a decisive geopolitical factor. Nevertheless, by the 1590s, the imperial economy started to crack under the weight of the "invincible armada" (285). In November 1596, Philip suspended payments to the bankers. The Crown was in bankruptcy, and the mining business started repeating its cycle of crisis and polemics around the mita introduced by Toledo. This crisis marked the end of Philip II's imperial dreams (287). The "invincible armada" was defeated in 1588. In Mexico, the combined effect of epidemic disease and mining work produced a demographic catastrophe in 1545–1546 and then again in 1576–1579. The crisis of the imperial economy of the 1590s widened the gap between the economies of Spain and the Americas, and the English and the Dutch started to take advantage of the situation. By 1615 the Spanish supply of silver was continuing to decrease and the Crown was still spending money and building an unsustainable external debt.

The decade between 1573 and 1582 was one of opulent splendor for Potosí (González Casasnovas, 55), but from 1590 to 1630 the contradictions of the Toledan regime exploded. On the one hand there was a centrifugal process that tended to rupture and destroy the indigenous community through the actions of private interests (*encomenderos, corregidores,* and merchants), and on the other a centripetal process, animated by the stubborn insistence of those in power on maintaining the indigenous labor force as the ultimate source of economic exploitation in the colony (González Casasnovas, 54). In the 1590s the production of silver stagnated as a result of the structural problems faced by the extraction of metals in Potosí. There were no longer any discussions around the problem of the mita, with the exception of the discussion around the *rescate*. The rescate was the practice of buying metals stolen from miners themselves or any other person that was not the owner of the mine. José de Acosta wanted to forget the empire's violent origins, employing a certain use of force and coercion in order to sustain the economic needs of evangelization. In Potosí, mining was becoming more and more dependent on the flow of an indigenous labor force that decreased with every passing month. These problems reignited the polemics around the mita. After all, Toledo could not exorcise the specter of polemics and discordance, and the system was functioning poorly.

Viceroy Cañete promulgated the *Ordenanzas* in 1594, which tried to regulate the working conditions of the indigenous labor force (González Casasnovas, 63). In 1596 in Carangas and Paria there was a protest of local Andeans who could not meet the demands of both the mita and the textile industry (González Casasnovas, 64). Between 1597 and 1600 the Jesuit Antonio de Ayanz, Diego de Paz, and the Dominican priest Miguel de Monsalve all sent reports to Madrid where they revisited the legal and political foundations of the Toledan mita in order to perfect its mechanism and functioning and cure it of its contradictions. Private interests were still trying to maintain control over the indigenous labor force, while the Indians were trying to evade the impositions of tribute and mita.

In 1601 a royal decree ordered that Indians would work voluntarily in the *obrajes* (textile workshops), while others would be placed in *reducciones* (forced settlements) in Potosí for the common benefit of the Indians and the mines, giving them the opportunity to stay there if they wanted by assigning them to different parishes. The use of an indigenous labor force in the mines made visible the contradictions in the central axiom of Spanish colonization, the protection and conversion of the Amerindians. The Spanish Empire was supposed to be grounded in the voluntary and peaceful conversion of non-Christians while it was actually grounded in the extraction of riches. The voluntary and rational conversion and the free consent of the Indians was central to the Thomism of Las Casas, while the extraction of riches depended on Sepúlveda's appeal to the servile nature of the indigenous labor force. Elliott summarizes this contradiction: "Driven by the twin imperatives of its thirst for precious metals and its obligations towards its new Indian vassals, the Spanish Crown was interventionist from the beginning in its approach to the government of the Indies. It sought to mould the developing colonial society in accordance with its own aspirations, and its own high sense—fortified by university-trained jurists who had entered the royal service—of the all-commanding nature of its divinely ordained authority" (Elliott, 130). The weight of these contradictions was part of a complex geopolitical context marked by the perception of the decline of the Spanish Empire. The rebellion in the Low Countries, Portugal, and Catalonia, commercial competition, pirate attacks, and the drop in the extraction of silver were part of the international scenario. As More points, out "by the end of the century, it became increasingly difficult to contradict the perception among Spaniards and non Spaniards alike that the empire was spiraling downwards" (36).

In 1609 another royal decree tried to find a solution to the impractical aspects of the 1601 decree. The 1609 decree maintained the forced indigenous labor for the mines while the authorities tried to convince the Indians, whose natural inclination was to refuse to work, to change their ways and customs (González Casasnovas, 69). In 1601 Miguel Agia wrote a commentary on the Decree of 1601 titled "Treatise containing three serious perspectives on law," which was published in Lima in 1604. Agia employed Aristotelian Thomist philosophy with the purpose of justifying the need to compel the workers to work in the mines: "because for them there is nothing more odious than working, even if it is for themselves, on top of the fact that the Spaniard and the Indian are diametrically opposed: because the Indian by nature has no avarice, and the Spaniard is extremely avaricious, the Indian phlegmatic and the Spaniard choleric, the Indian humble and the Spaniard arrogant, the Indian deliberate in everything he does and the Spaniard hasty in everything he wants, one the friend of order and the other the enemy of service. And ultimately dissimilar in condition, life, and customs" (Agia, *Tratado*, 56).[1] Agia represents the clearest defense of the Toledan mita using the typical Sepúlvedian arguments about the Indians' natural inclination to leisure dressed in Aristotelian and Thomist language.

The polemics around the impact of the mita in the indigenous world increased after the reception of the decree of 1609. The viceroy convoked a "junta" in order to find a solution to the problem and royal officials, *azogueros* (mine owners), and religious intellectuals met. Jesuits drew up a report in March 1610 titled "On the mita of Potosí and the reduction of the realm," in which they basically repeated the arguments used by Matienzo and Acosta (González Casasnovas, 76). They demanded a census of the population and a strict demographic control, but the magistrates who would be in charge of overseeing the situation distorted the facts in order to accommodate their interests and the interests of the azogueros.

Following a movement of contraction and expansion, of destructuration and restructuration, the output of silver that had already been falling since 1595 reached another crisis in 1620. This crisis was the first since the introduction of amalgamation. There were clashes and conflicts among the Basque miners, who wanted to monopolize the exploitation of the richer veins, and the rest of the Spanish and the creoles. This episode was called the "Basque-Vicuña War" and was material for Arzáns de Orsúa y Vela's monumental *Historia de la Villa Imperial de Potosí*.[2]

In 1634 after Viceroy Chinchón's attempt to end the corruption around the mita, Sebastian Sandoval y Guzmán's *Pretensiones de la Villa Imperial de Potosí* was published in Madrid, which defended the azogueros' interests. Consequently, the document led to the conviction that to oppose the mita was not to oppose the particular interests of the azogueros but, rather, to interfere with the Crown's interests. In fact, in 1636 there was another royal decree that compelled the viceroy to give *indios de repartimiento* (forced indigenous labor) to the azogueros. As a result, the Crown showed a strong commitment to the azogueros after the publication of Sandoval's *Pretensiones*, which employed the old Toledan argument that, without Potosí, both mining and the viceregal economy would collapse.

In 1624 the Consejo de Indias decided on the preparation of a manuscript that would collect the totality of the laws of the Indies. This task was completed by Antonio de León Pinelo (1590–1660), historian and chronicler of Indies, and author of *Paraíso en el Nuevo Mundo*, a monumental baroque oeuvre that was not published until the twentieth century. Juan de Solórzano Pereira (1575–1655) contributed to drafting the final manuscript, which would be published in 1681 without the consent of either of the two authors. The *Recopilación de las leyes de los reynos de las Indias* (1681) gathered the multiple writings that circulated across the Atlantic. The *Recopilación* was a rationalization of the ideology of Spanish imperialism that had remained unchanged since the sixteenth century despite the complex geopolitical situation (More, 38).

Juan de Solórzano Pereira wrote what is likely the strongest apologia for the Spanish Empire, *Política indiana* (1648), a translation of his own original Latin work, *De Indiarum iure* (1629–1639). Solórzano studied law at the University of Salamanca in 1587, graduating in Justinian Roman Law and Canonical Law in May 1599 from the school of Salamanca. His work reflects a profound knowledge of civil law and the canonic law of the School of Bartolo de Sassaferato, Roman jurisprudence, and Scholastic philosophy. The multiple volumes of *De Indiarum iure* written between 1629 and 1639 are a systematic commentary on the *Recopilación*.[3]

The success of *De Indiarum iure* led Solórzano to translate it into Spanish in 1648 with the purpose of its reaching a broader audience.[4] Written a century after the debates around the Spanish dominion, Solórzano's work became the greatest apology for the Spanish Empire ever written, in spite of the new complex geopolitical context (More, 36). By the end of the sixteenth century the supranational authority of the Spanish Empire was represented by the term *monarchia universalis* (Pagden, *Lords*, 43). Spaniards considered that the American territories had been incorporated either by contract or by just war, an idea that found its clearest expression in the concept of "composite monarchy, of which Castile was one member, although increasingly *primus inter pares*" (Elliott, *Atlantic*, 119). Despite referring to the colonies as an empire, these kingdoms "were always, in theory and generally in legal practice, a confederation of principalities held together in the person of a single king" (Pagden, *Spanish Imperialism*, 3). Nevertheless, this "theoretical principle of equality" among kingdoms "within one overarching monarchy" could not erase the historical fact that the Indies were incorporated by violent means (More, 33). Moreover, in spite of the language of equality among autonomous kingdoms, by the end of the sixteenth century, the centralizing effort of the Spaniards created a "state of the Indies," establishing "a system of government and control that might well be the envy of European monarchs struggling to impose their own authority on recalcitrant nobles, privileged corporations and obstreperous Estates close to home" (Elliot, 129). The Crown tried to maintain an absolute control over the colonies, yet it could only reign through negotiations with local agents. As Kamen points out, there were no autonomous local organs, and all decisions were made in Castile through the council of Indies (142). Nevertheless, the Crown established its control through a series of compromises that relied heavily on local agents such as the viceroy, Francisco de Toledo (142). The ideal of total control proved extremely problematic, since "the world's greatest empire of the sixteenth century, consequently, owed its survival to the virtual absence of direct control" (142). For

Solórzano, the unity of the Spanish Empire, the largest known in history, coexisted with a multiplicity of kingdoms, each of them being an autonomous and perfect community. The sum of autonomous kingdoms was no less complete than the totality of the globe whose circumference it was possible to travel without abandoning the dominions of Castile and Portugal. The kingdoms of the Indies coexisted with Castile, Aragon, Naples, and Portugal, each of them having their own ecclesiastical and secular institutions.

Solórzano's work represents the most perfect convergence of theoretical and practical interests, exemplified in the positions that he held in the Indies. In fact, Solórzano was judge in Lima (1609), and he was in charge of the mercury mine of Huancavélica, an experience that would contribute to his approach to the problem of mining in the Andes. Among the authors that most influenced Solórzano were Francisco de Toledo, Antonio de Herrera, and José de Acosta. Solórzano also invokes Roman jurisprudence and Acosta's arguments in order to justify the *reducciones* created by Toledo, and he cites Aquinas in order to justify the imposition of tributes and the pay of church tithes by the Indians. I argue that, if Vitoria represents the fissured foundations of the imperial network of the production of profit and Acosta a pragmatic and empirical materialization of Vitoria's metaphysical instrumentalism, then Solórzano represents the exhaustion of an administrative solution to the theological problems generated by the antinomies and irresolvable contradictions in the Spanish imperial politics.

Brading sustains that if Roman jurisprudence and absolutist principles occupied a central role in the tasks of sustaining the authority of the king, "the prestige of the monarchy was further buttressed by the Neoplatonic, Hermetic philosophy that flowed in the Baroque culture of the post-Tridentine Catholicism" (226). Brading reminds us that the Jesuit leader of the Neoplatonic school of thought was Juan Eusebio Neremberg (1595–1658), who saw in the will of the prince an image of divine omnipotence. The king was a sort of viceroy of God on earth, and this divinization of the figure of the king possessed strong resonances with the panegyric of Constantine made by Eusebio of Cesarea, "whom he described as the only true philospher, exercising perfect control over his passions, dwelling in communion with his Divine original, and acting as virtual intermediary between heaven and earth" (226). This Neoplatonic divinization would have a historic correlate in the succession from

Charles V to Philip II, which meant a transition between the warrior king and a sedentary monarch, "invested with sacral aura through his residence at the Escorial, a palace that was also monastery, church and dynastic sepulcher" (226). In Brading's words, "If Charles V was acclaimed a new David constantly engaged in battle, Philip II could only figure as another Solomon, an exemplar of wisdom and justice" (226).

Although teleological Christianization played a decisive role, "the seventeenth century focus was not on expansion through conversion but rather on the conservation of an empire that was beginning to show signs of fragility" (More, 37). As Pagden explains, apologists of the empire came to realize that conservation of the empire was more important than conquest because conservation required "prudence and wisdom, virtues superior to force" (Pagden, *Lords*, 111). Kamen also sees a shift in perspective when he examines the impact of Toledo's reforms in regard to the politics of the Spanish Crown. The Spanish Crown could dispense with transcendent justifications such as the papal bulls, because "from now on, only the Crown decided" (195). The Crown was the ruler of American kingdoms and the true owner of the wealth, lands, and properties of Peru and, in turn, had to defend the natives and "devise laws for their conservation" (195). In sum, "there was henceforth to be one sole empire, ruled over one sole authority, the Crown of Castile" (195). In Solórzano's words: "Indeed, as Anyone knows, it is better to prevent ends than to find beginnings, and it requires no less decency to conserve the Realms than to acquire them, and that those who guard and defend the prudence and caution of the Princes who rule them remain firm and durable" (*Política indiana*, 1:8). Commenting on this passage, More concludes that, instead of reexamining the origin and validity of the Spanish Empire, Solórzano recommends "the practical and urgent task of sorting through and organizing the colonial legislation that had accumulated chaotically over the past century" (More, 38).

According to Pagden, Vitoria's natural law had lost its power of persuasion when Solórzano offered his own take on the debates around dominion (*Spanish Imperialism*, 33). Vitoria's concerns about inventing titles that could not be easily appropriated by the Lutherans was replaced by Solórzano's attempt to "preserve the force of the civil law and to strengthen the power of the Crown" (33). The worst mistake Vitoria could have made was to use the vocabulary of natural rights, which only placed limits on the bulls of Donation issued by Alexander VI (33). Pagden also stresses that, for Solórzano, Ferdinand and Isabella acted in good faith when the pope gave them dominion over the Americas (33). As a result, Solórzano attempted to ground Spanish dominion in civil law, sustained by the will of the legislator, in perfect accordance with the principles of Roman jurisprudence (34). The ultimate source of legitimacy was the principle *Quod principi placuit legis habet vigorem* (What pleases the ruler has the force of law). The core of Solórzano's argument, according to Pagden, was that, even if the Crown came to be in possession of the Indies only by believing it had a legitimate right, it could still claim dominion by subsequent occupation (34). It was a retrospective justification of the conquest.

For Pagden, grounding the law in the will of the prince was essential to Antonio de Léon Pinelo, the lawyer in charge of collecting the existing laws dealing with the Indies, and to Solórzano's project. Since the maxim "what pleases the ruler has the force of law" is what separates Solórzano from Vitoria, it is useful to examine Aquinas's position with respect to this maxim. Aquinas examines this problem in Article 1, *Summa Theologica* IaIIae.90, titled "Whether law is something belonging to reason," about the essence of the law. Aquinas interprets the principle "what pleases the prince has the force of law" as an objection to the ultimate rational nature of the law and natural order. In Objection 3 we read: "Moreover, law moves those subjects to it to act rightly. But, properly speaking, it is the function of will to move someone to act, as is shown by what has been said above. Therefore law belongs not to reason, but to will; and this is in accordance with what the Jurist says: "what pleases the prince has the force of law" (*Political Writings*, 76–77).[5] The principle "what pleases the prince has the force of law" is associated with will and not with reason. Aquinas wants to adapt this principle to the rational ground of law because reason is "the guiding principle of human acts, as is shown by what has been said above, for it pertains to reason to direct to an end" (78). Aquinas answers the objection by arguing:

> Reason receives its power of moving from the will, as stated above. For it is because someone first wills an end that his reason then proceeds to issue commands concerning the things which are directed to that end. But in order that what it commands may have the character of law, the will itself must act in accord to the rule of reason. And it is in this way that we are to understand that the will of the prince has the force of law: otherwise, the

will of the prince would have more the character of iniquity than of law. (*Political Writings*, 78)

As a result, in Aquinas's political theology, for the will of the prince to have the force of law it has to act according to reason and therefore direct actions and people to the right end. Vitoria comments on this passage stating: "Aquinas replies that law is a function of reason because giving commands issues from reason, and law is a kind of rule or measure for human activity, the word *lex* being derived from the *verg ligare* "to bind, oblige" (*Political Writings*, 155). Vitoria also comments on Aquinas's reply by saying that in it "this is all confirmed, because all doctors agree that a good act is one which conforms to law, while an evil action is one which does not. But what conforms or fails to conform to law also conforms or fails to conform to reason" (157). Aquinas interprets the maxim within his own teleological, instrumentalist conception of the law as that which directs things, actions, and people to their proper end. I understand that Pagden places too much emphasis in the division between the will of the monarch and the natural law. Yet the synthetic and totalizing character of Aquinas's metaphysical instrumentalism was able to incorporate the Justinian Code by subordinating it to his teleological instrumentalism. It is evident that Solórzano rejects the limits that reason places on the imperial enterprise and not the alliance between reason and the will of the monarch. As we will see below, this will prove to be of central importance in Solórzano's attempts to find a solution to the impasses generated ever since by the mita. Both defenders and attackers of the mita framed their arguments in the language of Thomist philosophy.

Pagden offers an interpretation of Solórzano centered on the force of law borne by the will of the prince, based on the time lapsed since the acquisition of the Indies. Pagden quotes a passage where Solórzano claims that time can convert tyrannies into legitimate monarchies:

> Each reason in itself, and all together in the fact, as it consists of the histories of these conquests, and as they are sufficient in law to consolidate and strengthen the security of our kings: as very serious authors admit and recognize as they examine them one by one. Forcefully concluding from them, and proving with efficacious and irrefutable examples, that even when they overcome less perfect or legitimately acquired dominions, they are sufficient to ratify it and purge its defects: especially when the peoples

in possession do not object, and they have intervened over the long course of time with which *even tyranny becomes a perfect, legitimate Monarchy, as happened with the Romans and with others among the best the World has known and is now common law and assented to by all peoples.* (Solórzano, Política indiana, 1:111–12; my emphasis)

For Pagden, the time that passed since the conquest appears to be a sufficient reason, and questioning Spanish dominion is the product of mere envy (*Spanish Imperialism*, 35). Within Roman jurisprudence, the civil law grounded in the will of the sovereign "was sufficiently compelling in itself to make further inquiries into its legitimacy unnecessary" (34). Roman law "allowed for long-term *de facto* occupation of a particular thing (*praescriptio longi temporis*) to be recognized *de iure* a case of *dominium*" (Pagden, *Lords*, 89). By appealing to time as the foundation of rights of dominion, Solórzano is using the "language of Roman jurisprudence," which justifies Spanish dominion in a "subsequent occupation" (Pagden, *Spanish Imperialism*, 34). Pagden summarizes this retroactive justification in an elegant and succinct manner: "Time, the Spaniards' historical presence, is then the sufficient condition of dominium, for it is the objective condition that confers legal rights; and, in the end, it is legal, not natural rights, that are under debate" (34). I think that it is necessary to nuance this statement by examining Solórzano's explanation of this idea of time as a basis for rights.

Nevertheless, time does not ground dominion under any circumstances. There are conditions that are explicitly stated in the following passage:

> And as that great Master of all Baldo wrote, saying: *time, of which much has passed, causes us to judge as if we never have been because it is now erased and consumed by its contrary use, and we must accept and adapt to this. Because it is not very important to know now if it were more just for Pompeyo to rule than Caesar. Rather to live where we find ourselves, without going about scrutinizing the principles and origins of times in which men could not find any other reason or conviction than will or God's permission, which allows us to observe what we see happening in our own century, which gives us our customs and laws and in which gives our life, sustenance, and essence.* (Solórzano, Política indiana, 1:110; original emphasis)

In this passage, Solórzano provides a retroactive justification that accommodates any past event to the governing present. Time legitimizes only on the condition that we treat it *as if* it had not passed since it is erased and consumed in its present "contrary use." The violent past must be transformed into its opposite, the legal present. Paradoxically, in order for time to ground dominion, it is necessary to disavow time, to treat it *as if* it did not pass. By treating the illegal past as if it had not happened, the present retroactively posits its own legality. It is futile to inquire if a ruler has more legitimacy than another because once we look into the principles and roots of dominion we only find God's will. It is impossible to find a solid ground of power within history, which only presents contingent facts. Conserving the kingdom, conserving the present state of things, accommodating the past to the present, hiding the passage of time and the illegal origin of the empire, all are parts of the same operation: to retroactively ground the law from above, from the end, from its transcendent finality, which is God's will. In order for the positive law to have force of law, it is necessary to ground it not in events that happened in time, but outside time, since only the will of God commands in an absolute way. The difference between Vitoria and Solórzano is that the former provides a prospective justification of the empire using the principle of the subordination of matter to form and means to an end, while the latter provides a retroactive justification of the empire, employing the same axioms of metaphysical instrumentalism after the failure of Vitoria's project. It is a retroactive foundation that ends up privileging Roman jurisprudence because all the polemics around the dominion proved that nothing can provide an unshakeable ground for the law without presupposing the *unconditional force of law of a divine command*.

The use of the language of natural law collapsed under the weight of its own irreducible contradictions, and the only way out of the deadlock was to admit the violent origins of Spanish rule and turn these apparent obstacles into conditions of possibility by retroactively grounding the will of the prince in time. By getting rid of the question of the origin, Solórzano can simply focus on the conservation and administration of the means. The failure of natural law to provide an ultimate ground for the empire was due to the impasses inherent in metaphysical instrumentalism. Vitoria's instrumentalism depended on the view that ends were the most necessary causes, and it was possible to ground the Spanish Empire in valid titles. The debate between Ginés de Sepúlveda and Bartolomé de Las Casas proved that the principle of natural subordination of the material means to the formal ends could be used both to justify and to deny the Spanish dominion. José de Acosta thought that the conquest was violent and illegal, but he also believed it was still possible to use Vitoria's arguments to justify the conquest. Finally, Solórzano admits, with Acosta, that the origin was violent, and the task is to conserve the means. Yet, in order to conserve the means, Solórzano will have to reintroduce the unconditional commands of the ends. In this sense, Solórzano represents the exhaustion of the language of natural order and dialectical inversion that radicalizes Vitoria's metaphysical instrumentalism. Vitoria's rational instrumentalism ends up being materialized in the administrative voluntarism of Solórzano. Now the deadlock that was intrinsic to Vitoria's imperial reason is redoubled within Solórzano's corpus in his examination of the reasons against and in favor of the mita, the rotational system of compulsory labor.

ARGUMENTS IN FAVOR OF AND AGAINST THE MITA

Solórzano's examination of the arguments in favor of and against the mita is interesting when seen within the frame of Aquinas's metaphysical instrumentalism. This examination repeats the polemics between Las Casas and Sepúlveda. But before we examine the arguments for the mita in the context of mining, it is necessary to analyze the arguments gathered by Solórzano around the problem of its use for agriculture. After presenting the arguments against the use of tributary labor in agriculture, he explains that such "main foundations" of this doctrine "cannot be denied" to be "in [themselves] very just and pious, and commendable" (1:170). But the practical problem of this doctrine is that "whenever a rigorous attempt to execute it has been attempted" it has produced serious difficulties (170). Solórzano explains further that there have always been professors of jurisprudence and theology, experts on political government, who closely examined the nature of the Indians and their lands and dispositions and still supported the mita (170). Such experts dare to affirm that these personal services principally concern the cause of public utility and cannot be removed without causing great harm to the kingdom and the Indians themselves (170). The mita does not contradict the "rules and reasons of right" because the workers receive a "competent" wage and do not receive any harm in their "persons or possessions," rigorously regulating their days of rest even though they are being obliged to work against their will (170).

In order to reinforce his argument, Solórzano in Chapter 6, sections 6–8, introduces the idea of "Republic" as a "mystic body." Moreover, it is not a coincidence that he appeals to Aquinas in order to use the classic corporatist example of society as a harmonious totality: "and Saint Thomas says: then there will be a perfect and well-governed city, when the citizens sometimes help one another, and each one carries out his duty, and acts dutifully in every task asked of him, also availing himself of this example, the plot of the human body of all authors is very frequently for the mystic, or the politis of the republic" (1:171). According to the doctrines of Plato and Aristotle, the republic has only one body made up of different "professions" composed of many members who cooperate and help each other carrying each other's weight (171). Shepherds, peasants, and manual workers who devote themselves to mechanical arts are all "unavoidable and necessary" each in their own area as the apostle Saint Paul declares (171). For this reason, nobody can be excused when they are commanded to work in agriculture or in any other job or "responsibilities necessary to the Republic" (171).

Like Acosta, Solórzano appeals to the authority of Aquinas, who explains that a "perfect community" is well governed when the citizens cooperate among themselves and each of them fulfills his function, like limbs of a human body (171). He argues that it is not unfair if the Indians are obliged and compelled to perform tasks "for which, because of their condition and nature, they are more apt than the Spaniards (172). At the same time, the Spaniards have to govern and help them using their industry and ingenuity (172). Such division is justified in that "nature gave them more robust and vigorous bodies for labor, and less understanding or ability, providing them more with tin than gold in this way, they are those employed in it, like the others, to whom more is allowed, in governing them, and in other fuctions and utilities of civil life" (172). Solórzano adds that they are "coarse and rude men, raised on little reason, and characterized by almost the same nature as in a number of animals, and that for this reason we can boil them for their limited abilities" (172). The difficulties and efforts that are suffered by those who are more qualified for manual labor than for intellectual labor are what the ancients called a "necessary evil" (174). These obstacles have to be faced by the ruler who directs the republic to its proper end, in such a way that "those who govern the Republic must imitate skilled pilots who are not deterred by any impediment or headwind which may arise" (174). The good pilots are those who, "turning the sails, adjust and endure until another, more favorable [wind] comes along, and arrive at the desired port toward which they steer" (174). At the same time, "no one can doubt that with the direction and assistance of the Spaniards in as many offices and ministries as are exercised they have become more able and industrious in them" (175).

The indigenous people could not reach the achievements of the Spaniards due to their "limited abilities," profiting also from the salaries that they receive in order to buy their tributes or *tasas* for their sustenance (175). The greatest benefit received by the workers is that they are instructed in the Christian faith, without which they would return to their drunkenness, idolatry, and other vices to which they would be given if they lived idly (175). Facing such profits "it is not much, nor can it be surprising, that in return for such goods those who receive them make of them something useful or comforting" (175). The "reciprocal obligation" commands that "wise men, just by being that, should instruct, direct, and improve the ignorant with their science and likewise those who receive this instruction should return what they can according to their quality and capacity" (176). This is what Aquinas says "when he teaches us that, considered absolutely, it does not naturally make sense that the ignorant serve or are enslaved more than others" (176). Their reason for being such is "therefore [of] some use insofar as it is important that the ignorant are ruled by the wise and that the wise are assisted and served by the ignorant" (176). For Aquinas, everything that exists has an ulterior purpose (176).

As we can see, Solórzano is employing the classical arguments of the imperial school of thought epitomized by Sepúlveda's use of the principle of the natural subordination of matter to form and means to an end. Solórzano also repeats Aquinas's examples where there is a clear transposition of technique to nature and politics (that is, the example of the pilots). In *De regimine principum,* Aquinas affirms "that the government of a king is like the Divine government and that such government may be compared to the steering of a ship. . . . To govern is to guide what is governed in a suitable fashion to its proper end" (*Political Writings,* 39). Everything has a function and a purpose and the indigenous people have to serve the ultimate utility and common good of the colonial society. Briefly put, like Sepúlveda, Acosta, and Matienzo, Solórzano appeal to Aquinas's principle of the natural subordination of matter to form and means to an end in order to justify colonial coercion and the material practices of the empire. Once again, Solórzano justifies the personal service of the Indians, employing arguments borrowed from Aquinas's metaphysical instrumentalism.

ARGUMENTS IN FAVOR OF THE USE OF THE MITA FOR MINING

Solórzano examines the reasons in favor of the coerced labor in Chapter 15, Book 2, titled "On the service of the mines and the refining of their metals. And if it is licit to force the Indians into involuntary labor: Presenting the usual reasons and foundations that can be used to favor the affirmative response." He examines the reasons against the mita in Chapter 16, titled "On the same service of the mines, in which are presented the foundations of the negative response." The attempt to put an end to the polemics appears in Chapter 17, titled "On that which corresponds to thinking well about the resolution of the aforementioned question, and to expect that God will increase in other ways the treasures that are reduced by providing relief to the Indians. And on what human means can serve to licitly and securely acquire them" (1:262).

Solórzano begins his examination of the arguments in favor of the mita in an enigmatic way saying "the material that I attempt to present in this chapter is no less profound than the same mines which concern it, nor less laborious and obscured by contradictory opinions and royal charters in which can be found the serious reasons and foundations on which these are usually supported" (262). The attempt to find "reasons and foundations" faces the continuous threat of ungrounding, of an infinite regress of an unsolvable problem. Solórzano uses the metaphor of the vertigo produced by the mines to exemplify the attempt to examine the antagonism and opposition of the ideological field. Behind each argument there is another argument, and behind each cavern another cavern. The dark empty space of the mines threatens the rationality and legitimacy of the production of riches. Solórzano explains his intention of examining the problems as if they were veins of metals, proceeding like workers who find rich veins and work inside the mine (262).

THE EXHAUSTION OF NATURAL SUBORDINATION | 301

The first veins found by Solórzano, which are his continuous source of inspiration, are Matienzo and Acosta, who maintain that it is "just and legal to give the Indians the mita to work them and to benefit from the metals that they take from it and to oblige them although they do not want to be in this service (262). According to both authors, the Indians can be obliged because they receive a wage in exchange for their contributions, the conditions of work are regulated, and they are treated justly. One of the arguments in favor of the use of enforced Indian labor in mining is that this method is already being used in agricultural work:

> I consider in the first place that if agriculture (because it needs the labor) and industry of men is so essential, useful, and necessary for the earth to yield the fruits with which we sustain ourselves, it is permissible and must be understood as legal to conscript Indians for forced labor, as I dealt with at length in Chapter 9. It seems that we should not prohibit the extraction and refining of the metals, which amounts to self-preservation itself, and which had to be sought out, and which Mother Nature does not yield if the industriousness and ambition of men does not take them, as Pliny and others gravely declared. (1:262)

Solórzano states that the distribution of Indians is licit in agriculture based on the public need and utility of the service. The argument depends on the preexistence of natural ends of the state as perfect community. If the use of an involuntary labor force in agriculture for the "union and conservation of Spain" is legitimate, then mining is even more necessary than agriculture and should be justified by invoking the same end. Moreover, if agriculture is "useful and necessary" for the conservation of the social body, mining surpasses agriculture in utility since *beneficio* means "taking care" of metals, and metals are not given by nature but by "industry" or artificial means.

Therefore, "beneficio" as technique is useful and necessary for the "beneficio" and "care" of the social body. The weight of the proof resides in that artifacts and techniques are necessary for social life. Vitoria used this argument in order to state that humans are imperfect and incomplete animals that lack the supplements of nature, replacing them with artifacts in order to fulfill their needs. The validity of the social obligation is derived from the transposition of technique to nature. Also, José Luis Capoche argued that the mita was necessary for fulfilling the demands of the amalgamation technique, hiding the fact that the Indians refused to work in the mines not because of their lack of inclination to work but because of the lack of economic motivation or interest.

Once technique acquires the character of legal or moral obligation because of the way it sustains and benefits the kingdom, Solórzano proceeds to explain that the two republics (the Republic of Indians and the Republic of Spaniards) need each other, each of them fulfilling its proper obligation by occupying its proper place in the natural and rational hierarchy: "And delivered or produced as they are, they yield such utility, and they are deemed necessary, as with agriculture and its fruits, for their sustenance and conservation, and for those kingdoms, and of the two republics that, now mixed, constitute Spaniards and Indians: these latter would appear to be suffering or at least would suffer great harm, and much would break in the spiritual doctrine, government, and material protection of the same Indians if we were to be without their labor" (1:262).

Since both republics are mixed yet united, "the Indians should not and cannot be denied it [work in the mines], as, now mixed with us, we make one body, and we must help to sustain, conserve, and defend it as much as posible" (1:264). This argument reproduces ideas from Acosta, who argued in favor of the mita based on the unity of the social body. Indigenous people and Spaniards are part of the same hierarchical structure in the same way that limbs must be subordinated to the head of a statue. The Indians sustain the kingdom with their work, and their absence would be equivalent to a total breakdown of the system. Kamen's thesis that the source of power and weakness of the empire was its dependence on the cooperation of agents who were outside its control is useful for highlighting the geopolitical scope of the argument. Labor is outside of Spanish control, therefore it must be coerced in order to avoid a total breakdown of the unity of the Indies and Spain.

The necessary role of technique and amalgamation makes possible the codependence of the parts, grounding the colonial division of labor, according to which the prime matter must obey the dictates of the form. Briefly put, there is a paradox that emerges from Acosta's hylomorphic account of mining, because even though the two republics need to remain united, one of them is obliged to work sustaining the kingdom and the entire imperial economy. Although the perfecting form is supposed to be the source and origin of the conservation of imperfect matter, the imperfect matter of indigenous work and metals sustains both republics. Such dependence becomes ambivalent, paradoxical, and contradictory once Solórzano describes the singular and contingent perils and violence involved in mining, making visible the abysmal fissure in the attempt to ground the mita in indemonstrable and self-evident principles.

Solórzano continues, explaining that these reasons are summarized by José de Acosta, who, in a letter to the viceroy of Peru in 1595, recommends that he take into consideration the *"labor and benefit of the discovered mines and to ensure that other new mines are sought out and excavated: as the riches of the earth are the principal nerve for the conservation and the resulting prosperity itself of these kingdoms"* (1:262; original italics). Solórzano associates Acosta's defense of mining practice with Gregorio Agrícola, who in *De re metallica*, "responding to the arguments of those who discount them or gainsay seeking, excavating, and refining them" proves that "in metals there is food, clothing" (263). Moreover, the social life of the colony depends on precious metals to such an extent that, if they are absent, "we will even come to feel it in religion and in the spiritual instruction of the Indians" (263). The reciprocal presupposition between mining and religion can be seen in the "ardent zeal and care" that the monarchy employs in searching for these treasures, that it "does not depend on their greed, as in other aforementioned places" (263).

Solórzano insists there can be no doubt that, if people do not travel to the Indies, there will be no tributes or royal *quinto*, and the Crown "would suffer considerably" (263). These tributes defend and conserve the "provinces," making possible the labor of "archbishops, bishops, parish priests, religious, missionaries, and other ministers occupied with the conversion and instruction of the Indians" (263). In the absence of precious metals, "initiating the well-being and the health of their souls and bodies would continue to cease, and they would return to the vomit of their idolatry and bestial customs that by the infinite mercy of God they were leaving behind" (263). Solórzano is merely summarizing all the arguments employed by the imperial school of thought and the circle of Toledo: precious metals are what sustain the kingdom. Precious metals are the point around which all the temporal and spiritual power of the Indies is organized, and riches are the material support that joins the two kingdoms separated by the ocean. The centrality of precious metals does not make the Spaniards guilty of "being motivated by the lure of the mines" since God's Divine Providence wanted these riches, "which are usually the ruin of other mortals, to help to cause the salvation and conversion of [the Indians]" (263).

Solórzano's text points out not only the hylomorphic aspect of these arguments but also the prophetic teleological structure that understands the Indies as an available resource for the empire. Mining is an ambivalent thing, a mere means, something good in the hands of the Spaniards and bad in the hands of their enemies. Solórzano summarizes Acosta's position, saying that the justice and legality of the Indians' personal service is grounded in the "public utilities and necessities and the universal well-being of the entire kingdom" (264). The empire's ultimate end commands that the Indians "serve" in the mines, doing what is legally permitted to demand of them so long as it is deemed "useful" and "necessary" (264). The extraction of metals makes possible "the union and conservation of Spain and the Indies and, even better said, the entire expanse of the monarchy: and the defense and exaltation of the Holy Catholic Faith" (264).

Furthering his inquiry into the arguments in favor of coerced labor, Solórzano declares that "the universal well-being of the entire Kingdom ... embraces and encloses within itself everything and anything that can be required to deem it useful and necessary according to the doctrine of the authors who address it" (264). To the necessary character of the means Solórzano adds the indispensable character of indigenous labor, since the Indians are the only people who have this practical knowledge, and "no other could do more or better in this occupation than those who have always been the most capable and necessary." Experience shows that "neither Spanish nor Negros are fit for it" (264). The necessity of the "common utility" first makes precious metals indispensable, then techniques of amalgamation, and ultimately the cooperation of indigenous work itself. As we can observe, the complementarity between mita and amalgamation depends on an imperial metaphysics that divides the mastering perfect form and the servile imperfect matter.

Solórzano adds that following the steps of "wise men," it is possible to affirm that "the implementation and compulsion of Indians into the service of these mines and their metals seems neither reprehensible nor horrible" (266). Vassals have always served kings and republics. The Athenians, for instance, employed millions of chained slaves, and the Macedonians paid tributes to Alexander the Great and the Romans with the abundant material "that only one mine yielded to them" (265–66).

Following this line of thought, he adds that the Incas and the Aztecs commanded their subjects to work in the mines guided by less noble final causes than the creation of coins and wealth. The Incas and "Motezumas," before the arrival of the Spaniards, "tyrannized these provinces of Peru and those of New Spain and customarily employed [their vassals] in the labor of minerals" (267). The only purpose for which they used metal was to "paint or color themselves with their vermilion," which employed an infinite number of Indians, using them like slaves and appealing to absolute sovereign will (267).

Another argument employed by the defenders of the mita is that the "horror and duress" of this labor can be excused "based on its usefulness and its indispensable necessity for the union and conservation of the kingdoms of Spain and the Indies and of the Indians themselves" (267). This is the doctrine of theologians and jurists who insist that is not possible for the Spanish to be negligent, especially "if the risks and dangers are not extremely evident and inevitable, although it happens that some get sick or die from these causes" (267). The unity and conservation of the whole justifies the death or illness of a few indigenous workers.

Moreover, he who possesses a technique, skill, or special knowledge is obliged to employ this technique for the benefit of the social body even if his life is in danger. For instance, a medical doctor can be forced to use his knowledge even in the case of a life-threatening disease. Such an obligation is similar to the one "that the members of the republic have to help each other in imitation of those of the human body" (268). The members of the body "must expose themselves to whatever danger in order to save and defend the head, and because public health is the supreme law of laws" (268). The primacy of the head presupposes the hierarchical division between mental and manual labor, between mastering form and passive matter. The subordination of the part to the whole is inextricably joined to the subordination of matter to form, even in life-threatening circumstances proper to the work in the mines.

ARGUMENTS AGAINST THE MITA

Solórzano also presented arguments against the mita, in Chapter 16, "On the same service in the mines, in which the foundations of its opposition are brought forth." He begins this chapter saying that the dominant opinion about the legality of the mita has a solid basis, but that pious men, like the Jesuit Francisco de Coello, are of contrary opinion (274). For this reason, although the arguments in favor of coerced labor in the mines are completely consistent, the arguments against such practice are also meritorious. As a matter of fact, the labor in the mines and the extraction and beneficio of metals were each considered a "menial task, even worse than menial, and for this reason the Romans did not engage in it, instead assigning it to criminals, miscreants, and men of humble and low conditions and fortune, and considered it to be a quite serious punishment, more so than death" (274). The extraction and refinement of metals is full of misfortunes described in the "infinite texts" that abound regarding how criminals were "condemned to metal" and how they were branded on the forehead with fire (274). For all practical purposes Roman law considered these criminals not only slaves but also "dead" to such an extent that if they were exonerated of such punishment they were considered "resurrected" (274).

Therefore, the extraction and beneficio of metals is not worthy of free men with legal rights. Only delinquents and outlaws should work in such terrible conditions. Obliging workers to labor in the mines is a death sentence: "And since no precept that mandates or advises this coercion of the Indians into the mines and its dangers can be found, it seems that there is no benefit, and it must be admitted (but not permitted, as it is not permitted) to order them to be killed: this supports the right to kill someone or take him and put him in a place where he will die or will be killed" (275). The perfect state capable of issuing valid laws has to employ people outside the law because working in the mine is a death sentence, and giving indigenous workers to the azogueros is the same as giving them a right to kill, a prerogative otherwise limited to civil power.

Besides, positive law indicates that humans are not obliged to work in such inhuman conditions. Again, Solórzano quotes Aquinas, for whom nobody is obliged to do the impossible: "Arduous or extremely dangerous and difficult things do not fall under the precept of positive law, which never obliges doing the impossible or putting oneself in mortal danger, but rather some natural or Divine obligation that mandates it accidentally concurs with this precept: as according to Saint Thomas: the common school of Theologians and Canonical Experts resolves it" (1:275). Although forced labor is necessary for the conservation of the social body, it is not legal to kill free vassals of the Crown.

Solórzano continues explaining that "when the Sacred Scripture means to exaggerate the most difficult tasks, as in hyperbole, it compares them to what happens in the collection and extraction of metals" (276). Both in sacred scripture and in ancient texts about mining, images of the separation of metal from the earth abound in order to signify the life-threatening character of this work. For this reason, poets "attributed to the Iron Age the invention and beginning of digging in the *mines*: signifying that those who had the audacity to undertake such a thing had hearts as hard as iron" (276; original italics). Such audacity consisted of "daring to disembowel Mother Earth to extract the metals that God had hidden away in her harshest and deepest parts" (276). And God hides these metals because he knows that "their debased use would result in many evils and much harm to mortals, as Ovid, Seneca the Tragic, and the Philosopher Horatio, as well as other authors, have gravely and elegantly declared and cautioned" (276). These poets "do not stop making exhortations against and cursing such an evil invention and faulting the avarice of the men in this place" (276). Among them, Thomas More declares that the "appreciation and estimation" of gold and silver is equal to the difficulty of finding it, and that "for the other purposes of human life, iron is much more to our benefit" (276). Solórzano also mentions the "exaggerations" of Acosta, who "relates or describes with such grave and elegant words the works and dangers that occur in it, and of which he says just wanting to recount them causes horror; and still he surpasses the efforts to recount them by Pliny, Seneca, and Cassiodorus, as well as the others I have cited here" (277).

The horrors and difficulties of mining are nothing but signs of the prohibition against penetrating the earth in order to appropriate its hidden precious metals. Such prohibition is inextricably related to the origin ("beginning and invention") of the practice of mining, which transgresses the natural limit imposed by God. Mining is the transgression of the limit between the natural and the artificial. The horrors and difficulties of mining appear as obstacles, guided by the arbitrary decision to appropriate that which cannot be appropriated, yet they sustain the imperial enterprise by becoming the source of value of precious metals as a measure of the value of the rest of reality. The transgression of the limit between the artificial and the natural is materialized in Acosta's exaggerations, especially when depicting the horrific and abysmal space of the mines. The end, the value of precious metals, is separated and hidden (*escondido*), in order to remain valuable and desirable, and for this reason mining is hard and painful work. The whole argument against compelling the indigenous to work is grounded in the pains and horrors that result from enjoying the use and exchange values that are the result of the absent and postponed character of the precious metals.

Among the multiple horrors that abound in the interior of the mines, Solórzano counts the illnesses contracted by workers who are exposed to the mercury:

> They especially tell of the damages and illnesses that are contracted in the quicksilver mills, just as I experienced in those at Huancavélica, where I was inspector and governor from the year 1616 to 1619. Whose very dust causes a great slaughter of those who dig the mines, that here we call *mine sickness*; and the fumes from the same quicksilver quickly penetrate to the marrow the bodies of those who boil and refine it, weakening all the limbs, causing constant tremors in them; although by fortune some are of a robust temperament, few do not die within four years, according to Matiolo and Bisciola, and before them Pliny, St. Isidore, Dioscorides, and others. (277; original italics)

The extraction of riches depends on a technique that produces silicosis or "mine illness." The materialization of the principle of the subordination of matter to form and means to an end is the destruction of the life of the worker. The "mine illness" is the reverse of the "perfect community" and its imperial metaphysics. The limit that separates metals from human industry is materialized in the exhalations generated by the process of production of metals. As a result, the paradox of mining is that the body must obey the head, but the more it obeys the head the more it destroys its members. The weakening and destruction of the social body is the result of the commands given by the head, producing the toxic remainder of an ambivalent technique that is the benefit of amalgamation.

Solórzano continues, explaining how the interior of the mines produces an experience of death and darkness: "Indeed, almost all the *mines*, whatever they may be, Pliny and other Authors who have written about this teach: the climes and locations are harsh and sterile, the odors and exhalations intolerable, the air pestilent and scarce, no light at all, thus they are always filled with perpetual night; and the candles they use to help banish it cause great difficulty with their smoke" (277). To the lengthy list of material conditions of work, such as the sterility of the lands of the mines, the smell, the deadly fumes, the lack of air, and the lack of light that produces a perpetual darkness, Solórzano adds the supernatural realm: "And the worst of it is that in many are seen ghosts and very frightful apparitions of the demons themselves, called subterraneans, that appear to have been posted as guards of the treasures" (278). Solórzano is referring to the stories of how demons cause physical harm to the miners in order to guard the precious metals. In this way, God wanted the extraction of metals to be painful because humans desire what is "rare and difficult" (278). These underground specters interrupt and obstruct the appropriation of metals.

Another argument against the mita is that the service provided by the indigenous workers is different "from that of those condemned to metal in the time of the Romans" (278). During the Roman Empire, "these latter served as slaves, without being able to acquire anything for themselves or leave their misfortune, and were regarded as dead" (278). Nevertheless, "our Indians are always free and earn their own wages and salaries, and they move around their mitas or shifts which makes their work more tolerable" (278). Solórzano immediately retorts that "they have this freedom more in name than in fact because one cannot truly have that which is forced upon on, as Quintillian says" (278). They do not act freely because they are "compelled to serve for earnings and foreign commodities" (278). A payment can hardly be a wage if it cannot augment the fortune of the workers, compensating the dangers to which they are subjected (278). As a matter of fact, indigenous workers are treated worse than slaves because the mine owners "do not lose their money" and do not care if the Indians die because "they are able to conscript others in their place" (278).

The use of this argument shows how, although receiving a wage is the correlate of freedom, indigenous workers are worse off than slaves since their freedom is merely nominal. As a matter of fact, the wage of the *mitayos* barely covered their basic needs, and as a result they ended up contracting debts in order to comply with the mita. While the slave has some value because the owner needs money in order to buy more slaves, the life of the mitayo does not have any monetary value, because it can be immediately replaced. The use of mitayos involved the possibility of using a resource that is always available, dehumanizing the workers to a point that is worse than slavery. Unlike the precious metals they extract, the mitayos do not have any value, being reduced to an available raw matter that can be mobilized at will. The mixture of paid and coerced labor turns the worker into a person who is less than a slave.

Besides, Indians should not be "condemned to metal" because they did not commit any crime. On the contrary, their meek and humble nature and the services they provide "make it so that any grace they have is deserved, and thus it is forcibly that they are sat down to be seen punished and beaten without having committed any crime against that which the law provides" (279). Besides, the cause for ceasing to oblige the indigenous peoples to pay a service to the encomenderos was in the excesses and grievances that were worthy of slaves and not of free vassals (280). But the distribution of Indians in the mines "opens the door" for treatment that was worse than that received by the encomenderos, since experience shows how the miners oppress and punish the indigenous workers (280). Briefly put, Solórzano seems to be conscious that Toledo's reforms tried to put an end to the abuses of the encomenderos but they caused more problems than solutions, increasing the abuses and the oppression. The introduction of compulsory yet paid labor ended up dehumanizing indigenous peoples, making their work valueless.

Many times the Indians carry animals, women, offspring, tools, food, and drinks, "and upon leaving, their loads are much more burdensome still, as they carry the metals they have mined on their shoulders and many times enveloped in the folds of their own clothing, because they are not given even a bag or sack for them" (281). They carry this heavy load along "dark and murky twists and turns with shortness or infirmity of breath" (281). The dangers turn the mine into a trap where accidents are waiting to happen. Making a mistake "takes them more quickly to the depths and to death along with the same load" (281). Literally, the heavy loads imposed on the workers produce accidents that increase the list of evils that destroy the support that produces value and riches. After comparing mining with pearl diving, Solórzano points out that such works and misfortunes can be compared with those of prisoners described by authors who "equate in law the punishment of life imprisonment and the punishment of metal and that all punishments contain a kind of servitude or slavery" (283). Allowing the mita, "it does not appear that the end, or intention, the advocates of the opposing position want to defend is achieved: to wit, that these and other kingdoms are preserved" (283). Although the defenders of the mita appeal to unconditional ends, the mita does not achieve this end. The unity of the kingdom is threatened by "the great harm that has come to the Indians through this work" (283).

Acosta and others insinuate that conserving the kingdom would be achieved if they would free the Indians of this obligation, since "keeping them in a state in which they are completely spent and falling down, or in briefer terms, this mystical body on the feet of which we establish and consolidate ourselves" (283). In other words, the argument against the mita is that the arguments in favor of it fail to achieve their own ends. The structure of the argument is homologous to that of God's double injunction to produce value and to refrain from enjoying the secrets of nature. The application of the principle of natural subordination to the end excludes the end itself, turning the condition of possibility of the conservation of the kingdom into an obstacle, making visible the inconsistency of the principle itself: there is no mining without the kingdom, but there is no kingdom with mining either.

Potosí mining is the paradoxical object produced by the principle of natural subordination: it is both the necessary means and the reminder of the inconsistency of the metaphysical grounds of the mystic body of society. It is an obstacle engendered by instrumentalist metaphysics as the core of imperial ideology. Compulsory labor in the mines is the result of a blackmail proper to the imperial command that obliges the choice between mining and chaos. Nevertheless, chaos is the result of the attempt to impose an order on reality. The forced choice between order and catastrophe yields to a dialectical contradiction: compulsory labor is simultaneously the condition and the obstacle of the subsistence of the empire, and the conservation of the kingdoms. The arguments against the mita are not in diametrical opposition to the arguments in favor of the mita because they involve a shift of perspectives according to which what was experienced as a condition of possibility is now experienced as an obstacle created by the first perspective. Allowing the mita is counterproductive: it is grounded on the principle of the subordination of matter to form and means to an end (the conservation of the kingdom), but it does not fulfill the end, destroying it instead.

By building an empire upon forced labor, the Spaniards built an empire with clay feet: "Although we may have them (as we do have them) as such, and we determine to be made of clay, as Daniel says, that which supported the statue of gold, silver, and bronze, which Nebuchadnezzar saw, and that they were broken, everything came down and was made into dust, as the same Daniel says" (283–84). The Spanish Empire has feet of clay because it depends on a contradictory and immoral practice. The image of an empire with feet of clay is the materialization of the inner contradiction in the principle of the natural subordination of means to an end. Elevating the means to the status of an end in itself equals destroying the means and losing the end. The ultimate end of the ruler is to promote the "common good" of the "perfect community," yet mining technique destroys the material basis of the "perfect community" itself.

As a result, Solórzano, the author of the most radical defense of the Spanish Empire, is conscious of the contradictions engendered by the technical deployment of the classical corporate society, the perfect community, when he says that the sovereign makes the worst mistake when destroying his source of riches (284). Those who create laws and write about them know that the preservation of the people is a good that is incomparable to the "greatest treasures" of the kingdom. Those who think that the labor in the mines and the extraction and benefit of metals contribute to conserving the religion and faith also err (285). There is no evangelization without people to evangelize (285). In other words, the violence employed when obliging Indians to work is not the moderate violence of the gentle yoke of Christianity, the purification and mortification of the soul, but the complete annihilation of the body: "Because Christ himself, the Redeemer and our lord and true Author of the Gospel that we preach to those who call to and are moved to him, promised them through Saint Matthew that if they are overworked and burdened he will provide relief and respite, and if they carry his yoke, because it is as tender and smooth as that which they are made to carry, they will have solace and peace in their souls" (285).

Attracting indigenous people to the faith is incompatible with throwing them to the mines. Using an almost Lascasian tone, Solórzano condemns the violence done to the Indians as an obstacle to their conversion: "Elsewhere, David said that they should enjoy and look upon the extent of their tenderness, and how blessed are those who trust and believe in him. All of which will be difficult to persuade the Indians of if they see this oppression" (285). It is not possible to introduce them to the "meditation and contemplation of Faith," because excessive work hinders the understanding and tiredness weakens human nature (285). The scandal is exacerbated by the injustice of the distribution of roles in the mystic body of the perfect community: "Especially, seeing the Indians who place on their shoulders all this weight, in which we say the survival of the kingdom consists, without asking others to lift a finger, not even the load, as it is they who make use of it" (285). The Indians serve as support of the economy and the Spaniards benefit from their work without moving "a finger" to ease their burden. Indigenous labor is the condition of possibility of the extraction of riches, and it is a job that it is not fit for human beings, in spite of the fact that they are the only ones who know how to do this job.

The reasons in favor of the need of tributary work are based in the belief that without metals there is no temporal or spiritual kingdom. But the destructive violence of mining will also result in the lack of precious metals, leading to their relapse into idolatry, chaos, and catastrophe. The arguments in favor of the mita are based on the instrumentalism implicit in the principle of subordination, while the arguments against the mita highlight the dysfunctionality of the system, showing how the means do not lead to the alleged end as invoked by the first group. The arguments for and against the mita both repeat the structure of the debate between Las Casas and Sepúlveda: the principle of subordination is sufficient to demand the unconditional subordination of indigenous people because they are imperfect matter that must be subordinated to the perfect community, but the rigorous application of the principle contradicts such subordination as the Indians are not only the means but the end as well. This aporia or double bind is not an accidental feature of imperial reason but a necessary and structural effect.

THE SOLUTION TO THE IMPASSE

In Chapter 17, Solórzano looks for a way out of this impasse, which is to make a political decision that is both practical and administrative and which clings to technical means:

> These are the reasons and most substantial foundations that are offered to my limited understanding and which can be considered on the one hand, and on the other, on the question we continue dealing with of the Indians' forced personal service in the mines (I content myself with having proposed them; their ultimate and precise resolution is yet to come and will be issued by a superior and more accurate judgment), supposing that, although some

royal charters have mandated or permitted that it continue for now, and this is what is being practiced, these same charters admit the doubts of the matter and demonstrate a desire to relieve the Indians, always the urgent and present necessities that are found today in the Monarchy of Spain. (290)

This passage alludes to the 1634 charters that allowed the use of forced indigenous labor despite doubts about its ethical grounds. The charters also expressed the doubts of the Crown. Moreover, the same laws that commanded continuing with the practice were the laws that protected the Indians. The arguments invoked against the mita, expressing the need to protect the Indians, were used as arguments in favor of the mita. Solórzano's dictum is: "For now the distribution will continue" (289).

Solórzano explicitly states that to continue the practice of the mita while searching for other legal means does not mean that the imperial enterprise should stop: "commerce with Spain should not cease for this reason, nor should the propagation of the faith, or support for the wars, as Father José de Acosta advises" (290). Gold and silver are not everything, "the Indies have many other beneficial and attractive things" (290). The good functioning of the mines cannot rely exclusively on coerced labor. Even if the mines produce fewer precious metals, "God will allow this to liberate from all doubts, to work in all great tasks and with better results that those that have been attempted in the past, as Acosta himself also promises" (290). Solórzano insists that the basis of a society cannot be riches or military power, which cannot be ends in themselves, but only religious piety, which can end up producing riches and power as a by-product. Solórzano appears to be relatively conscious of the way the means threaten to become autonomous and substitute the transcendent ends. As a result, his instrumentalism remains disavowed, and divine will is presented as the only possible source of sovereign will and rational ends: "Only true piety is sufficient for the princes to be able to save everything; just as, on the contrary, without it, they will benefit from neither the armies, horses, halberdiers, abundant arms and swords, batallions of soldiers (however numerous they may be), nor gold nor silver nor have any other apparatus, regardless of its value" (290–91). The means appear to be insufficient for conserving the empire. Roman Catholic piety is the only principle and end capable of bonding the perfect community. As a result, amalgamation and indigenous labor are instrumental means for conserving the kingdom, but they cannot be ends in themselves.

The solution to the problem of the mita is more mining: "As will happen here, yielding more metals and of more law in the already-discovered mines, and bringing about this relief for the Indians so that they themselves discover others for us; because we know from evidence . . . that they have knowledge of many very rich and abundant [mines], and that by fleeing from the labors they suffer in them they will no longer reveal them to us" (291). Solórzano's solution continues within the terms of the aporia: the mines are simultaneously the problem and the solution, and the only way to find a solution to the problem is to create more problems, that is, to discover new mines. The counter-arguments against the reduction of the Indians to the status of instrumental means are invoked as arguments in favor of the necessity of the end. No matter how hard Solórzano tries to regulate the technical means by subordinating them to higher principles, the whole logic of power remains trapped in instrumentalist logic. If the possibility of discovering new mines is the only remedy, the problem persists since the mita causes the flight of the Indians, who also refuse to point out the location of the mines.

Solórzano does not simply condemn the Indians who flee as "we Spaniards also did in ancient times, as did the Balearics, the Oriental Indians, and other Nations . . . that for the same reasons abhorred gold more than the plague" (291). Although he does not condemn the fleeing Indians, he notes with concern that, "in Peru, there are Indians who superstitiously believe that their Inca must be resurrected, and for him they guard all the rich mines of which they have knowledge, and through neither pleas nor threats nor punishments is there a single one who wants to reveal them to the Spaniards" (291). Solórzano is alluding to the myth of the return of the Inca, which would bring only disorder and chaos. The worst catastrophe would be the possible restitution of sovereignty and mineral riches to the Indians. In other words, the threat of indigenous insubordination, the specter of Las Casas's doctrine, demands a political decision to postpone any decision to get rid of the mita, filling the gap produced by the infinite proliferation of reasons and questions.

Solórzano does not hesitate to use an apocalyptic tone that demonizes the indigenous, blaming them for their own misfortunes, since "In this place they act like the demons that are said to guard covetously similar treasures, not so much to take advantage of them as to reserve them for the Anti-Christ, the son of perdition to whom many refer, with which they will make much war at the time of his coming" (291). The demons that separate the miner from enjoying the forbidden fruits of the earth and appear in the arguments against the mita return under the form of the Indians keeping their treasures for their own illicit ends, reinforcing the need to defend the unity of the two republics against the return of the Inca. Like Vitoria, who was not able to reach a decision on the servile nature of the Indians, Solórzano oscillates between arguments in favor of and against the mita, only in order to cut the aporia by formally transforming the arguments against the mita into arguments in favor of maintaining the practice while searching for new solutions. The decision to postpone the abolition of an evidently dysfunctional mita/amalgamation complex is inseparable from his formal decision to stick to the principle of natural subordination, despite acknowledging the lack of origin and ground of the natural law for guaranteeing the legality of the mita.

To the imminent threat of the return of evil indigenous practices, embodied by the restitution of sovereignty and wealth to the Incas, Solórzano opposes the correct use of reason guided by Catholic piety: "In whatever manner it may be, the true and prudent reason of state is to watch and to aspire only to what is licit, and nothing has ever come out advantageously by subordinating divine precepts and respect to human interests" (291). The "true and prudent reason of state" needs to be guided by "divine precepts and respects" based on the authority of the prince and in the metaphysical principles that steer human law to its transcendent end. The only solution is to remain pious since God breeds metals in the earth and decides who is worthy of his favor and who is not: "In Peru it has happened that, having discovered the rich silver mine of Vilcabamba and inducing conflict between the discoverers, they were ordered by Justice to install gates and locks while it was resolved, and when they were later opened, no sign or trace of any metal could be found" (292).

Solórzano states clearly that he wants to question not the validity of the ends but, rather, the morality of the means: "But my intention for this is not to say that they should stop prospecting and excavating the mines and its metals which I know well (and have already noted in another chapter) God has bred so that men can take advantage of them and help themselves, and more so in times of urgent necessity. I only want to say, and I do say, that they should be excavated with licit and fair ways and means on which we can found and secure the abundance procured from them and which the Holy Spirit promised in the Proverbs" (292–93). He explicitly says that he does not pretend to delegitimize the labor in the mines since his aim is only to Christianize it. The only way to Christianize the means is to subordinate them to the end.

The rules of law and theology exist in order to treat slaves humanely, "without harshly punishing them or exposing them to obvious risk and danger to life" (293). Now, the rights that Spaniards have over slaves are not the same as the rights they have over Indians because the slaves "are like the dead, or animals, and we can make use of them with little harm for our advantage and comfort, even though they are exposed to some danger" (293). There are people who think that it is legitimate to kill them, "and that in such a way they are compeled to subordinate their health and life to those of their masters" (293). Therefore, only voluntary work is legitimate and not oppressive "because, although the work is taxing and seems impossible to compensate with any wage no matter how large, volunteering dispels any hint of oppression or injury, according to the law" (294).

Solórzano also thinks that it would be licit to hire voluntary Indians: "It will also be more licit and tolerable to hire Indians who voluntarily want to direct themselves to it, or Spaniards, or Free Blacks, Mestizos, or Mulatos, who are abundant in the Indies, and paying them well there will always be some who will apply themselves to this work, as they do in Germany and other provinces and has been seen in the Indians who by their own will hire themselves out or work collectively in Potosí" (293). Although the payment cannot remunerate Indians for the work, it would not be oppressive to hire voluntary Indians in the mines, as was the custom in Potosí.

Solórzano admits that the work in the mines cannot be remunerated because of the huge gap beween the suffering produced in the mines and the money employed to compensate for those sufferings. The solution of the problem of the abyssal distance between the payment and the material conditions in the mines is transforming the latter into a technology of power capable of redirecting whatever is outside the law to the common good. Solórzano gives a list of people who should be obliged to work in the mines instead of the Indians:

> And it would be no less convenient to condemn wicked men and criminals to this work, be they Spaniards, Indians, free Blacks or slaves, Mestizos, or Mulatos, sending them to the Judges who, instead of other punishments, impose this one when the severity of their crimes merit it; also as the afore-

mentioned charters have already ordered, and which we declared to be common among the Romans and other well-governed Nations; and we see that the punishment of the stockade, which is compared to [the punishment of forced labor in the mines], is mandated by the collected laws of our Kingdom; and which Thomas More, in his *Utopia*, greatly approves of, saying that it is the best and most useful way that can be found to punish crime. (294)

The solution erases racial differences by drawing a clear line between those who are inside the law and those who are outside it.

Introducing this custom of punishing criminals in the mines "would serve as a restraint, as in them not so many crimes can be committed" (294). Such measures would be useful for the "general reformation of customs, and those who commit sin would pay for it here, seeing this punishment as merciful if it were deserving of punishment by death, and remaining slaves to it (as the law thus calls them)" (294). Condemning those sentenced to death in the mines would also absolve their crimes "with public service, and they will clean the earth of bad seeds and [allow] the Indians to rest and increase so they can help in other tasks that would not be so laborious" (295).[6]

Moreover, Solórzano thinks that it would be useful to commute the sentence in work that serves the common good (295). The measure of converting mines into jails would set an example: "to see criminals endure such protracted labors was still for others a valuable lesson" (295). Solórzano dismisses the arguments against this possible measure, saying that "the usual criticism against this is not of much substance; suffice it to say, no miner will want to be sent men of such poor quality to work in this task" (295). He argues that this would be a problem already faced by the miners, since "today they are sent innocents, and they buy and maintain slaves, which gives rise to many difficulties" (295). The Spaniards must be satisfied—even if there is less gain with the prison/mining complex than with "the involuntary labor of the Indians"—and hope that in the future the mines will yield more "advantageous results" (296). In this way they will prevent the Indians from "carrying off all the work of the mining of Gold and Silver, which they need and share less than the Spaniards and other idle Nations, and from viciously and prodigiously spending the majority of what was mined with so much pain and labor" (296). Substituting other licit means for the mita is the only consistent solution to a policy that seeks the common good of the vassals: "Besides the fact that many kingdoms and provinces are found that have neither gold nor silver mines, or if they have them they do not mine them because they place a greater value on the health and preservation of the vassals, which is the most precious treasure princes can possess, as has already been said" (297). The preservation of the vassals has to be the direct goal of the prince. The accumulation of money has to accommodate to the common good of the vassals: "if he cannot measure and accomodate his things and they want to draw it all out in money, while they will have more they will need more, without all the rivers having to bear gold, it can be sufficient to slake the thirst of their insatiable greed" (298). Once gold becomes an end in itself, it becomes the value against which all other values distintegrate.

Solórzano's solution makes visible the inner inconsistencies of instrumentalism: by remaining stubbornly attached to the destructive means, Spaniards deprive instrumentalism of its higher purpose. Such is the paradox of Potosí mining and the mita/amalgamation complex. The persistence of this practice is a sign of the inconsistencies of the entire colonial enterprise, which means a great deal of trouble, but its absence means total catastrophe, because a dysfunctional kernel sustains and keeps together the changing and expanding network of production, circulation, and credit. To conclude, what appear to be arguments against the mita are arguments to keep practicing mining by other means. First, Solórzano says that it is necessary to keep practicing the mita while they also look to substitute it with other practices that would alleviate the Indians' situation. The necessity to keep practicing the mita is based on the decision to not stop exploiting the mines. Also, the Crown's doubts express the intention of protecting its vassals. Second, the arguments against the mita, which appeal to its destructive character, are employed as arguments in favor of practicing mining by other means, substituting the mita with a system for punishing criminals. If the arguments against the mita invoke how the means contradict the ends of the empire, which are civilization and the conversion of the natives, then the transformation of mines into a penitentiary system would be consistent with the ends of the empire, because it would be employing people outside the law as a labor force. Third, establishing a parallel with Vitoria is also significant for showing how the arguments against the mita become the arguments for continuing to mine by other means. After contemplating the possibility of not being able to use valid titles for justifying the Spanish dominion, Vitoria states that commerce and extraction of the abundance of precious metals should not cease.

In the same vein, Solórzano states that stopping the mita should not mean stopping commerce and the extraction of metals because mining would have to continue. After the failure of Vitoria's natural law and the historical experience of the destructive effects of mining, Solórzano reintroduces the unconditional character of the end in order to conserve the means while improving them. Finally, Solórzano invokes Christian piety, continually invoked by those opposed to the mita, in order to strengthen his own position and to evade the impasse. In order for the extraction of metals to be effectively subordinated to the means, the profit has to be produced only as unintended side effects of the end. The question arises: What is the meaning of the reintroduction of the unconditional end when cutting the impasse between arguments against and in favor of the mita? The compulsive return of the unconditional end means that instrumentalism, in order to remain functional, has to remain metaphysical. Concretely, as soon as the Crown's decisions are directly motivated by the accumulation of riches, the whole ideological justification becomes self-defeating. As soon as the subjects of ideology become aware that the invoked ends, civilization and conversion, are justification of the means, the ideological effect is null. The point of invoking unconditional piety to regulate profit is that the Crown can achieve temporal profit only if it directs its actions to the transcendent end, the common good of the vassals. Although the purpose of religion is to justify the temporal profit enabled by Christian piety, for ideology to keep its hypnotizing power, profit must appear as an unintended side effect of the end. Solórzano's message is that profit should not be aimed for directly but, rather, as an inadvertent after-effect of the transcendent end.

Solórzano provides us modern readers with the authentic form of the imperial paradox: what matters is to continue pursuing the transcendent ends, even if artificial violence lay at the origin. Imperial subjects have to act as if their decisions were solidly founded on unconditional transcendent ends. What matters is the consistency of the imperial ideology because once the means explicitly substitute for the ends, the whole enterprise becomes self-defeating. This is exactly how the transposition of technique to nature works in metaphysical instrumentalism: the origin and the end are meant to justify the artificial manipulation of nature and people. The natural end was from the beginning the means of the means.

Thus far, I have considered how the presuppositions of metaphysical instrumentalism served as a support to justify the exploitation of the indigenous labor force described in the work of Solórzano. Metaphysical instrumentalism works also in the *Política indiana* to justify the possession and technological manipulation of metals by conceiving them in terms of plastic and mutable entities that reproduce themselves like living beings. The attribution of life to metals in metallurgy is compatible with their technological exploitation and its transformation into money.

THE LIFE OF METALS

In the sixth book of the *Política indiana* where Solórzano discusses the royal hacienda, which means the properties of the Crown in the Indies, he devotes the first chapter to the riches extracted from the mines. The mines of the Indies are comparable to Ophir and Tarsis, the places from which the king Solomon obtained his riches (301).[7] These riches are "inexhaustible" and fall over Europe like rain, "because they come from the East and West Indies that are laden with such treasures and that God has provided as a perpetual harvest, because what is collected and brought home in a year seems to have no use other than to predict, prepare, and promise what comes next" (Solórzano, *Política indiana*, 4:301). There is an implicit comparison between mining and harvesting, according to which the mines provide inexhaustible riches because metals grow in the mines in the same way as plants grow in the fields. Solórzano provides an anecdote that links the capacity of precious metals to self-generate with the envy that other nations have towards the Crown:

> Now, when I try to print this, a sensible paper has come into my hands, written, so it seems, by some minister or secretary of the king of Denmark, which, responding to some poorly founded reasoning with which the representative of France present at the Münster congress attempted to diminish the glory and power of Spain and cause its downfall, says: That she is such, that she can arrogate or attribute to herself that of the poet, who said, speaking of Niobe: *I am greater than the injuries of fortune because, although much is taken away, I am left with more*. And notice of this was taken later

because of the newly arrived galleons at the same congress, in which more funds came to our king than France and Sweden and all their kingdoms could yield to them in ten years. (4:301)

Later in the text, Solórzano proceeds to examine the alchemical belief in the capacity of metal to self-generate:

> But leaving this for those who would have the ability and position to remedy it and all that can be said of the nature, generation, differences, and properties of metals for those who have written particular treatises on this subject, to which many attribute a vegetative soul, as with plants, and affirm that, like plants, metals grow while more is extracted and that gold usually increases if it is buried and that copper sprouts if it is sown as if it were from others, Cardano, Monardes, and Juan Barlerio tediously attempt to prove, and speaking particularly of gold, copper, and other metals from our Indies, Pedro Mexía, Simon Mayolo, and Tomás Porcacho, refer to some laws and authors of our jurisprudence that through these causes usually place and count mines and their metals among the fruits, which is very important and dignified to inform about for many reasons. (4:302)

After declaring himself not to be a specialist on the subject matter, he quotes authorities such as Girolamo Cardano (1501–1576) and Nicolás Monardes (1493–1588), who sustain that metals literarlly have vegetative souls that allow them to grow again in the bowels of the earth. Solórzano declares this to be an important subject because of its legal consequences, since metals are included as a category of "fruits." Before explaining the importance of considering metals as fruits, it is instructive to bring to the discussion the work of Álvaro Alonso Barba, author of the *Arte de los metales* published in 1640.

Álvaro Barba was a Spanish priest and metallurgist who lived in Peru from 1604 to 1650. He was the inventor of a method of hot amalgamation in copper cauldrons designed to improve the yield of silver ore mined in Potosí. In the seventeenth and eighteenth centuries his book received seventeen translations and six reeditions. His *Arte de los metales* is one of the most important metallurgical treatises of the Renaissance.[8] Barba addresses the problem of the self-generation of metals in his Chapter 18, titled "On the generation of metals." Since metals are hidden in the bowels of the earth, it is no wonder that they invite all kinds of speculation about their origin: "It is no wonder that there has been such a diversity of opinions about the subject of metals among people who are permitted them; it seems that it was with a peculiar providence the Author of nature wanted to hide them in the dark profundity in which they grow and the painful harshness in which they are enclosed in order to place some sort of obstacle against human ambition" (30). Like Solórzano when appealing to classic literature to explain how mining transgresses a limit imposed by God, Barba states that God intended to hinder human appropriation and transformation of metals, in order to ground their value by means of scarcity. Before offering any empirical description, Barba briefly mentions the theological-philosophical tradition, employing the standard teleological and hylomorphic motifs common to Neoplatonism and Aristotelianism:

> Those who have been granted the name of *philosopher*, by understanding the knowledge of causes, leaving raw material as the most remote beginning of metals, as it is with all other corporeal things in the world, note another cause, albeit also remote, that consists on the one hand of a certain humid and unctuous exhalation and on the other of a piece of viscous and oily earth, the confluence of which results in a material which is not only of metal but also of stone; because if aridness prevails stones are generated, and if there is more than abundant humidity it will become metal. (30)

Barba differentiates between the matter of metals and the prime matter, the pure passive potency deprived of any substantial form. The matter of metals is "a humid and unctuous exhalation," a reference to Aristotle's theory of exhalation as it appears in the *Meteorology*.[9] When looking for the immediate origin of the self-generating power of metal, Barba anticipates that gold is the end of nature: "From the varying temperament and purity of the aforementioned material originates the diversity of metals, from which comes the purest end of all and what nature principally intended, gold" (31–32). The immediate matter of metals is subordinated to an end, which is gold.

Barba attacks those who think that metals did not change since the time God created them: "Many of the masses, avoiding difficult discourses, say that from the beginning of the world God cultivated the metals to the way they are found in seams today" (32). Such a thesis is the result of the intellectual laziness of those who try to avoid difficult discourses (32). The self-generating power of metals cannot be reduced to the power of God or the celestial spheres. The philosophers who ignore this fact "make grievances against nature, baselessly denying the *productive virtue* it has on all other terrestrial things" (32).

Barba thinks that attributing all productive properties to God expropriates nature's productive virtue. This productive virtue will be the form of metals, which manipulates their immediate matter. This matter, which is the immediate cause of the generation of metals, is *vitriol*, or metal sulfate: "this raw or fundamental material of the generation of metals is vitriol" (32). If the productive virtue is the causal form of metals, vitriol is the material cause. Vitriol engenders sulfur and mercury, which are the two components out of which metals are created: "This vitriol, through the heat of subterranean fire and celestial attraction, gives off two kinds of smoke or vapor, one earthy, fine, and unctuous, and somewhat distributed, which the philosophers call sulfur, because it resembles it in its qualities; the other is damp, aqueous, viscous, and mixed with fine earth, which is the proximate matter of mercury" (32). Following the alchemical doctrine, Barba maintains that sulfur and mercury are the material cause of metals. He provides more information about the formal cause of metals in Chapter 20 where he writes: "Other than the heavens, which attend the generation of all things as universal cause, as with metals, the efficacy of another proximate cause is necessary, that impresses virtue upon them, opens them in their own matter; because the qualities of the elements by themselves are not sufficient nor are they determined by the production of a certain kind of mixture but rather insofar as they are directed by another particular virtue, as is manifestly seen in animals" (35).

The heavens and universal causes are insufficient to explain the efficacy of proximate causes. The proximate cause is a force impressed on matter. He argues that the elements by themselves are incapable of producing mixed entities such as metals. Only the productive virtue, form, can direct elements to the production of their end. Barba identifies form and function by showing how the productive virtue transforms the elements into its own instruments: "Thus, this proximate cause or mineral virtue uses the elemental qualities as instruments, especially heat and cold, in the generation of metals: with heat it uniformly mixes the earthen with the damp, which is the material that composes them: it boils it and assimilates and thickens it, and with cold it hardens it and crystallizes it in the form of metal, more or less perfect, according to the greater or lesser purity found in the material's present disposition" (35). The form, or productive virtue, "uses" the qualities of the elements as its instrument. Barba employs a series of verbs such as "assimilate," "thicken," "harden," and "crystallize" to illustrate the way the form acts over its matter. The final product is more or less perfect depending on the disposition of the matter.

As we can observe, Barba transposes technique to nature by conceiving nature as analogous to metallurgy. Material nature subordinates the elements and the imperfect matter to the creation of precious metals, which are the most perfect end of nature. If nature has its own productive virtue that operates over its material in the same way metallurgy does, then it enjoys a certain autonomy with respect to the first, remote efficient cause of everything, which is God. This certain autonomy is the capacity of metals to self-generate outside of external divine causation. Metals' capacity for self-generation is the result of the transfer of instrumentalism to nature itself. The capacity of the metals to grow and reproduce is explained by appealing to a now immanent form that operates over matter by instrumentalizing the elements.

Far from contradicting instrumentalism, the living capacity of metals presupposes technical manipulation. The best example of self-generation of metals is Potosí itself: "Many conclude, and we all at least see, that the stones were left within the mines years ago, because they had no silver, they were later taken out with it so continuously and abundantly that it can be attributed to nothing other than the perpetual self-generation of silver" (32). Potosí is the best example of the self-generating power of metals. Barba appeals to his experience in Potosí in order to argue in favor of the self-generating power of metals that certainly makes possible the imperial enterprise by enabling the continous and never-ending appropriation of metal. The extraction of the royal quinto is perfectly compatible with the life of metals as a never-ending source of profit.[10] The living character of metals is the outcome of metaphysical instrumentalism and assumes the earth to be a never-ending resource of riches. After considering how the life of metals is compatible with metaphysical instrumentalism and the imperial politics of extraction of the royal quinto, let us return to Solórzano's text in order to explain the legal consequences of considering metals as the fruit of the earth.

Before explaining the legal importance of such a thesis, Solórzano proceeds to give an etymological explanation of the word "metal":

> What is offered to me to say what belongs to my intention is that this word "metal" is Greek, and some say that it was taken from a verb that in the same language means to scrutinize or to search; others, along with Pliny, say that the natural property that is experienced in the veins of metals, once one is found and discovered another is found nearby; but it is in everyone's interest, generally taking this name, to comprehend and embrace *whatever material is usually extracted from the bowels of the earth*, be it gold, silver, mercury, copper, iron, lead, tin, sulfur, alum, salt, lime, gypsum, clay, slate, stones, quarries of all types. (Solórzano, *Política indiana*, 4:302)

In Greek, "metal" means to "scrutinize" or to "search," a reference to the activity of looking for veins. Solórzano is not just providing an etymology of the word. He is establishing a parallelism between the experience of mining (which always finds veins close to other veins) and the nature of metals. The jurist seems to be less interested in making pronouncements on the capacity of self-generation of metals than in providing their definition. He is interested in defining metals as any matter that comes from the bowels of the earth, meaning that they are the fruits of the earth as he expressed when discussing the self-generating capacity of metals. If Solórzano is ready to accept that metals are self-generating it is because he wants to consider them as fruits of the earth. And the fruits of the subsoil belong to the Crown: "The most common of which is what the miners from where it is extracted call *Regalías* [Royalties], which is to say, goods belonging to the Kings and supreme Lords of the Provinces where they are found and by themselves and incorporated by law and custom in their patrimony and the Royal Crown, and which are now found and discovered in public places, now in lands and possessions of private persons" (4:303). The *regalías* are the objects of property of the Crown. The mines are *regalías* regardless of whether they are spaces that belong to private individuals. Therefore the fruits of the earth, metals, are properties of the Crown:

To such an extent that although they claim and demonstrate that they possess such lands and their terms for a particular grant and concession from the same Princes, however general the words they used may have been, they will not make use of nor take advantage of this to acquire and gain for themselves the mines in which they are discovered, if that is not especially stated and expressed in the said grant, as many regulations of the common law of the Kingdom have decreed and declared in which all commentaries proceed and expand greatly on this point. (4:303)

Metals are property of the Crown, who grants the usufruct to particulars by means of grants (*mercedes*).[11] Solórzano contemplates the idea of the capacity of metals to reproduce themselves only because this is a sign that they are the fruits of the bowels of the earth, which belongs to the Crown.

On the one hand, the beliefs about the self-generation of metals is the result of the transfer of technique to nature by attributing to the natural productive virtue a capacity to generate metals by instrumentalizing the primal elements. On the other hand, this same is the result of God placing metals in the subsoil for the king, who then grants them to particular miners in exchange for the royal quinto. José de Acosta emphasized how metals did not have true interior life but only resembled life. Within Acosta's teleological design, the life of metals was the life of an imperfect matter that was subordinated to the perfect form, placed in the bowels of the earth to serve the interests of the universal monarchy.

Solórzano not only finds alchemical ideas such as the capacity of metals to reproduce themselves as perfectly compatible with the teleological design, but he also tolerates the practice of alchemy, under the form of amalgamation by mercury. For this reason, after stating that metals are "engendered" from mercury and sulfur, he recommends tolerance of the alchemical arts:

> And thus the Art of Chemistry or of the Alchemists should not be prohibited, as with the mixture of these and the other ingredients that they employ and the concurrence of the elements, attempt to extract gold and silver because this is not to change the substance of things, which only God can do but, rather, to imitate or to make use of these and other wondrous secrets, including warding off glamor or the evil eye and other medicinal effects. In

consideration of this, Father Acosta piously and wisely concludes: "In all the oddities that this metal contains, the Author of its nature deserves to be glorified, since cultivated nature so readily obeys his hidden laws." (4:313)

Solórzano, who was in charge of the mines of Huancavélica, is conscious of the importance of amalgamation by mercury for refining silver and producing money. Like Aquinas, he thinks that alchemy cannot change the substance of things, since it is only capable of aiding nature by imitating it. Therefore, alchemy does not contravene the natural order of things, being perfectly compatible with the imperial enterprise. Aquinas follows Alberto Magno, for whom it was not possible to transmute species, since neither humans nor demons emerge through the imposition of new forms over matter.[12] Art can only replicate nature but it cannot perfect it. Aquinas's instrumentalism is still metaphysical, and only God has agency over his creation since only he can create a substantial form. Art can only impose accidental forms, and it is not capable of altering substantial forms: "This form is, rather, an accidental form. All artificial forms are accidental forms. For art works only on what has already been put into existence by nature" (in Bobik, *Aquinas on Matter and Form*, 10). For Aristotle and Aquinas, while nature has an inner capacity to move itself, a technical artifact cannot move or reproduce itself since it lacks the inner capacity to change (Aristotle, *Physics* 2.192b16, in *Basic Works*, 236). When commenting on this passage, Aquinas explains "but things which are not from nature, such as a bed and clothing and like things, which are spoken of in this way because they are from art, have in themselves no principle of mutation, except *per accidens*, insofar as the matter and substance of artificial are natural things" (*Commentary on Aristotle's Physics*, 75). Aquinas assumes that artisans operate by imitating nature, and that is possible to use their language to describe natural processes. But this does not mean that artisans can alter or perfect natural order. Ironically, artisans are subordinated to a natural order that is described in terms borrowed from artisanal manipulation. The limits Aquinas puts on artificial transformation result from the preeminence of God, who also stands in relation to nature in the same way the artisan stands in relation to the product of his art. The incapacity of the artisan to bypass the Divine Artisan can help us see how, for Solórzano, the arguments invoked by Aquinas against alchemy are not arguments for prohibiting it but for *practicing* it. The limits to technical manipulation are also its conditions of possibility since it is not only possible but necessary to practice amalgamation in order to extract more metals.

Solórzano considered the belief on the self-generating capacity of metals to be fully compatible with regarding them fruits from the subsoil, which belongs to the Crown. Solórzano thinks that alchemical beliefs and practices are compatible with the imperial project because they are consistent with metaphysical instrumentalism, which serves as an ideological frame for considering them as a "standing reserve" available to the Crown. To conclude, metaphysical instrumentalism presupposes instrumental manipulation not only because the justification of the means are the true aim of attributing transcendent ends to natural order but also because it reduces craftsmanship to the status of a neutral and available device that can be controlled by an efficient cause. There is an intrinsic, substantial, and not merely accidental relation between the self-generation of metals and the consideration of them as an infinite source of riches. Moreover, if Solórzano is capable of accepting this idea, it is not only because it is consistent with instrumentalist metaphysics but because it is its axiomatic consequences.

By moving the emphasis from remote to proximate causes, from transcendent forms to the productive virtue, the instrumental subordination of matter becomes immanent to nature. The form impressed over matter moves from heaven to the earth but still subordinates matter to heaven in the last instance. The goal of nature becomes the production of gold. As a result, nature itself becomes a machine for producing infinite riches, analogous to the capacity of money to reproduce itself. Yet the most important paradox is that, by employing technical language to ground the capacity of self-generating of metals, metaphysical instrumentalism produces a convergence between Andean and Spanish beliefs. Solórzano's incursion in the self-generating capacity of metals invites an examination of the relations between Spanish instrumental metaphysics and Amerindian beliefs regarding the capacity of metals to grow inside the mines.

There was a convergence of indigenous and Spanish ideas about metal. Since the life of metals is not an exclusive patrimony of the Spaniards but also an essential part of indigenous beliefs both in the sixteenth century and in the present, it is useful to discuss the different views on this subject offered by contemporary academic production. An analysis of the way these works conceptualize Andean beliefs on the self-reproductive capacity of metals will show how such beliefs become readable retrospectively through their mediation by metaphysical instrumentalism. Andean vitalism has been an object of attention of disciplines such as anthropology and history of science on the one hand, and decolonial theories such as Walter Mignolo's local thinking on the other. I argue that decolonial theory tends to romanticize Andean vitalism by conceiving it as some pure exteriority or alterity that resists Eurocentric appropriation. My understanding is that this theoretical posture runs the risk of reifying indigenous beliefs under the form of a thing-in-itself, a reserve of locality that dwells beyond the limits of Eurocentric thinking.

The problem with such a critique of Eurocentrism is that it does not go far enough, since critics of Eurocentrism refuse to mediate these indigenous beliefs with the actual forces of technological and capitalist expansion. To avoid the hypostatization of the life of metals into a reserve of pure resistance, it is useful to bring to the discussion the works of contemporary ethno-history, history of science, and material culture that examine Andean vitalism within different contexts. These contexts go from the confluence of Spanish and indigenous beliefs to Andean metalworking and the ritual uses of money. By examining these contemporary academic works, it is possible to see how Andean vitalism is read retroactively through metaphysical instrumentalism, which was exhausted in Solórzano's work.

ANDEAN SELF-GENERATION OF METALS

Many scholars have investigated the pre-Hispanic cultures of Peru's beliefs in the germinal, vital, or genetic power immanent in metals. While some scholars emphasize absolute exteriority and the disjunction between Andean and European beliefs, others see convergence and synthesis. Most of them point out that sacredness was not necessarily reflected in its external form but embedded in the material thing itself. Many scholars thus end up opposing Eurocentric technological expansion to the Andean belief in the living power of material beings. The Argentine philosopher Rodolfo Kusch opposes Western metaphysical and technological thinking to the "seminal" thinking of the indigenous peoples of the Andes; likewise, Walter Mignolo opposes Acosta's Western rationality to the Aymara vitalist universe. Although he interprets Andean cultural concepts such as *sallqa* (nature) and *huaca* (deities) as parts of cosmologies structured by colonial power relations, he opposes notions such as the "form of life" as a distinctive singularity of the Andean earth to José Acosta's rational and Aristotelian universe (Mignolo, "Commentary," 494).

I understand that it is not legitimate to romanticize the belief in the self-generating power of metals by elevating it to the status of some alterity that is exterior to the imperial metaphysics and its instrumentalist reduction of nature to raw matter. On the contrary, the relation between living metals and objects of manipulation is a dialectical one: the conflict is irreducible, yet the opposites are inseparable. For instance, the idea of a life proper to matter and its pan-animist cosmology are not an exclusive patrimony of Amerindian cultures that turn any opposition between Spanish and indigenous beliefs into an abstract dichotomy. As a matter of fact, during the Renaissance, authors such as Marsilio Ficino and Girolamo Cardano used the stoic notion of force (*pneuma*), with vitalistic and empiricist implications, which offers an image of nature as an organic whole. Spanish writers such as Acosta, Solórzano, and Barba also asserted that metals had some semblance of life and a certain power to grow in the earth.

In his seminal work, Jean Berthelot pointed out the equivalence between minerals and the fruits of the earth. The term *mama* meant the principle or origin that generated silver, and it had connotations of singularity and reproductive capacity. For this reason, Andeans venerated golden *mamas* as huacas bridging the division between agriculture and mining. Comparing Hispanic beliefs such as Barba's with the description of indigenous beliefs present in the texts of Cristóbal de Albornoz and Bernabé Cobo, Carmen Salazar-Soler sees a certain convergence between the two cultures. Albornoz describes the *mamas* as huacas:

> There are other kinds of huacas that they revere and serve with much care, which are from the first fruits of what they harvested from some land that was not sown. They choose the most beautiful fruit and they guard it, and in its likeness they made others from different stones or from gold or silver, such as an ear of corn or a potato, and they call them *mamacara* and *mamapapa*: and they do the same with the other fruits and in this form of all the minerals of gold or silver or mercury that they have long ago discovered. They have chosen the most beautiful stones from the metals, and they have guarded and continue to guard them, and they cut off pieces from them, calling them the mothers of the mines. (Albornoz, 165)

Another author who portrays the pre-Hispanic belief in the *mamas* as huacas is Bernabé Cobo, who writes that "Those who went to the mines worshipped the hills there and the mines themselves, which they call *Coya*, beseeching them to give them their metal, and in order to get what they requested they held a night-long vigil, drinking and dancing in reverence of the aforementioned hills. Furthermore, they worshipped the metals, which they call *mama*, and the stones from these metals, called *corpas*, and kissed them and conducted ceremonies with them" (Cobo, 166). Following Berthelot, Salazar-Soler explains that the term *mama* had two meanings. On the one hand, it meant "image" or "imitation" of the first mythic product, which is *saramama*, mother of corn, and *corimama*, mother of gold. On the other hand, *mama* meant that the germ is the producer that creates silver called *mama sara*, mother maize, or *mama cori*, mother gold (Salazar-Soler, "Encuentro," 245). In both cases, the fruit, *mama*, is simultaneously originary (first, initial), original (singular, extraordinary), and originator (fertile, reproductive) (245).[13]

Salazar-Soler draws important consequences from the presence of what she calls this "embryologist" theory of metals in the Hispanic and pre-Hispanic world. First, she considers such convergence a proof that the beliefs in the living power of the *mamas* did not constitute an obstacle to technical progress but, on the contrary, allowed the incorporation of the indigenous population into the modern world of the mines ("Encuentro," 248). Second, confronting the "occidental ideas" of metaphysics imbued with pre-Hispanic ideas makes it possible for us to question the dichotomy between rational and irrational knowledge (248). Third, she asserts that the Andean mines were a "melting pot" that allowed for the mediation and exchange of ideas about matter and mining (249). Finally, she argues that the concept of huaca was inseparable from mineralogical, topographical, and geological knowledge implicit in its morphological description (Salazar-Soler, "Las huacas," 254). Therefore, the huaca had a religious dimension that was not "incompatible with technical knowledge" (254; my translation). This dimension was immediately grasped by the Spaniards who used this kind of knowledge to find mines and treasures, propitiating their exploration, surveying, and exploitation (255). Where Mignolo sees ideals of harmony and plenitude that obstructs the reduction of earth as a living organism to a natural resource, Salazar-Soler sees the possibility of integration for the indigenous peoples into the modern world of the mines. Kendall Brown also argues that believing in cosmological harmony and the organic living unity of nature helped indigenous miners to cope with the disruptive nature of mining since colonial times.

As a matter of fact, attributing to metals the capacity to proliferate and regenerate in a maternal earth can also enhance the fantasy of the endless availability of metals. Moreover, the life of metals coexists with refining techniques such as amalgamation, which treats precious metals as an amorphous material capable of acquiring the universal form of pure exchange value. The link between Andean beliefs in the life of metals and the different meanings of money can be seen in the work of Olivia Harris, where she discusses money, its meaning, its sources, and northern Potosí. Scholars who want to present Andean beliefs as irreducible to Western frames can end up retroactively reading Andean beliefs against the background of instrumentalist metaphysics. These acts of taking for granted metaphysical presuppositions show how, far from being an obstacle to technocapitalism, Andean beliefs are a local condition of possibility. Andean beliefs about money appear to be fully fetishist since they attribute to metals a life of their own, which then endows things and artifacts, the products of human hands, with a life of their own. To endow metals and artifacts with lives of their own parallels endowing money with the capacity to create more money.

In a text titled "The Sources and Meaning of Money: Beyond the Market Paradigm in an *Ayllu* of Northern Potosí," Olivia Harris examines the Andean "ontological conceptualizations of money" among some rural Laymi men in order to explain certain remnants from colonial times that help the economy to function in the present. Although the text is centered on the uses of money among Laymi people, it also focuses on beliefs that were central to mining, such as the belief in the self-generating power of metals. One central claim of Harris's work is that the signifying functions of money in Andean culture cannot be reduced to either the liberal or the Romantic discourses on money. The irreducible element is the Andean belief in the fertility of metals guarded by the *supay* or devils of the mine. Her work shows how the two main factors that structured the meaning of money in the Andean region are tribute payment and the well-known Andean proximity between mining and agriculture. Although Europeans introduced money as a means of exchange, exchange "does not exhaust, or even dominate, its meanings" (Harris, "Money," 305). Although she attempts to demonstrate that the Andean meaning of money is irreducible to exchange value, Harris also points out that its "fetishistic quality derived in good measure from the way it appears to produce wealth—and thus the power to acquire everything—in and of itself" (305).

In the twentieth century, the Laymi use money for purchasing goods and for "festive expenditure" (coca leaf, cigarettes, rum, and so on) (306). Harris explains that these uses of money cannot be relegated to the individual sphere, and this becomes evident in the "festive expenses," "whose purpose is to acquire the necessary materials to communicate with the divine sources of power and so to reproduce the *ayllu*" (306). Harris declares that "money as such seems to have a largely neutral value" (307). The use of money in the ayllu economy is limited to "practical" uses that cannot be reduced to exchange or to any other cultural or ideological reasons.[14] In other words, money appears to be an instrument, a neutral device for exchanging goods. Harris also explains how "money is therefore not seen per se as alien or threatening" (308). When looking for examples of the neutrality of money, Harris turns her gaze upon the "ceremonial gifts as *arkhu*" in community feasts (307). The first thing that I want to emphasize is that such a notion of money as a practical neutral device is inseparable from an instrumentalist metaphysics where money is seen as a means to an end that remains subordinated to the controlling subject in the same way that the efficient cause remains external to the passive matter imposing forms on it to produce a final object (I will come back to this point).

Despite claiming that exchange does not exhaust the Andean meaning of money, Harris proceeds to explain that Indians justify profit by appealing to the well-deserved compensation for traveling costs. In other words, the costs of transportation, travel, and circulation justify the appropriation of profit as surplus value. Harris quickly adds that "in reality, profit is objectively visible only when a circuit of exchange begins and ends with the money form; money is unique in that it alone both initiates and completes a circuit of exchange" (309). She explains that surplus can be calculated by converting what is exchanged by barter into monetary equivalents. Nevertheless, she denies that the Laymi engage in this kind of calculation, explaining that they refer to profit as money "giving birth" (*wawachi*) to more money. Giving birth to money means performing a "valuable social service" (309). Profit is the sign of the fertility of money understood not merely as exchange but also as planting in order to fulfill

THE EXHAUSTION OF NATURAL SUBORDINATION | 335

real needs (309). The fertility of money is a concept associated with circulation, the creation of riches, and the fulfillment of social needs.

Yet fertility is not only associated with profit but also with debt. Harris illustrates the relation between fertility and debt by explaining that the Aymara in the ayllus of northern Potosí use the word "manure" (*wanu*) to refer to debt. Based on Bertonio's translation of "debt" and "loan" as *wanu*, Harris concludes that "in the metaphorical association of debt and credit with manure we can detect a vision of circulation itself—or rather delayed circulation—as a fertilizing force" (309). The idea of metals growing like plants inside the mine is associated with a "delayed circulation" as being born out of credit, making evident the fact that, in Harris's words, "for them circulation and production are part of a single process and that we should be wary of separating them" (310). Since money flows outside the ayllu, the latter is not a self-contained unit but is "embedded in wider circuits of reproduction" (310). It is important to pay special attention to Harris's analysis of metals as growing, quasi-alive beings, because the idea contains an analogy between this self-reproductive power and the totality of economic relations as a whole. In other words, the circulation of money producing money is transposed to nature itself.

Since Harris's main interest is the examination of the meaning of money, she insists that the "primary use" of money is "purchasing ritual materials" in order to counteract the money that flows outside the ayllu, maintaining the "productive and reproductive" cycle alive and working (310). The productive and reproductive side of money is apparently irreducible to exchange because the Laymi think of sacrifices offered to the mine and the earth in terms of "feeding." In other words, "humans feed the sacred beings so that they will in turn provide food for human society" (310). To illustrate the "primary" use of money, which is associated to mining and rituals performed in order to enhance the fertility of growing metals, Harris analyzes the celebration of the carnival de Oruro and Llallagua with its commemoration of the Virgin of the Mineshaft, where performers wear masks as devils.

These devils represent the *supay* or Tio who rules the underground world and is the owner of the mines. The abundance or scarcity of metals depends on the *supay*'s capricious and contingent will, which also decides on the life or death of the miners. All mines have images of the Tio, and miners offer him cigarettes, alcohol, and sacrifices. August is the "month of devils," when minerals buried on the underground "rise to the surface," revealing hidden treasures. Harris adds that "people say that to possess the treasure one must make a human sacrifice to the devils" (311). The association between mining and money has a ritual priority in the month of August, since it is a part of a cycle where all the material goods increase (315). Harris observes that the identification between money, mining, and fertility corresponds to the role of the mine of Potosí in promoting silver over gold in the European monetary system. I understand that an important consequence we can draw is that the fertility of the mines was inseparable from the *support of worldwide capitalism in the sixteenth century, making possible the emergence of what Kamen sees as an imperial network of credit, commerce, and production of profit.* The process of conversion of raw metal into money took place after the establishment of the mint, and after the beginning of mining in Potosí.

Harris's point about the irreducible character of this primary use of money should be understood against the background of Michael Taussig's *The Devil and Commodity Fetishism in South America*. In this work, Taussig opposes "use-value," based on gift exchange and reciprocity in indigenous culture, to capitalist "exchange-value," where profit is identified with money producing money. The libations of the Bolivian miners are attempts to preserve a certain sense of harmony, preserving use value against exchange value. The pact with the devil "exploits the antichrist to redeem the mode of production of use values and to wrest it from the alienation of means from ends under capitalism" (96). As a result, Taussig sees both exchange value and wage labor as an "unnatural" distortion of the "natural economy" of use values and ayllu reciprocity. Harris explicitly rejects this view, insisting on the irreducibility of "feeding" to exchange. In Harris's words, "Feeding is the proper behavior of human beings: you expect that other human beings will feed you; offering food is not an act of exchange. Humans must feed the devils to assuage their voracious hunger; but this does not guarantee these humans food or prosperity" (Harris, 325). In other words, the Laymi believe that by feeding the owner of the mine, they can calm the voracious hunger for surplus value, but without the guarantee of prosperity provided by the external measure of exchange value. It is possible to interpret Harris's argument against Taussig's in terms of an important aspect of capitalist ideology. As a matter of fact, her argument about the irreducibility of the logic of the gift implicit in feeding to the logic of exchange appears to be associated with the lack of any preexisting guarantee that the real needs of the ayllu will be fulfilled.

Going back to the association between money, mining, and fertility, Harris illustrates how the unity of the spheres of production and circulation is similar to the unity between the mining and agriculture, the Tio and the fertility of the *pachamama*. The best example of such unity is the belief in the *mamas* as living metals already registered in the sixteenth century by the extirpator of idolatry, Cristóbal de Albornoz. Harris analyzes the specific ways in which the Laymi explicitly identify money and fertility in their libations, which are directed to three forms of abundance: *phaqhara, llallawa,* and *phaxïma*. Phaqhara, which means "flower," is used to increase the flocks and livestock. Llallawa, the name of strange-shaped tubers or maize cobs, means increasing the harvest. Phaxsïma, which means "like the moon," ensures the fertility of the mines, "which not only produce money in the sense of minerals but also the source of the markets, of urban consumers cut off from the process of food production, and monetary wealth" (315). Libations ensure enough currency to buy things to reproduce the wealth of the ayllu (315). Like potatoes or maize, money is the result of the fertility of the earth that makes possible the reproduction of the flocks. As a result, the purpose of offering food is to enter into communication with the source "the shadowy domain of the devils or *saxra*" (315). As one can observe, all these uses consider nature as if it produced money by itself.

One of the most valuable assets of this seminal article is the way in which Harris introduces further dualisms that add more layers to the idea of money as emerging out of fertility. After identifying circulation with fertility, she introduces two new distinctions, one between the raw and the minted and the other between past and present origins of money. According to the first division between the raw and the minted, raw money metal comes from the underground realm of devils while minted money is guaranteed by the state, which stamps it, transforming it into coins, and regulates its circulation. In other words, the source of metals, which is the earth (*pachamama*) is associated with the underground realm of devils, while the state apparatuses that secure the functioning of the social order by imposing taxes are associated with God. God's sphere maintains order and good government, while the devil's sphere is chaotic and unpredictable since it is the source of both abundance and misfortune (316).

Although Indians use money for economic exchanges, they may also consider it a threat to social bonds, "in Laymi concepts today, money relates primarily to the core values of prosperity and reproduction on the one hand and the state on the other" (316). Money is a product of both nature and the law, but "these two sources are antithetical to each other, money itself has a dual, ambivalent character" (316). Harris makes an interesting analogy between Andean thought and Western economic thought by reminding the reader that there are two theories concerning the origin of money. She explains that, according to the state theory, money was created and guaranteed by a political system and functioned as a sign of the mutual obligations of the state and its citizens; according to the commodity theory, the functions of money depend on its intrinsic value. The duality between the raw, associated with the underground world of chaotic forces, and the minted, associated with God as law and order, is present also in the ritual practices. The libations offered to the devils are made of raw metal, fragments of gold, silver leaf, and other minerals, while saints do not receive raw metal but only worked metal (316). As hinted at above, if raw money comes from the underground ruled by the devil, coins are like mirrors stamped by the state and illuminated by the sun and God (318). While raw metals are offered to the Devil of the mines, coins are offered to God and the saints, in the same way as the ayllus offer tribute to the state. Harris interprets these dualisms not as rigid dichotomies but as being highly ambivalent, as in the cultural concept of "I," according to which opposites are complementary and independent yet essential parts of a harmonic whole.[15]

The second distinction introduced by Harris is the one between past and present money. The Incas did not use metals as money, and the cultural importance of gold and silver (the "sweat of the sun and tears of the moon") lies in the Inca cult. The Inca controlled the production of those same metals that were used by the mining communities to pay tribute to the state. In Europe, metallurgy was employed in warfare, transportation, and agriculture, while in the New World societies "metals *served mainly symbolic functions, communicating power, status, and religious beliefs*" (319). Harris quotes Heather Lechtman, who is an important reference in the study of pre-Hispanic Andean mining and metal working. A brief detour through this important vitalist aspect of Andean metalworking in Lechtman's work and in Carolyn Dean's study of the roles of stones in Inca culture will help to clarify Harris's link between the notion of the fertility of money and past beliefs. In "Andean Value Systems and the Development of Prehistoric Metallurgy," Lechtman explains that the basis of Andean metalworking, which she calls "technology of essence," is the incorporation of the essential ingredient, the gold or silver, "into the very body of the object" (30). Letchman points out that color was the most important property of precious metals, and she ties the metalworking techniques used by the Andeans with the idea of the "essence" of the object as being invisible. The essence, the object's surface appearance, "must also be inside it" (30). For an object to be an object, it must contain "essential quality, even if the essence is only minimally present" (30). Metals have the appearance of gold and silver, concretely their colors and their reflectivity, "even if they incorporated very little of these precious metals in their structure" (15). Moreover, "a large proportion of the gold and silver-*looking* objects we have from the Andes are not made of the pure metals" (15). Objects "contain relatively small amounts of gold or of silver" (15). Lechtman sees an inseparable relation between the Inca ideology and the expression of power through colors and this technology of essence where power remains invisible. Lechtman relies heavily on both Regina Harrison and Gerald Taylor, ethnohistorians of the Andes, who illuminated the use of the Quechua term *camay*, the act of giving life spirit into an animate object. In Lechtman's words: "*Camay* refers to the domain of the material and the concrete, to the domain of people and of natural and cultural objects they fashion and use. Perhaps the notion of 'technological essence'—of the visually apprehended aspect of an object as revealing its inner structure—are related to these fundamental Andean concepts of the divine animation of all material things" (33).[16]

From my perspective, the most interesting aspect of Lechtman's vast work is the way she relates Andean vitalist visions with concrete technological styles. In doing so, she ties the notion of "technological essence" as technique for dividing between the interior and the surface of the object with the Andean notion of camay as the act of giving spirit to an object, and this is useful for understanding the sacred character of matter itself. The most recent systematic study of this idea in the mineral world is Carolyn Dean's *A Culture of Stone*, which focuses on the role of rocks in Inca material culture. Contemporary attempts to categorize the "essence" of mineral things are blinded by what the things "look like" (4). Like Lechtman, Dean introduces a difference between essence and appearance that is central to her approach, since "from an Inca perspective, what the eye perceives (a thing's surface appearance) was important, but nearly always less significant than what the mind conceives (a thing's substance or essence)" (4). The distinction between appearance and essence parallels the distinction between the processual character of working with the substance and the emphasis on the end product. In Dean's words, "as a corollary, process—an emphasis on working with the substance of a thing—was often valued over the end product, its 'finished' appearance" (4). The Incas thought that "sacredness was embedded in the material of the thing, rather than in its form" (4). Essence could embody a whole array of hosts and it "was not necessarily reflected in its external form" (4). The shape or visible form is different from the essence, which "was transubstantial, and so its significance was independent of form" (5). While appearances are subject to generation and corruption, the essence is stable and permanent (6). Stones are the expression of such stability because they are "life immobilized" (6). More important, rocks are not just endowed with life force but are "potentially animate, transmutable, powerful, and sentient" (6). Incas did not worship rocks as rocks, since they "recognized as something beyond mineral composites, rocks that were revered or had symbolic import because they were at once rocks and more than rocks" (5). To summarize this position, Andean matter has a substantiality of its own, which is independent of the visible form.

Combining the insights of Harris, Lechtman, and Dean, it is possible to assert that within the Andean framework, the mineral world has an intrinsic value, and metals have a substantial essence beyond their appearance of immobilized life, which is fertility as a "potentially animate" source of all value. The "technology of essence" makes it possible to reveal the inner camay as an essence that contains more than meets the eye. Nevertheless, the distinction between essence and appearance is common to all the metaphysical traditions from Plato to Kant going through Aquinas. As a matter of fact, Western metaphysics has been blinded by a fetishistic fascination with the content hidden behind the form, the substance behind the veil of appearances. Both Lecthman and Dean invert the hierarchy between content and appearance by identifying form with changeable and mutable appearance and matter with substance, which subsists behind the changeable qualities of the mineral world. What is undeniable is the fact that in order to read the specificity of Andean beliefs these authors invoke a philosophical vocabulary of

Western metaphysics to dissociate them from it. Briefly put, they invert Platonic metaphysics while conserving its distinctions.

They privilege matter and substantial content over form and appearances, but they still think that indigenous peoples believe in something beyond appearances. Nevertheless, there is more proximity between this procedure and Aquinas's than they would be ready to accept. It is an inverted Thomism that still strips all material entities of their determination by means of an abstraction. It is an abstraction because it supposes that there is something beyond or behind determinations, which are understood as appearances. This matter is precisely the prime matter, the ultimate identical matter behind all appearances or concrete empirical determinations. This idea inverts Thomism by dispossessing it of its substantiality in order to attribute it to matter. Behind the veil of appearances there is a full substance identical to itself, free of conflicts and contradictions, permanently present.

This idea may resonate with other inversions that took place in Western philosophy such as in the case of Baruch Spinoza who identifies God and substance. I am not interested in attributing a philosophical position to the Andean people, nor am I in a position to challenge the accuracy of Lechtman's and Dean's readings. My aim is more modest, and it consists in highlighting how Andean animism becomes readable only as a retroactive effect of Aquinas's metaphysical instrumentalism itself, since, in order to find some specificity these authors have to appeal to a philosophical vocabulary. Andean animism is the inversion of many Western presuppositions only because it is the result of a colonial context. The reading of the self-generation of the mineral world offered by Lechtman and Dean allows us to problematize the reading of Walter Mignolo, who reads it in terms of a locality irreducible to the categories created by Eurocentrism and capitalist and technological expansion. Since Lechtman's and Dean's interpretations of Andean culture implicitly interpret the living power of the mineral world against the background of the distinction between matter and form, it is possible to affirm that this vitalism is readable retroactively through the inversions of instrumentalist metaphyscis.

Returning to Harris's texts, she associates the notion of past money with the self-generation of metals understood as an immanent essence beyond the world of appearances. Past money is the value generated in the underground world, while present money is the money that circulates in the surface and is stamped by the state. In order to illustrate this division, she introduces the difference between past and present as embedded in the cultural concepts of *illa*, associated with treasure, and *chullpa*, associated with the monumental tombs of pre-Inca rulers. Illa are amulets associated with secrecy, since people outside the social group are forbidden to see these objects. The translations of *illa* in some Quechua dictionaries make explicit the secrecy that surrounds the illa. Harris quotes Diego González Holguín who, in the *Vocabulario* of 1608, defines *illa* as "whatever is very old and hidden away." The word *chullpa* refers to the monumental tombs of the rulers of the pre-Inca population. Chullpa money, like the illa (treasure), belongs to the underground dark world of the dead and devils (321). This buried treasure is also connected to the topic of the Incan gold that went underground with the murder of the last Inca, and which will return to the surface with the return of the Inca, who will then subvert the present order of things. The Andean notion of illa is directly associated with the hidden self-production of life, a foreclosed and forgotten source of value, whose return threatens to invert the order of things.

Harris also notes that chullpa money, then, is also a primary source of money, but it does not refer to coins as currency but, rather, to the intrinsic value of the money stuff itself, "treasure." Moreover, chullpa money is directly linked to the veins hidden in the dark underground of the earth. Finally, there is a connection between chullpa money and the "buried treasure" guarded by the devils that return to the surface in the month of August. The "generous aid of the Tio," the "unpredictable source of riches" makes possible the extraction of the "intrinsic value" of money hidden in the underground (321). Harris distinguishes present-day money from past money: "the way that Laymi distinguish present-day money from ancient money, then, suggests not a historical comparison between the present and the past as a means of conceptualizing the ontological origins of money" (321). In other words, in Andean thought, the "ontological origins of money" follow a sequence: there is a passage from the most ancient past underground treasure guarded by the devil (chullpa, illa, raw metals), to the money stamped by the state, making possible the reproduction of the ayllu and the fulfillment of the needs of the whole.

Once again, Harris elegantly ties the series of dualisms—raw versus minted, devil versus God, chullpa money versus Inca money, ancient versus more ancient past—back to the fertility of metals, by means of the notion of *yanantin* as complementary opposites that contribute to the harmonic functioning of the whole: "Both it seems are necessary to ensure the functioning of the economy" (321). The article ends saying: "Chullpa money, Inca money: the first is associated with the fertility of metal beneath the earth's surface, while the second, engraved with the 'heat of the prince' and the insignia of the state, constitutes the mysterious sacred source of actual currency. Andean cultures harness the powers of the past in this way: remnants of a bygone era are put to work for the living and for the reproduction of the world" (322).

According to Harris, the dichotomies between the raw and the minted, the infernal underground world (where metals reproduce themselves) and the surface (where God and the state guarantee the validity of money) work as opposites that keep up the functioning and the reproduction of the harmonic whole. Yet, these opposites are not symmetrical because the self-generation of metals is asymmetrically subordinated to the production of metals and to the state. Let us recall that in Solórzano, the Crown was the owner of the subsoil and it had rights to a fifth of its fruits. Besides, during many centuries the relation between the ayllus and the state was mediated by tribute. Nowadays, when the vice president of Bolivia, Álvaro García Linera distinguishes between extractivist and non-extractivist capitalism to defend the ownership of the subsoil by the state, he argues that the surplus generated by the subsoil can be used in order to be redistributed to the ayllus. Briefly said, the life of metals is compatible with the state's ownership as well as with technological extractivism. The living character of metals is not an obstacle but a condition of possibility of its technological and capitalist manipulation.

In bringing Harris into the discussion about the self-generation of metals, my purpose is to distance this idea from that of a pure exteriority that resists Eurocentric appropriation. Harris's reading shows how the self-generation of metals participates fully in the circulation of money and sustains the economy as a whole. The problem is that Harris forecloses the generation of surplus value (calculation of benefits) by introducing the notion of giving birth to money, which appears to be subordinated to the satisfaction of needs and providing services. She dissociates the reproduction of the whole from the search for surplus value by restricting the self-generation of metals to the satisfaction of needs. Nevertheless, it is necessary to keep in mind that capitalist ideology justifies itself by asserting that generating money out of money is the most efficient way to satisfy needs, when in reality the satisfaction of needs, the creation of use value, is strictly subordinated to the exchange value and the generation of profit. Harris associates the underground self-generation of metals to the debt of living beings towards the earth and characterizes this debt in terms of a delayed circulation.

It is necessary to take an extra step and realize that this is also the way capitalism works. Capitalism also thrives for future credit because it only produces use value on the promise that it will become value to come, more value in the form of profit, surplus value. The sphere of production is retroactively valued by the sphere of circulation and the generation of more value, which valorizes production itself. This value is always promised and delayed, and it is only retroactively actualized because it depends on a leap of faith, the promise of obtaining more value. Despite the attempts to conceptualize money in instrumental terms, attributing to it the function of feeding the earth and satisfying human needs, the generation of metal becomes an end in itself.

Now, if circulation, production, and reproduction are intertwined cycles of a process where money produces more money, it is possible to infer that Harris's description of the cultural meaning of money for the Andeans appears to be formally subsumed within capitalism under the form of money as measure of everything else. If, in the underground world of the mines, money is an entity endowed with a life of its own that also reproduces itself, then nature itself *functions like capital*, where investors have to sacrifice some money to get more money. We also saw how Harris also separated her view from Taussig's by opposing feeding to exchange and stressing the role of the Tio as an unpredictable source of riches. The argument is that, when offering a libation, receiving something is return is not guaranteed as in the act of commodity exchange. Nevertheless, there is a parallel with the act of investing money in a commodity with use value in order to transform it into more value. In commodity exchange, there is no guarantee that buying or producing a commodity with use value will produce surplus value at the end of the process. In the three volumes of *Capital*, the three processes of production, circulation, and reproduction are retroactively signified by a certain leap of faith in the final production of surplus value. The ultimate capitalist fantasy is that of the "speculative" self-generating substance of wealth that fulfills the satisfaction of needs in the most rational and efficient way. In reality, capital does not engender itself but exploits the "living labor" of the workers by producing surplus value. Exploitation does not take place exclusively in the factory, since capital is capable of valorizing all spheres of social life. The goal-oriented stance of the primordial use of money is subsumed in the self-propelling circulation of capital, resignifiying the instrumental role of money itself. As a result, Andean animism becomes retroactively readable through capitalist ideology.

CONCLUSION

The exhaustion of metaphysical instrumentalism in Solórzano was the result of a long history of failed attempts to find a coherent solution to the impasses already present in Vitoria's efforts to ground the Spanish dominion. The impasses in Vitoria result from grounding the imperial enterprise in natural law and reason. He considers reason to be the only way to provide accountability for the Spanish Empire. Therefore, he turns to Aquinas's metaphysics in order to find an unshakable ground for the empire. He proposes the right to trade, commerce, and communication as the first valid title of the Spanish Empire. As a matter of fact, the importance of this title is such that commerce and taxation also appear in Vitoria's conclusion where he unequivocally states that business, extraction of metals, and tribute should not cease even if the Spaniards lack valid titles for justifying the conquest. If the Indians blocked commerce, the Spaniards could consider this an injury that could then be avenged through war. The right to declare war is a prerogative of the civil power, which has four causes and is grounded in the principle of the natural subordination of matter to form and means to an end. The final cause, which is the most important cause, is the common good. The efficient cause is God, who is the author of the divine design and the natural order. The formal cause is the force of law, and the material cause is the commonwealth. *Imperium*, the capacity of the sovereign power to command, is solidly grounded in metaphysical instrumentalism, which derives its power of persuasion from the transposition of technique to nature. Political mastery is like artificial mastery. Natural subordination is modeled on artificial subordination. Ultimately, God stands in relation to creation in the same way that the artisan stands in relation to the product of his art. Aquinas conceives material nature as a manipulable stock because it presupposes instrumental manipulation. This inner contradiction of conflating the natural and the artificial in order to enforce hierarchical subordination would produce irreducible aporias that would determine the future discourse on Spanish dominion.

The basis of metaphysical instrumentalism, and of Western metaphysics, is thus a metaphysics of handiwork, of making by hand, that is endowed with a life of its own by the fetishistic transference of the properties of the craftsman to God. According to this metaphysics, the agent possesses an idea that is then imprinted on available material in order to produce a final product that preexists in the agent's mind. In their empirical activity, imperial subjects presuppose material nature in general and precious metals in particular as available manipulable stock. Empirical activity is possible because metaphysical instrumentalism presupposes a world given in advance.

For this reason, I have argued that metaphysical instrumentalism is ideological, since in it there is a split between knowledge and practice that does not mask reality but, rather, structures reality in advance. The imperial subject thinks he is guided by a higher end, while in practice the same subject acts as if the technical means were ends in themselves. The divorce between theory and practice, the irreducible contradiction between the two, is redoubled in theory itself: metaphysics borrows its teleological and hylomorphic schema from technique only to subordinate manufacture to a superior origin and end. Now the aim of this procedure was, from the very beginning, to justify the means. The end was the means of the means. Although the real stake of natural right is the defense of temporal profit (the circulation of commodities and the extraction of riches), this gain has to appear as a by-product, a secondary effect. As soon as the subject realizes that the real aim is the enjoyment of the fruits of mining, the whole ideological enterprise is undermined by disbelief.

Nevertheless, before falling into disbelief or producing a crisis of sovereign foundation, the imperial ideology laid down by Vitoria starts to show its first fissures, which are visible in the debate in Valladolid in 1550–1551 between Bartolomé de las Casas and Ginés de Sepúlveda. Sepúlveda grounds imperial subjection and the Spanish dominion, that is, the enjoyment of the fruits of mining and of exploiting the natives' labor in the mines, in the principle of the natural subordination of the imperfect to the perfect. Sepúlveda represents a radicalization of Vitoria and the most perfect example of the imperial consequences of the hylomorphic doctrine: indigenous people are considered an imperfect matter that needs to be molded by the imposition of superior cultural forms. Sepúlveda is following Aquinas, for whom the sovereign commands are to the subordinate administrators what the plan of the architect or designer is to the inferior craftsmen, manual workers, and manufacturing crafts. When Sepúlveda employs the principle of the natural subordination of matter to form, he assumes this subordination of inferior crafts to superior crafts. Aquinas's instrumentalist division between rulers and ruled has strong geopolitical implications, since Aquinas himself writes in his commentary to Aristotle's *Politics* that foreigners are weak and fit to be slaves. Despite Sepúlveda's insistence that the subordination of the imperfect Indians to the perfect Spaniards is for their own benefit, what remains implied is that Indians are living tools, an available manipulable stock that can be subordinated to the ends of the empire.

Las Casas deconstructs Sepúlveda's hylomorphism by taking Aquinas's doctrine literally; since the imposition of form over matter is carried out in order to make one differentiated entity, it is impossible to apply the form of one material entity over another. The geopolitical consequences of Las Casas's interpretation of the hylomorphic doctrine are important: Indians are already a formed matter, a self-sufficient community that can repel force with force and defend their territory. Moreover, Las Casas declares that the Indian commonwealth constitutes the efficient cause of civil power, as opposed to Vitoria, who conceived the commonwealth as the material cause. As a result, Indians could reject the papal donation and Spanish rule based on their own political agency. The impasses of instrumentalism result from a fissure that was already present in Vitoria: there is an irreducible contradiction between conceiving of the Indians as imperfect matter and potential living tools and considering them perfect and autonomous communities that could resist the Spanish rule.

José de Acosta represents the most practical, utilitarian, and technical deployment of Vitoria's and Aquinas's metaphysics. He reactivates Vitoria's arguments about the right of free circulation, but he also acknowledges that the empire had a violent origin, although he believes that the empire is a political fact and that questioning it would endanger the project of evangelization. Not only does Acosta employ Aristotelian and Thomist principles of causality to explain the empirical particularities of the New World, but he also employs them to justify the Spanish dominion and Spanish enjoyment of the fruits of mining. While Sepúlveda employs the principle of the natural subordination of imperfect matter to the imperfect form to say that the Indians are natural servants, Acosta employs the same principle in order to argue that the role of metals is to be used and transformed into instruments by humans, "the inferior serving the superior." Nature is humans' handmaiden, and God placed the metals in the New World in order to favor the universal monarchy.

According to this view, metals are an imperfect matter that is subordinate to humans by many degrees. They have only an instrumental value, and they have to be subordinated to transcendent ends—in this case, the project of evangelization. Nevertheless, in Acosta there is another internal deadlock. While he considers precious metals to be imperfect matter that has to serve superior ends, he also considers money to be an instrument that is virtually all-encompassing, since it can buy any other useful thing. The impasse in Acosta's extremely pragmatic metaphysical instrumentalism is ultimately the fissure between the use value and the exchange value of metals. Once metals are considered as money (exchange value), they subordinate everything else to their power, becoming ends in themselves, triggering the perverse drive to accumulate riches for the sake of accumulating riches. And Acosta knows very well that the source of riches is the inhuman underworld of the mines of Potosí. He knows it very well, but he keeps invoking natural subordination because he acknowledges that the whole political and evangelizing enterprise depends on mining. In Acosta, it becomes even more evident that metaphysical instrumentalism is invoked to justify the control over the means. Even if the empire had an illegitimate origin, it was nevertheless a political fact, and the economic network had to be maintained for the sake of the salvation of the natives.

Both arguing for the centrality of mining and invoking providential design were part of the ideology of the circle of Francisco de Toledo's supporters and lieutenants, and this is evident in the *Anónimo de Yucay*. Francisco de Toledo's radical reforms in Peru, such as the introduction of the mita, the introduction of amgalgamation, and the *reducciones*, were inspired by Juan de Matienzo and then defended by José Luis Capoche. All these authors invoked the Indians' inferior nature, revitalizing Sepúlveda's arguments, although now in service of the forced mobilization of Indians to work in the mines in exchange for a wage. They also were aware of the political consequences of the economic crisis caused by the decline of silver production, and they did not refrain from using an apocalyptic tone in order to justify the introduction of these radical measures. By painting a gloomy picture of Potosí and thus of an empire on the edge of chaos, these Toledan apologists saw in amalgamation the only source of salvation. Since the labor of amalgamation required massive numbers of Indians (the same Indians who were declared naturally inferior by invoking Aristotelian and Thomist metaphysics), these apologists appealed to the pre-Hispanic policy of the mita in order to supply the mines and the amalgamation mills with enough workers. An essential part in the justification of this forced mobilization was that it was both expedient but also absolutely necessary, since it was demanded by the new technology. Although the Toledan reforms still were framed within the providentialist design of metaphysical instrumentalism, there was a shift in perspective where the need to administer the empire and to conserve mining practices started to substitute for the search for an ultimate and unshakable origin or ground of the empire.

Once the official ideology shifted from invoking transcendent origins and ends and started invoking expediency and some sort of economic and technical necessity, we witness the presence of two different logics, which appear intertwined yet each one irreducible to the other. On the one hand, they can be read as following a logic of power and domination, which can be characterized in Heideggerian, Foucauldian, or Frankfurt School terms. Whether we read them as signs of the history of the metaphysics that ends with enframing (Heidegger), attempts to regulate and administer a population (Foucault), or incipient signs of an Enlightened instrumental rationality (Adorno), these texts point to some need to stabilize a material chaos by imposing forms according to a design.

Stabilization is different from excess, the drive to pursue surplus-value, the self-movement of money producing more money—which is not only a political but also an economic logic. For instance, José Luis Capoche only wants to give an air of necessity to amalgamation and the mita by employing a rhetoric of catastrophe that emphasizes the possibility of economic and political crisis because of a short circuit in the circulation of money and commodities. This logic is not just one of domination but also a logic of excess and capitalist deterritorialization, the disruptive power of the cycle M-C-M´ mobilized by the never-ending production and appropriation of an excess—ever more precious metals, ever more money. This excess, which had been kept under control by Aquinas, for instance, when he stated that excessive accumulation of money is a perversion or deviation, was now the principle and the rule. Mining could only survive as its own excess, as a destructive drive, an elevation of the means to the category of ends, indifferent to the concern for the well-being of the people that was invoked in order to justify it.

These two logics, the logic of taming the excess and the logic of elevating this excess to the character of a principle, are the two logics of modernity that remain incompatible, since one cannot be translated into the other.[17] However, they are the two sides of the same coin, two irreducible aspects of imperial expansion. My thesis is that these two aspects of modernity arise both historically and dialectically from the inherent contradictions of metaphysical instrumentalism visible only in a colonial context. It is necessary to redouble the gap between the logic of domination and the logic of capitalist deterritorialization within metaphysical instrumentalism. The irreducible gap between these two logics is the result of the inner impasses of Aquinas's metaphysical instrumentalism, which generated a monster that it could not control, an instrument that subordinates all instruments to itself and therefore undermines its own metaphysical presuppositions. Money was an instrument that escaped imperial control, and the indigenous worker was its objective correlate, since the empire was absolutely dependent on money. The matter of the empire becomes a material flux beyond imperial control. The limit of metaphysical instrumentalism, its dysfunctional inner core, is a result of instrumentalism itself, which ends up being deprived of its transcendent ends.

Solórzano Pereira's *Política indiana* was the greatest apology for the Spanish Empire ever written, and it represents the exhaustion of metaphysical instrumentalism. After the long history of the failure to ground the empire in natural law, Solórzano ends up arguing that it is necessary to forget about the search for origins and instead find a solution to the practical problems that threaten the empire. Nevertheless, as the examination of the problem of the mita makes evident, the impasses of instrumentalism reach their acme because the arguments against the mita are also invoked as arguments to keep practicing it, since Solórzano reintroduces the iron necessity of the Christian end in order to exorcise the evils associated with the underground world of mining. The solution to the problem of mining was more mines and more money, on condition that the pious Catholics considered this profit to be a by-product of the transcendent end, and not an end in itself. The belief was that new mines could be found, because, among other reasons, metals reproduced themselves in the earth. This belief, shared by both Spaniards and Andeans, could be used to justify the never-ending extraction and exploitation of nature and the appropriation of money by the Crown. Once we take this into consideration, we can see how Andean vitalism, far from being a reserve of pure alterity that resists Eurocentrism, is actually the other side of instrumentalism, which becomes more apparent once metaphysical instrumentalism exhausts its possibilities and is deprived of a higher end. In other words, Andean vitalism becomes readable only retroactively through the impasses of metaphysical instrumentalism.

In this book I have argued that technology is not a neutral device that can be impartially examined through empirical observation. Instrumental manipulation is inseparable from a metaphysical frame that assigns it a principle and an end. Yet metaphysics presupposes instrumentalism because it transposes instrumental manipulation to the order of nature. In short, imperial command presupposes instrumental manipulation. Such a transposition from technique to nature engenders a circular contradiction in which the means of the end are also the means of the means themselves: the whole ideological enterprise consists of justifying the means. The instrumental means acquire a life of their own when they are elevated to the status of an end in itself. The idea of material nature as an external and passive resource able to be manipulated by an agent and the idea of nature as a harmonic totality that reproduces itself are two sides of the same ideological coin. Therefore, the self-generating power of metals is not an exterior point of resistance irreducible to Eurocentrism but, rather, the internal effect of a colonial situation shaped by metaphysical instrumentalism itself. Imperial technology ambiguously integrates the exteriority of indigenous beliefs.

Such ambiguity is visible in the twofold character of Andean animism. On the one hand, vitalism was not an obstacle to technological capitalism but, instead, the condition of possibility of the integration of indigenous beliefs to modernity. On the other hand, vitalism was also associated with an organic view of nature that is inseparable from the return of the Inca and a project of anticolonialist emancipation. The dialectical relationship between Andean vitalism and metaphysical instrumentalism makes evident that the history of technology is not the history of European achievements or the triumph of the arduous work of empirical observation or rigorous reasoning over a theological past. With the colonial structure emerges a dysfunctional kernel inherent in modernity that is both an obstacle and the condition of possibility of technological capitalism: an excess of life inseparable from the imperfection of matter. As a result, Andean animism is simultaneously a dysfunctional excess of the privation of matter that is read retroactively through the paradoxes of metaphysical instrumentalism. Correlative to privation is the excess of life with no functional value, a life that escaped colonization. Imperial metaphysical instrumentalism and its disavowal are two sides of the same coin because colonization can only be sustained by a remainder that eludes it, thereby making it failed and incomplete.

The excess of life—the drive to repeat, regenerate, return to a lost past—is not only the product of colonization, the dysfunctional excess generated by metaphysical instrumentalism, but also the sign of a revolutionary potential that antagonizes colonization. Andean vitalism emerges from the interstices between communities, as an ambivalent "both-and," a symptom of the asymmetrical relations of power inseparable from colonial antagonism. This excess of life is repetition, regeneration, and also revolution. For this reason, the decolonial perspective that tries to validate Andean animism by assigning it the role of a local knowledge that is absolutely different to Eurocentric science fails to grasp the dialectical relation between imperialism and Andean vitalism. It fails to see the double and paradoxical co-constitution according to which indigenous beliefs and practices internalize European frames and fails to see how this internalization generates an excess that is an external remainder of colonization. The decolonial option that tries to make a critique of the European bias of techno-capitalism cannot help but fall short of its aspirations because it ignores that the deterritorializing force of capitalism is its universal power. As a result, the only way to make a radical critique of colonialism that is relevant today is by unearthing the metaphysical presuppositions behind instrumentalism. Both colonial heritage and the anticolonial struggles of the present can be read as mediated by the inner impasses of imperial metaphysics. The two logics of modernity, technological domination and capitalist deterritorialization, emerge from the inconsistencies of metaphysical instrumentalism and are still at work today.

NOTES

INTRODUCTION: *IMPERIUM*, METAPHYSICAL INSTRUMENTALISM, AND POTOSÍ MINING

1. For a history of mining in Potosí, see Ignacio González Casasnovas, *Las dudas de la Corona*; P. J. Bakewell, *Miners of the Red Mountain*; Jeffrey A. Cole, *The Potosí Mita, 1573–1700*; Kendall W. Brown, *A History of Mining in Latin America*; Enrique Tandeter, *Coercion and Market*; Nicholas A. Robins, *Mercury, Mining, and Empire*.
2. See Ralph Bauer, *The Cultural Geography of Colonial American Literatures*; Jorge Cañizares-Esguerra, *Nature, Empire, and Nation*; Antonio Barrera-Osorio, *Experiencing Nature*. Cañizares-Esguerra (*Nature, Empire, and Nation*) shows how the pragmatic and empirical observation of nature practiced by Iberian imperial science was an antecedent of Anglo-Saxon empiricism traditionally associated with the name of Francis Bacon (1561–1626) and the scientific revolution in Europe. The argument here is that the scientific, technological revolution, inseparable from the emergence of the modern/colonial world, was indebted not only to Francis Bacon and John Locke but also to what is perceived as the enemy of modernity, Aquinas's Scholasticism. Against the common view of Scholasticism as a useless abstraction destined to be superseded by empiricist, utilitarian,

and pragmatic ideologies, in my view Scholastic metaphysics is oriented toward instrumental manipulation and imperial expansion. Here we study both the imperial ideology implied in natural right and the concrete methods of mining and metallurgy described in some key texts that belong to what D. A. Brading calls the Spanish "imperial school of thought" (*The First America*, 50). Imperial ideology ends up depending on the indigenous labor and the infernal underground space inside the mines, which is the exact reverse of the moral and philosophical grounds of the civilizing and evangelizing mission of the Spanish Empire.

3. See Rolena Adorno, *The Polemics of Possession in Spanish American Narrative*; Anthony Pagden, *Lords of All the World*; Anthony Pagden, *Spanish Imperialism and the Political Imagination*.

4. The importance of the School of Salamanca for modernity resides not only in its being the first to theorize about international rights and economics but also for the emergence of the "modern ontology of the subject." Martin Heidegger pays particular attention to the role of Francisco Suárez on the formation of modern metaphysics and Descartes's subjectivist turn. Along this same line, Dennis des Chene's three volumes on the influence of late Scholasticism on the dualist subjectivism of Descartes are an invaluable contribution. The presence of late Scholasticism in the first attempt to theorize the economy of imperial Spain in the sixteenth century is the focus of Elvira Vilches's *New World Gold*, where she explores the ways in which the precious metal of the Indies had such a destabilizing effect that it caused many to question the more traditional ideas about the value of money. Vilches's book analyzes "writings on money," paying particular attention to the School of Salamanca, arguing that the reflections of late Scholasticism resulted in "the reversal of centuries of Scholastic thinking in economic matters" (184).

5. See Santiago Orrego Sánchez, *La actualidad del ser en la "primera escuela" de Salamanca*; Miguel-Anxo Pena González, *De la primera a la segunda "Escuela de Salamanca"*; Cirilo Flórez Miguel, Maximiliano Hernández Marcos, and Roberto Albares Albares, eds., *La primera escuela de Salamanca (1406–1516)*; Ángel Poncela González, *La Escuela de Salamanca*; Virginia Aspe Armella and Idoya Zorroza, *Francisco de Vitoria en la Escuela de Salamanca y su proyección en Nueva España*; Miguel-Anxo Pena González, *La Escuela de Salamanca*; André Azevedo Alves and J. M. Moreira, *The Salamanca School*; Elvira Vilches, *New World Gold*.

6. A medieval philosophical doctrine, nominalism sustains that only individuals exist and universal concepts are nothing but abstract and vague fictions. It is opposed to "realism," according to which universals are really existing essences. Platonism, for instance, is a form of realism since it stipulates the existence of a world of ideas. There are forms of realist Aristotelianism (for instance, Thomism) according to which real things contain really existing essences. Michael Allen Gillespie shows how modernity is inseparable from a "nominalist revolution" that began with William of Ockham. In this sense, nominalism is the ontological presupposition of the Protestant reformation, Machiavellianism, and liberalism. Michael A. Gillespie, *The Theological Origins of Modernity*.
7. According to the doctrine of participation, things "share" different degrees of perfection. Participation implies certain distributions of qualities and determinations that introduce different degrees or hierarchies in the chain of being. See Emmanuel K. Cosmas, *Participation as the Fundamental Basis of Analogy in St. Thomas Aquinas'* Metaphysics.
8. The centrality of Scholasticism did not go unnoticed in contemporary colonial studies. In *The Fall of Natural Man*, Anthony Pagden examines the first attempts to theorize the ontological nature of the American Indians through the lenses of Thomas Aquinas. In *Lords of All the Worlds*, Pagden reconstructs the continuity between the idea of *imperium* as sovereignty and expansion in the Middle Ages and modernity through the Spanish universal monarchy. Nicolás Wey Gómez reconstructs the deep presuppositions of the theory of place in the cartographic decisions of Columbus, a gesture that is absolutely novel. In his *Tropics of Empire*, he argues that the notion of place was "a crucial cosmological concept in the intellectual tradition that witnessed the encounter between Europeans and the native people of the Americas" (Gómez, 48). I understand that the most important contribution of this monumental book is that the centrality of it refutes the modern separation between the spheres of knowledge, nature, and politics since "in a cosmos conceived as working artifact, place was a fundamental category for understanding the behavior of *all* physical bodies, including the besouled bodies of humans, beasts, and plants. For the writers considered here, concepts of place—no matter how implicitly—pervaded *and* joined seemingly remote fields of knowledge, such as physics and politics" (66). Moreover, Scholastic concepts of place anticipated European colonialism, since they were employed to discuss problems of extraterritorial rule,

because natural concepts "were thought to assign nations their unique positions in a hierarchy of polities as well as their roles in teleological history (69).

9. The "substantial form" is that by virtue of which amorphous raw stuff becomes substance, something that subsists and persists in itself. Like Platonic ideas, Thomist forms are preexistent ideas that enjoy certain superiority with respect to matter. Yet, unlike Platonic forms, Aristotelian-Thomist forms do not dwell in a separate world beyond the material one because they are inseparable from the given composite of matter and form. The doctrine of the composition of matter and form is known as "hylomorphism" and is a central axiom of Aristotelian and Thomist physics. The central aspect of hylomorphism is to explain the unity of the physical entity as a result of the imposition of a preexisting form over an amorphous matter. Forms are exemplary patterns or ideas that determine the belonging of an amorphous stuff to one or another "universal" species or kind. Form provides ontological consistency to matter. Exemplary forms are like blueprints or prototypes that *subordinate* the amorphous passive potential of matter that is deprived of any actual existence. Like matter, "substantial form" is also a concept oriented to instrumental manipulation. Hylomorphism is also inseparable from teleology because forms are also the end of matter. Returning to the example of the statue, the form is the idea that the artist has in mind before imposing it on matter. The idea works as a preexisting end that has a mandatory power.

10. Prime matter is not bronze but the raw amorphous matter out of which *this* or *that* exists. In Aquinas's metaphysics, reality is stratified or structured according to different levels of "perfection" that range from prime matter to stones, from stones to plants, from plants to animals, from animals to humans, from human to angels, from angels to the perfect being of the Prime Mover. The more "perfect" (autonomous, self-sufficient) something is, the farther it is from the imperfect, "prime matter." Therefore, prime matter is imperfect and deprived of any consistency, and God is the ultimate perfection because he is the actual *being* that does not need anything else to exist. The concept of prime matter is the result of abstracting all the peculiar qualities of material beings and attributing them to formed matter.

11. It could be objected that this is not taking into consideration the distinction between poesis and praxis. In Book 6, Chapter 4, *Nicomachean Ethics* 1140a, Aristotle writes, "making and acting are different . . . so that the reasoned

state of capacity to act is different from the reasoned state or capacity to make" (*Basic Works*, 1025). Also, in Book 1, Chapter 1, *Nicomachean Ethics* 1094a3–6, Aristotle writes that everybody desires something good as its end, "but a certain difference is found among ends; some are activities, others are products apart from the activities that produce them" (935). While acting is an end in itself, producing the end is the product. Acting is superior than making because inferior arts are subordinated to superior arts. What this objection, which is based in a modern confusion, misses is the problem of how, in political subjection, the agent stands in respect to itself (as in the case of acting) but how it stands in relation to the other (in the case of making).

12. The agent, or statesman, can be self-sufficient with respect to itself because its acting is an end in itself. But it stands in relation to the population in the same way that the artist stands in relation to a work of art: "just as a weaver or a shipbuilder or any other artisan must have the material proper for his work . . . so the statesman or any other artisan must have the material suited to him. First among the materials required is the population" (*Nicomachean Ethics*, 1325b41–1326a8, in *Basic Works*, 1283). Moreover, the vocabulary of instrumental manipulation that is presupposed in the analogical relation in Aquinas between the dominating principle and ruled matter shows that the principle of natural subordination is intrinsically technical.

13. See, for instance, Robert C. Scharff and Val Dusek: "Most philosophers of technology would probably agree that, for good or (at least as often) for ill, Martin Heidegger's interpretation of technology, its meaning in Western history, and its role in contemporary human affairs is still the single most influential position in the field" (*Philosophy of Technology*, 247).

14. Heidegger's history of metaphysics as technological domination—also known as "history of being"—has many versions. I am following Reiner Schürmann's version, which presents this history of being as the progressive deconstruction of the founding principles of metaphysics, which ends up with the exhaustion of metaphysics embodied in technological mastery over the earth. See Reiner Schürmann, *Heidegger on Being and Acting*. See also Michael E. Zimmerman, *Heidegger's Confrontation with Modernity*, 166–90; Arthur Bradley, *Originary Technicity*, 68–93.

15. See Martin Heidegger, *Nietzsche*.

16. This does not mean that being under the spell of ideology can be reduced to mere and simple "false consciousness" about the real, material, and em-

pirical conditions. Ideology shapes empirical and material reality. For this reason, Marx considers that it is not possible to understand material conditions without first showing how ideological fictions mold reality itself. In other words, ideological mystification is a misrecognition that does not simply fall under the category of false consciousness because it shapes social reality itself.

17. See Paul M. Sweezy, *The Transition from Feudalism to Capitalism*; Immanuel Wallerstein, *The Modern World-System I*.
18. As Kojin Karatani puts it, "Money carries the right to be exchanged for commodities, but commodities do not have the right to be exchanged for money. Moreover, if a commodity fails to sell (if it cannot be exchanged for money), not only does it have no value, but it also has no use value. It is simply waste to be discarded. This is why Marx called the question of whether a commodity can be exchanged for money the "fatal leap" (*salto mortale*). Our rationalist-miser capitalist who wants to propagate money through the process money-commodity-money (M-C-M´) must venture the fatal leap: commodity-money (C-M´)" (*Structure of World History*, 95).
19. In Karatani's words, "Far from being materialistic, the capitalist system is an idealistic world based on credit. It is for precisely this reason that it always harbors the possibility of crisis (*Structure of World History*, 4).
20. See Slavoj Žižek, *The Parallax View*, 297–98.
21. See Juan Van Kessel and Dionisio Condori Cruz, *Criar la vida*.
22. When analyzing the notions of "good living" or "living in harmony" (*sumac kawasy* in Quechua and *sumac kamaña* in Aymara), Walter Mignolo writes: "'To live in harmony' has been undermined by a different philosophy of life, 'living to work and to develop,' wherein 'development' becomes the goal of life and is at the service of development. 'To live in harmony' means that the goal is *crianza* (nurturing), rather than development, and the re-generation of life, rather than the re-cycling of industrial product" (*The Darker Side of Western Modernity*, 313). Not only does this argument use the language of metaphysical instrumentalism by distinguishing between means and ends, but it also assumes a binary distinction between regeneration and recycling. However, in capitalism, self-valorization is an endless circular movement of expanding self-reproduction whose goal is not to produce more things but to regenerate value itself, exploiting the pure drive of life. Capitalism hijacks the excess of life inherent in life's regeneration by translating this excess into an urge to reproduce itself in an infinite circulation.

CHAPTER 1: GROUNDING THE EMPIRE

1. See Harry Cross, "South American Bullion Production and Export 1550–1750."
2. For the importance of the School of Salamanca, see Quentin Skinner, *Foundations*, 2:135–73; J. A. Fernández-Santamaría, *The State, War and Peace*; and Richard Tuck, *Natural Rights Theories*.
3. See also Anthony Pagden, *The Fall of Natural Man*, 60–108.
4. Pagden, in his translation of Vitoria's *Political Writings*, defines the *relectio* as a lecture intended to "re-read" a problem that was formerly raised during the year's lectures. In the University of Salamanca the *relecciones* had a "time limit of two hours" (Vitoria, *Political Writings*, 380).
5. Vitoria's *On American Indians* (*Relectio de Indis*) was printed in 1557 and circulated as a manuscript inside and outside the University of Salamanca (Adorno, 109; Pagden, *Fall*).
6. The purpose of this critique was to ground the Spanish Empire in reasons that fell strictly within the domain of natural rights and the order of reason. Such a critique is an attempt to provide a retroactive justification of the imperial enterprise in order to strengthen the empire.
7. See Brian Tierney, *The Idea of Natural Rights*, 256–72, 290–301.
8. Annabel S. Brett, *Liberty, Right, and Nature*, 123–37.
9. Skinner, *Foundations*, 2:135–66, 175–79.
10. Tuck, *Natural Rights Theories*.
11. See Monica Brito Vieira, "Mare Liberum vs. Mare Clausum."
12. See Toy-Fung Tung, "Vitoria's Ideas of Supernatural and Natural Sovereignty."
13. See Skinner, *Foundations*, 2:140, 145; Tierney, *Rights*, 293–301; Brett, *Liberty*, 135–36n40; Pagden's translation (in Vitoria, *Political Writings*, 14n32); Martti Koskenniemi, "Empire and International Law," 15–16n43; Georg Cavallar, *The Rights of Strangers*, 108–12.
14. Skinner also explores the medieval foundations of Vitoria's thinking. See Skinner, *Foundations*, 2:148–52; also the essays by H. M. Höpfl, "Scholasticism in Quentin Skinner's *Foundations*," and Annabel Brett, "Scholastic Political Thought and the Modern Concept of State."
15. For a discussion of Vitoria in the context of Schmitt, see Orlando Bentancor, "Francisco de Vitoria, Carl Schmitt, and Originary Technicity," *Politica Comun* 5 (2014).
16. I am using Vitoria's *Political Writings*.

17. I will not analyze these titles because they have already been discussed extensively by Pagden and Adorno. See Pagden, *Fall*, 57–108; Adorno, *Polemics*, 109–13.
18. See Watson, *The Digest of Justinian*, 98.
19. After analyzing the first valid title Pagden concludes that "Vitoria had left the Castilian Crown with a slender claim to *dominium jurisdictionis* in America but no property rights whatsoever" (*Spanish Imperialism*, 22). Pagden also notes that Vitoria thinks war was justified only if the Indians caused injury, and if the Indians do not voluntarily accept Christian princes, the expeditions must cease (Pagden, *Spanish Imperialism*, 22; Vitoria, *Political Writings*, 291). Based on this quotation, Pagden concludes that at the end of the day all Vitoria had was a "starkly objective claim" based on the factual presence of the Spaniards and that it was not possible to abandon the Indies without detriment to the interests of the princes (*Spanish Imperialism*, 22). Pagden is consistent with the interpretation of natural right and Scholasticism as impractical and premodern while only Anglo-Saxon calculation of benefits seems to be modern. I think that Pagden misreads Vitoria's assertion about the possibility of ceasing the expeditions, because Vitoria is not asserting his own opinion but, rather, including a possible objection: if the seven titles are inapplicable, the expedition must cease. Against the possibility of ceasing the expeditions, Vitoria categorically asserts that trade must not cease because the Indians have a surplus of things that the Spaniards lack, and therefore, the Spaniards can claim the lands that are not occupied. Let us recall that the first valid title is grounded not only in the law of nations but also in natural law and therefore derives from eternal law (Vitoria, *Political Writings*, 292). Although the Spaniards cannot claim possession over the Amerindians' bodies or things, they are entitled to claim not only the things in Amerindian lands that have not yet been explicitly possessed but also a tribute that is grounded on the right to subject the barbarians to their legal jurisdiction. In my understanding, far from anything being thin claims for *dominium jurisdictionis* and the absence of property rights, Vitoria provides a more modern defense of the Spanish Empire. Vitoria's defense of the empire is a rational one that refuses to ground the dominion in the papal power. His defense is not thin but indirect and formalist. In order to achieve the subjection of the Amerindians it is necessary to go through a series of steps that involve commodity exchange and the exploitation of the land.
20. See Aristotle, *The Basic Works*, 140.

21. According to Padgen, Vitoria relies on Lactantius's *De opificio dei* and Cicero's *De Natura Deorum* in order to counter Epicurean atomism with the Stoic theory of design (Vitoria, *Political Writings*, 5n7). My argument is that the Providentialist design is also relying on Aquinas's extreme teleological interpretations of Aristotle. See also Christophe Grellard and Aurélien Robert, eds., *Atomism in Late Medieval Philosophy and Theology*; Catherine Wilson, *Epicureanism at the Origins of Modernity*.
22. See Lactantius, *De opificio dei* 3.1–2; Lucretius, *On the Nature of Things* 5.222–27.
23. Also in Article 1, *Summa Theologica* IaIIae 95, titled "Whether it was useful for certain laws to be established by men," Aquinas writes that some "discipline" is necessary to transform virtue into "perfection": "in the same way, we see that although man is helped to secure his needs—food and clothing, for example—a degree of effort is necessary also. He has the beginnings of what he needs from nature: that is, *reason and hands*; but his needs are not fully met by nature, as they are in other animals, to whom nature has given enough in the covering and food" (*Political Writings*, 127; my emphasis). In other words, although "hands and reason" are *given* by nature, human needs are not "fully met by nature." One could argue that in Thomist theology, man is naturally destined to denaturalization.
24. See Martin Heidegger, *The Principle of Reason*, 62, 63.
25. See Pagden, "Introduction," in Vitoria, *Political Writings* (xix–xx).
26. Pagden points out that the "single unifying concern" in Vitoria's work is to preserve the civil state from the arguments employed by heretics and schismatics to introduce disruption ("Introduction," xvii).
27. This is consistent with Chapter 1, *Metaphysics* 981a28–981b6, where Aristotle writes that to know something is to know the cause (*Basic Works*, 690). Master-workers know more than manual workers because they know the causes of things (690). Aristotle emphatically claims that "thus we view them as being wiser not in virtue of being able to act, but of having the theory for themselves and knowing the causes" (690). Aquinas comments on this passage, saying, "In order to understand this [superiority of the master artisan over the manual worker] we must note that architect means chief artist, from *techne* meaning chief, and *archos* meaning art. Further, since matter exists for the sake of the form, and ought to be such as to befit the form, the shipbuilder knows the reason why the wood should be shaped in some particular way; but those who prepare the wood do not know this. And in a similar way, since the completed ship exists in order to be

used, the one who uses the ship knows why it should have some particular form; for the form should be one that befits its use" (Aquinas, *Commentary on Aristotle's* Metaphysics, 9–10). In Aquinas's commentary we can clearly observe how the subordination of the inferior to the superior craftsman is central to the instrumentalist core of Aquinas's interpretation of Aristotle's metaphysics. The subordination of carpenter to the shipbuilder and to the navigator parallels the subordination of the material means to the final form, since matter exists for the sake of its form, which is dictated by the use.

28. See Ralph McInerny, *Aquinas and Analogy*.

29. Among Heidegger's most important contributions to the genealogy of the relation between metaphysics and technology is his analysis of the way technological modernity arises from the imperial reason: "The Greek *aletheuein*, to disclose the unconcealed, which in Aristotle still permeates the essence of *techne*, is transformed into the calculating self-directing of the *ratio. This determines for the future, as a consequence of a new transformation of the essence of truth, the technological character of modern, i.e., machine, technology.* And that has its origin in the originating realm out of which the imperial emerges. *The imperial springs forth from the essence of truth as correctness in the sense of directive self-adjusting guarantee of the security of domination.* The 'talking as true' of *ratio*, of *reor*, becomes a far-reaching and anticipatory security. *Ratio* becomes counting, calculating, calculus. Ratio is a self-adjustment to what is correct" (*Parmenides*, 50; my emphasis).

30. Another motivation for Vitoria's use of these metaphors could be his desire to reject the pacificism of Juan Luis Vives and the Erasmists, who follow the Augustinian tradition of condemning war as a necessary evil. This is not surprising, given the "military ethos" that pervades Spain's monarchical ideology. Juan Luis Vives questioned this ethos in *De Concordia et Discordia* (1529), which was dedicated to Charles V. Vives's imperial pacifism demonstrates that it was not a given that the art of war could be the only structuring metaphor for explaining the natural subordination of positive law to divine law. Vives believed that the military ethos belied an unfettered appetite for power and world domination (*libido dominandi*). Augustine had judged that the desire for domination and the search for military glory was a consequence of man's fall from grace. Vives further claimed that even intellectual and spiritual life was suffused with this spirit of conflict, which was visible in oratory competitions and Scholastic "disputatio."

31. See Pagden, *Fall*, 91–92.
32. See Nicolás Monardes, Modesto Bargalló, and Francisco Guerra, *Diálogo del hierro y de sus grandezas*.
33. It is impossible not to think about Ginés de Sepúlveda's arguments about the servile nature of the Amerindians. Their lack of arts and crafts was, for Sepúlveda, an unequivocal sign of a deficient human nature.
34. Adorno follows Henry Wagner and Rand Parish in suggesting that Vitoria's uncertainty was "colored by the pressure put on him, and that his last title, about which he felt uncertain, was elaborated to mollify the emperor" (112). Vitoria would accept the application of the principle of natural slavery "only if the good of the Indians, not the gain of the Spaniards" (112) were the object. Adorno also reminds the reader that, in his last dissertation, Vitoria concluded that the sole just cause for war was to repel injuries received and that the offenses suffered by the Indians had to be serious in order to provide a sufficient reason for returning the injury (112). The conclusions reached by Vitoria may have impacted Emperor Charles V's decision to ask the professor-priests (the *maestros religiosos*) to remain silent on the questions of dominion of the Indies (Adorno, 113).
35. These contradictions remain abstract and speculative (see Chapters 1 and 2 of this book) but become visible in the material dependence of the Spanish Empire on the network of material technologies (described in Chapters 3, 4, and 5).

CHAPTER 2: THE IMPASSES OF INSTRUMENTALISM

1. For the topic of the Requerimiento, see Patricia Seed, *Ceremonies of Possession in Europe's Conquest of the New World, 1492–1640*; Adorno, *Polemics of Possession*, 265–66; José Rabasa, *Writing Violence on the Northern Frontier*; Jon Beasley-Murray, *Posthegemony*.
2. Adorno writes: "The core of the conflict between Sepúlveda and Las Casas was the principle of separation versus unity; that is, do the two entities, one more perfect and one less so, refer to different subjects or to the same one?" (122).
3. All citations are of Sepúlveda's *Demócrates*; the translations were made by Alfred MacAdam.
4. For Pagden's classic distinction between imperium and dominium, see *Spanish Imperialism* where he distinguishes between *imperium*, as the sovereign right to rule over the natives, and *dominium*, as the right to profit

from land, labor, and minerals (Pagden, *Spanish Imperialism*, 13–35). For a discussion about the importance of analogical thinking in the School of Salamanca, see Jaime Brufau Prats, *La Escuela de Salamanca ante el descubrimiento del Nuevo Mundo* (Salamanca: Editorial San Esteban, 1989).

5. When commenting on this passage, Adorno explains that "this is the basic presupposition of Sepúlveda's thinking, and it is similar to Vitoria's reasoning on the same subject. For this reason, Sepúlveda places great confidence in Vitoria's implicit support for his position" (118).

6. According to Pagden, the conflict between Sepúlveda and the theologians is based on disciplinary boundaries that the latter transgresses in his literary discourse by transforming axioms into metaphors (*Fall*, 114–15). Pagden pays excessive attention to the literary aspect of Sepúlveda but does not pay any attention to the philosophical principles and presuppositions that guide his work. Faced with the impossibility of the law providing its own ground, Sepúlveda, like Aquinas and Vitoria before him, employs literary tropes to guarantee the natural and rational origin of the law. For Aquinas and Vitoria, the force of law must be grounded on certain, really existing, yet indemonstrable principles. Since these principles are indemonstrable, yet necessary for demonstrating everything else, they must be rendered intelligible through examples. Metaphysical instrumentalism is idealist, in the Platonic sense of the word, because the examples are always second degree copies of a more fundamental preexisting perfect original. Yet from a materialist perspective, these examples contain more than the exemplified absent original. In this sense, the exemplified thing, the commanding origin, borrows its own power from technique, because technique contains more power than the commanding principle. The literary character of the tropes is the reminder of the disavowed transposition of technique to ontology and politics.

7. This passage illustrates the fundamental metonymic shift in Sepúlveda's discourse: one part rules and dominates the whole as if it were itself the ground of the totality and its own being. One particular element introduces perfection in a dispersed multiple chaos. The centrality of this statement, which makes visible the imperial consequences of the principle of natural subordination, will become evident when this argument migrates to José de Acosta. In Acosta, one particular thing—metal—is transformed into the measure of everything else, becoming independent, autonomous, or (using the Scholastic vocabulary), perfect.

8. In Vitoria's words: "It is good that such men should be subordinate to oth-

ers, like children to their parents until they reach adulthood, and like a wife to her husband" (*Political Writings*, 251).
9. As Adorno notes, Vitoria did not confuse civil slavery with natural slavery, since Indians should not be enslaved but only subjected to the Spanish government (119).
10. For a discussion of the classical division between Scholasticism and humanism, see Katherine Elliot van Liere, "Humanism and Scholasticism in Sixteenth-Century Academe: Five Student Orations from the University of Salamanca," *Renaissance Quarterly* 53.1 (Spring 2000): 57–107.
11. This is a circular argument based on the metonymic function of power, the totality of the people who agree to obey natural law, while those who do not are simply left out of the totality of the people.
12. Pagden comments, "as Democrates concedes, they may not *be* monkeys or bears, but their mental faculties are still only mechanical ones much like those of bees and spiders" (*Fall*, 116). While this view in itself was inoffensive to his contemporaries, "what *was* new—and offensive—was the rhetorical mode Sepúlveda used to present the evidence for his contention that the Indians belonged to the category 'natural slave,' because, as he phrased it, they are 'barbarous and inhuman peoples abhorring civil life, customs and virtue" (116). Thus, given their natural inferiority, for Sepúlveda it was necessary to subject and correct Amerindians for their own benefit. They are neither animals nor men but, rather, *imperfect matter*, a pure indetermination, that is, the *subordinate* pole of natural subordination that must be dominated. While Pagden claims that Sepúlveda stretches the limits of his rhetoric to the point of concluding that Amerindians are indeed animals, this operation can also be understood as a result of his application of the principle of the subordination of matter to form and his understanding of the Amerindian as matter lacking determination.
13. For an analysis of the importance of Nicolás Monardes, see Antonio Barrera-Osorio, *Experiencing Nature*.
14. See Juan Ginés de Sepúlveda and Bartolomé de Las Casas, *Apología de Juan Ginés de Sepúlveda contra Fray Bartolomé de Las Casas y de Fray Bartolomé de Las Casas contra Juan Ginés de Sepúlveda*.
15. Aquinas also writes, "As stated above, those things to which a man is inclined by nature to the natural law, and among these things it is a property of man to be inclined to act according to reason. Now reason proceeds from general principles to particular conclusions, as is made clear at *Physics* 1.184a16. But speculative reasoning is concerned chiefly with necessary

things which cannot be otherwise than they are, and so the truth is found just as surely in its particular conclusions as it is in its general principles; whereas practical reasoning is concerned with contingent things belonging to human actions, and so although there is a certain necessity in its general principles, *the more we descend to particulars the more frequently we find exceptions*" (Aquinas 121; my emphasis).

16. Adorno interprets this division as one between "philosophical ideas" and a "legal tradition." While Vitoria and Sepúlveda used philosophical ideas to apply a natural order to regulate relations between people, Las Casas appealed to a legal tradition in order to ground the civil order founded on the voluntary consent of the Indian subjects (Adorno, 123).

17. Las Casas borrows his interpretation of free consent of people in terms of efficient cause from Baldus de Ubaldis, an Italian jurist who wrote in medieval Roman law. Adorno maintains that the Las Casas radical edge came from using the legal tradition against the more philosophical inclinations of Sepúlveda. Yet the fact that Baldus read Aristotelian principles into Roman texts makes this distinction more complicated. See James Gordley, *The Jurists: A Critical History* (Oxford: Oxford University Press, 2013).

CHAPTER 3: MASTERING NATURE

1. See Pagden, "A Programme for Comparative Ethnology," in *The Fall of Natural Man*; Sabine MacCormack, "The Mind of the Missionary: José de Acosta on Accommodation and Extirpation, circa 1590," in *Religion in the Andes*.
2. See Ivonne del Valle, "José de Acosta, Violence and Rhetoric"; Del Valle, "From José de Acosta to the Enlightenment"; Del Valle, "Jesuit Baroque"; and Claudio M. Burgaleta, *José de Acosta, S.J., 1540–1600*.
3. See Jorge Cañizares-Esguerra, *Nature, Empire, and Nation*.
4. See Antonio Barrera-Osorio, *Experiencing Nature*.
5. By "metaphysical instrumentalism," I mean the use of metaphysics in order to justify the instrumental manipulation of reality. Borrowing insights from both Heidegger and Marx, it is possible to argue that metaphysics is itself the result of transposing the realm of artisanal techniques to the realm of ontology, nature, and politics. In the attempt to justify the ultimate origin and end of nature and politics, Thomist philosophy appeals to the metaphors of an artificer (efficient cause) that imposes forms (Platonic ideas or patterns) over metal (raw prime matter) in order to produce a stat-

ue (perfect end). God stands in front of the creature in the same way that an artisan stands in front of his or her product. On the one hand, philosophy borrows its power of persuasion from technique. Natural mastery presupposes artificial mastery, preunderstanding nature as an available standing reserve ready to be corrected, molded, and improved. On the other hand, metaphysics disavows its own transposition by reducing technique to an example of ontology, a mere medium, neutral, instrumental device that requires an efficient cause to be set in motion and directed to a preexisting end, since artifacts "have no inner impulse to change" (Aristotle, *Basic Works*, 236).

6. All citations are of *De procuranda*; translations are by Bryan Green.
7. See Pagden, *Fall*, 99–104.
8. The metaphors of the statue, the house, and the ship are present in Aristotle's and Aquinas's illustrations of the four causes as well as in the way they explain how statesmen administer a city. See, for example, Aquinas's *Commentary on Aristotle's Metaphysics* (281–83). Technical manipulation is an essential part of Aristotle's explanation, in Chapter 4, Book 7, *Politics* 1325–28, of how the statesman works with a population: "As the weaver or shipbuilder or any other artisan must have the material proper for his work (and in proportion as this is better prepared, so will the result of his art be nobler), so the statesman or legislator must also have the materials suited to him. First, among the materials required by the statesman is the population: he will consider what should be the number and character of citizens, and then what should be the size and character of the country" (Aristotle, *Basic Works*, 1283).
9. See Vilches, *New World Gold*.
10. See Heidegger, *Principle of Reason*, 62–63.
11. It is important not to confuse the purpose of my own study, which is to analyze the scope of application of the principle of the natural subordination of matter to form in imperial ideology, with the story told by Arthur O. Lovejoy in *The Great Chain of Being*, where Lovejoy provides a genesis of the idea of the chain of being, a continuum of created beings in which all possible beings coexist, in the principle of "plenitude," the idea that the universe is "full" and does not lack anything. The principle of plenitude results from the fusion of Aristotle's notion of continuity with Plato's insistence on hierarchy and gradation. Lovejoy narrates the history of the principle of plenitude from its Greek origins, through the medieval and modern history of ideas, to the eighteenth century.

12. All references to the *Historia Natural y Moral de las Indians* are from the English publication, *Natural and Moral History of the Indies*, ed. Jane E. Mangan.
13. See Michel Foucault, *The Order of Things*.
14. Translation is by Andrew Ascherl.
15. In "The Forbidden Food: Francisco de Vitoria and José de Acosta on Cannibalism," Pagden explains that, for Thomists such as Vitoria and Acosta, cannibalism violates the principle of subordination of imperfect matter to perfect form because cannibals consume foods destined for a lower level of being.
16. One can read in the *Diccionario* by Frédérique Lángue and Carmen Salazar-Soler the following definition: "artificio mecánico para moler metales." See also this definition from García de Llanos: "Although the name *ingenio* encompasses many other things which concern it, properly speaking, an *ingenio* is the device with which metal is milled, the principal parts of which consist of: canl, nozzle, wheel, axle, shaft, carriages, bearings, castle, triangle, heads, chains, mallets, cams, lower waterwheel blades, sledgehammers, discs, and muller" (García de Llanos, *Diccionario*, 62). We can read an extensive enumeration of the different *ingenios* of Potosí in Jose Luis Capoche's *Relación* (80–102). The problematic importance of these "ingenios de molienda" for the amalgamation method is stressed by Pete Bakewell: "The necessity of building what were, for the time and place, complex refining plants was the main drawback of amalgamation. To maximize the efficiency of the refining process, through optimum contact of the mercury with particles of ore, fine trituration of the ore was required. To achieve this, mechanically driven stamp mills proved essential. To begin with, refiners in Potosí made mills turned by human power. They quickly progressed to bigger machines driven by mules or horses; and finally, though still rapidly, to others powered by vertical waterwheels" (*Miners of the Red Mountain*, 19).
17. According to Marx, what prevented Aristotle from finding out the truth of quality is that he lived in a society founded on the labor of slaves that naturalized the master-servant relation (*Capital*, 1:152). "The secret of the expression of value, namely, the equality and equivalence of all kinds of labor because and in so far as they are human labor in general, could not be deciphered until the concept of human equality had already acquired the permanence of a fixed popular opinion. This however becomes possible only in a society where the commodity form is the universal form of the

product of labor, hence the dominant social relation is the relation between men as possessors of commodities" (152).
18. See Cristobal Colón, *Textos y documentos completos*, 217–21, 286–87, 302.
19. This does not mean that being under the spell of ideology can be reduced to mere or simple "false consciousness" about the real, material, and empirical conditions. Empirical and material reality is shaped by ideology. For this reason, Marx considers that it is not possible to understand material conditions without first showing how ideological fictions mold reality itself. In other words, ideological mystification is a misrecognition that does not simply fall under the category of false consciousness because it shapes social reality itself.
20. Vilches finds the historical background of the notion of financial capital as entrepreneurial credit and speculation in the market in the Scholastic notion of "Cambio de Arte" as both financial tool and abstract value. *Cambios* is a concept that applied to commercial, finance, and credit operations as opposed to the direct exchange of commodities. In her analysis of Tomás de Mercado, she explains how, for Scholastics, the "formless nature of credit" manipulated by the voracity of the *cambistas*, caused a feeling of anxiety inseparable from "the subversive power of the market," a power alien to the monarchical state in the hand of foreign bankers (*New World Gold*, 145–208).
21. See David E. Mungello, *Curious Land: Jesuit Accommodation and the Origins of Sinology* (Stuttgart: Steiner, 1985).
22. For Žižek's formulation of Peter Sloterdijk's critique of cynical reason see *The Sublime Object of Ideology*.
23. See Peter Bakewell, "Technological Change in Potosí: The Silver Boom of the 1570s," 91–92.
24. See Kamen, *Empire*.

CHAPTER 4: FROM IMPERIAL REASON TO INSTRUMENTAL REASON

1. For an exploration of the colonial origins of the reason of state, see Irene Silverblatt's *Modern Inquisitions*, where she follows the path opened by Hannah Arendt, arguing that the modern belief in the rationality of the state, inseparable from racial thinking ant bureaucratic rationality and totalitarianism, has its origin in the inquisitorial practices in colonial Peru. The Spanish inquisition in Peru was an antecedent of the reason of state,

since the inquisitors justified their surveillance by appealing to "the spirit of reason and equity, as they deliberated the truth of colonial subjects" (23). Such practices and technologies of power were "shaped by the dictates of an emerging absolutist state" (23). Bureaucratic technique of extraction and registration of truth have an indissoluble relation with "state thinking" and "race thinking," becoming an antecedent of the modern nation-states in the nineteenth and twentieth centuries. Silverblatt broadens Arendt's argument by showing how the methods of colonial classification were an antecedent of the totalitarian and fascist models.

2. In a footnote, Kamen states that the theories of "empire" tended to write about imagined rather than realistic notions (537n138).
3. See Lewis Hanke, Susan Scafidi, and Peter Bakewell, eds., *The Spanish Struggle for Justice in the Conquest of America* (Dallas: Southern Methodist University Press, 2002).
4. Demetrio Ramos, "La crisis indiana y la Junta Magna de 1568."
5. See also Lewis Hanke, *Aristotle and the American Indians*, 86–87.
6. Charles A. Truxillo, *By the Sword and the Cross*, 80.
7. Geoffrey Parker, *The Grand Strategy of Philip II*, 8.
8. See Carlos Sempat Assadourian, "Acerca del cambio de la naturaleza del dominio sobre las Indias."
9. See also Ramos, "La crisis indiana," 26–27.
10. Isacio Pérez Fernández, *Bartolomé de Las Casas en el Perú*.
11. Carlos Sempat Assadourian, "Fray Alonso de Maldonado."
12. Merluzzi contends that Sánchez's report made a deep impact on Espinosa: see Manfredi Merluzzi, *Politica e governo nel Nuovo Mondo: Francisco de Toledo viceré del Peru (1569–1581)* (Roma: Carocci, 2003), 36.
13. At the most basic level I attribute this ambivalence to the inner impasses of metaphysical instrumentalism.
14. See Hanke, *History of Latin American Civilization*, 1:87–89; and John Hemming, *The Conquest of the Incas* (New York: Harcourt, 1970), 421.
15. All references to the *Anónimo* are from Pérez Fernández Isacio, *El anónimo de Yucay frente a Bartolomé de Las Casas: estudio y edición crítica del Parecer de Yucay, anónimo*, and the translations are by Andrew Ascherl.
16. In 1575 the archbishop Manuel Jerónimo de Loaisa, who asistió the junta, denounced the mistreatment and indigenous oppression, using a Lascasian rhetoric. See Pérez Fernández, *Bartolomé de las Casas en el Perú*, 559.
17. In Article 4, *Summa Theologica* IaIIae, Aquinas stated that, although the ultimate principles of metaphysics and natural law are universal and nec-

essary in all time and places, "the more we descend to particulars, the more frequently we find exceptions" (*Political Writings*, 121). Of course, these failures or exceptions were due to "certain specific obstacles," and not to the general principles themselves, "because in some men the reason is perverted by passion, or evil custom, or wicked disposition of nature" (122). In other words, Aquinas saw these exceptions as sources of evil or deviation that were external to metaphysics and not the result of the inner inconsistencies of the presumed hierarchical or rational order of being. The author of the *Anónimo* considers that Peru *is* the exception, the world upside-down.

18. See Fernando Montesinos and Francisco de Toledo, *Memorias antiguas historiales y políticas del Perú*.
19. See Bakewell, *Miners of the Red Mountain*, 18–26.
20. For the problem of the mita, see Bakewell, *Miners of the Red Mountain*; Cole, *The Potosí Mita*; González Casasnovas, *Las dudas de la corona*; Tandeter, *Coercion and Market*.
21. All page references are to Louis Capoche, *Relación general de la villa imperial de Potosí*. Translations by Andrew Ascherl.
22. Most of the sources on colonial mining point in direction of the Andean origin of the huayra.

    According to Mary Money de Álvarez, González Holguín says in *Vocabulario de la lengua general de todo el Perú* that the quechua *huayrachina/wayrachina* was the "oven in which metal is smelted" and also *huayrachina/wayrachina* were "the instruments of smelting any metal." The noun *huayrachina* takes its name from *huayra* meaning "wind" and *china* meaning "to make a fan" (*Oro y Plata* 60; González Holguín, *Vocabulario*, 194). In the *Diccionario y maneras de hablar que se usan en las minas* (1609) by García de Llanos, we read the following definition of the huayra: "*Huaira, Huairar*, and *Hiradores*. The correct pronunciation of these words is also with 'H' for the said reason and for this reason *huaira* means wind. And the Indians give the same name to some small ovens for smelting rich metals, because they put them high on the hills where they catch the wind better, which they use for the bellows. And smelting in this way is called *huayrar*, and the Indians that do it *huairadores*, of which there are many in Potosí that do nothing else but recover the metal that the Indians steal during their labors which is best for the purpose. These days there are not as many huayras as there used to be because not much rich metal remains. However, when many are discovered from the hill at night (which seem better), it is a sign that the work is going well. When there are usually more at this time they reach fifteen, more or less" (57).

23. García de Llanos explains in the *Diccionario* that the *tacana* is the most valuable metal and is refined by amalgamation (80). It is characterized by "brown" or "almost black spots," and its value increases with the darkness of its color. The name of this metal comes from *tacani*, which means "to hit" (*golpear*). Potosí, and other mines with more "foundation and stability," abound in *tacana*. García de Llanos also adds that "the mines that are always rich in metals never last but rather with the muteness they ordinarily have sometimes yield similar riches" (80–81).

24. Recent bibliography on Toledo's reforms undermines this historically dated construction of Toledo as the supreme organizer of Peru. For instance, Mumford demonstrates clearly in *Vertical Empire* how the General Resettlement owes a great deal to the "tyrannical" Incas.

25. Ideology cannot be reduced to an explicit effect of discourse, because the implicit core of ideology is silently grounded on a nondiscursive real kernel around which ideology is organized. The problem with reducing metaphysical instrumentalism to discourse analysis in a Foucauldian mode is that it fails to consider the crucial nondiscursive element upon whose disavowal every discursive ideological practice relies, namely, the silent presupposition, an unacknowledged "as if," the endless impersonal compulsion to accumulate more riches.

26. The concept of vertical archipelago appears for the first time in Murra's path-breaking study of the Chupaychu and Lupaqa kingdoms in the sixteenth century. According to Murra's hypothesis, the Inca Empire owed its success to a vast network of vertical control that connected multiple settlements along the eastern and western cordilleras with the central groups in the highlands. It was vertical because the political organization of the networks of different "ecological zones" was centered in the highlands. And it was an "archipelago" because it connected a variety of satellite islands in the fertile areas of the east, the Pacific coast, and the tropical frontier. The vertical archipelago was a strategy that allowed the Inca Empire to adapt to the environment by guarding against possible crop failures. Another advantage of the vertical archipelago was that it allowed the Incas to control different ecosystems through the massive method of mobilization of labor called mita. The mita was an important part of the principle of reciprocal exchange, because it was a countergift given to the Inca Empire under the form of rotational labor power used in public works such as irrigation

systems, agriculture and the construction of terraces. The mobilization of *mitmaqs* provided natural resources from different areas. For instance, if there was a shortage of potato in one area, then the maize provided by another area would make up for the lost crop. According to Murra, the vertical archipelago was not an Inca invention, because the Chavin civilization (800–200 B.C.) already employed this model by controlling different ecological zones. See John Murra, *La organización económica del Estado inca*.
27. See Thomas Ertman, *Birth of the Leviathan*; Magali Sarfatti Larson, *Spanish Bureaucratic-Patrimonialism*.
28. Graham Burchell, Colin Gordon, and Peter Miller, eds., *The Foucault Effect*.
29. Mumford notes that the Spanish theologians played an important role in propagating notions of governmentality in the Council of Trent (1545–1563); see *Vertical Empire*, 252n12.
30. See Žižek, *The Parallax View*, 297–98.

CHAPTER 5: THE EXHAUSTION OF NATURAL SUBORDINATION

1. All references in this chapter refer to the original language editions, and were translated by Andrew Ascherl.
2. See Alberto Crespo Rodas, *La guerra entre vicuñas y vascongados (Potosí 1622–1625)* (La Paz: Collección Popular, 1969). "Vicuña" was an informal term for non-Basque Spaniards in Alto Perú, which derived from their habit of wearing hats made of vicuña (an animal similar to llama or alpaca) skin.
3. The first volume of *De Indiarum iure* was censored by royal decree of the Crown on September 28, 1637. The second volume was sent to the king by José de Nápoles in a document dated March 12, 1638. The Roman curia condemned Solórzano's work, including it in the index of forbidden books in March 1646 because he defended the ecclesiastical power exercised by the king of Spain. The censorship of the work in Rome did not have any effect in Spain, which was not ready to cede any power in ecclesiastical matters. *De Indiarum iure* was received as a most complete juridical and theological defense of the rights of the Catholic monarchy over the Indies.
4. The *Política* is not a literal translation of the *De Indiarum iure* since there are parts missing from the original and new parts have been added. The edition I will analyze here was published in 1972 and is a reproduction of the text as corrected and annotated by Francisco Ramiro de Valenzuela

in the eighteenth century. This edition is accompanied by legal citations from the collected Laws of the Indies of 1680 (after Solórzano's death). The six volumes of the *Política* cover the totality of the topics addressed by historians, missionaries, and chroniclers, going from the juridical and theological grounds of the conquest and colonization to the singular details of the administrative and juridical organization of the natural riches of the Indies. Solórzano begins his first volume with the discovery and the titles of dominion. The second volume is devoted to the new vassals of the Spanish king, the indigenous peoples, whom he considered free yet in need of protection from the Crown because of their "miserable" and servile nature that needs to be corrected and disciplined by teaching them the use of money and iron. The third volume comments on and defends the old problem of the encomienda; the fourth volume discusses ecclesiastical matters such as the Patronato; the fifth volume studies the system of secular government of the Indies; and the sixth volume discusses the financing system (xxxiii).

5. See Watson, *The Digest of Justinian*, 98.
6. In Ley 20, Título 25, Libro 6 of the *Recopilación*, blacks, mestizos, and mulatos are condemned to the mines of Huancavélica.
7. See also Juan de Solórzano Pereira, *Política indiana* (Book 1, Chapter 6, no. 32), 1:301.
8. The purpose of this chapter is not to give an exegetical examination of Álvaro Alonso Barba's *Arte de los metales* or to examine the confluence between hermeticism, Neoplatonism, Aristotelianism, and Andean beliefs on the living nature of the mineral world, but to put it in relation with the productionist presuppositions of the imperial school of thought as examined in this book. In my own view, Barba represents an exhaustion of the ontological frame of metaphysical instrumentalism grounded in the principle of the natural subordination of matter to form and means to an end. For an interpretation that reads Barba's work as irreducible to metaphysical instrumentalism see Orlando Bentancor, "Matter, Form, and the Generation of Metals."
9. See Aristotle, *Meteorologica* 4.381b3–9. For the importance of this work in the premodern theories of the generation of metals, see John A. Norris, "The Mineral Exhalation Theory of Metallogenesis in Pre-modern Mineral Science," *Ambix* 53.1 (2006): 43–65; David E. Eichholz, "Aristotle's Theory of the Formation of Metals and Minerals," *Classical Quarterly* 43.3–4 (1949): 141–46.

10. In 1649 Barba wrote a letter to the king, in which he affirms both that his method of amalgamating with copper cauldrons is useful to get profit from the useless "debris," and that metals do grow in Potosí. As Barba himself puts it: "On this same basis one can also presume that in such copious minerals as those that many other veins, seams, and branches were found in the tunnels, or the few that, because there was not much law, then stayed to dig, and today due to mercury amalgamation were able to become rich. Besides that which in so many centuries nature will have produced and matured, thus, in conformity with good philosophy and experience, it is certain that metals grow in its veins" (quoted in Barnadas 152). Barba finishes the letter affirming his loyalty to the Spanish Crown, as well as his desire to continue augmenting the royal tribute: "To Your Highness, I request and plead that you admit the manifestation of the things that this writing contains and that paying attention to the special and evident intelligence I have acquired on the subject of metals in the more than forty-four years I have spent in these Indies, with continuous study and experiences that I have communicated without any interested convention in very great service to your Royal Personage, I augment the *quintos* and the public good, and also attending to the satisfaction that I will have deserved I have served for more than thirty-three years in the office of Priest without interruption and I have carried out other important tasks in the most serious postings of this Archbishop, Lipes Pacajes, Oruro, Potosí, and this city of La Plata, in view of its superior ministers and in which I have always tried to increase my achievements and not pretend to solicit awards, I ask Your Highness to arrange in reason what I have proposed and offered so that it will be of more use to the Royal service, that I am ready to execute that which I am assigned and that I give this my writing as my testimony and that which it will provide, on which I will receive mercy with justice as I ask, etc. (Barnadas 154).
11. Patricia Seed explains in *American Pentimento* (57–71) that the Iberian practice of the ownership of the subsoil by the Crown had Islamic origins and it was essential for the extraction of surplus value. Since metals belonged to God, who gave them to the Crown, to become rich off the mines individuals had to mobilize natives to perform the work inside the mines. This argument is very compelling since it ties a genealogy of Spanish jurisprudence to its economic consequences in the Americas.
12. See William R. Newman, *Promethean Ambitions*, 1–34.
13. See Berthelot, *Une Région minière des Andes péruviennes*, 120.

14. It is important to keep in mind that in capitalist ideology, such neutrality is precisely the essence of commodity fetishism, appearing as a universal representative of all commodities. Commodity fetishism is the abstraction of money from its material, contingent, and empirical determinations, presupposing an inherent universal empty character that makes it appropriate for representing things. Exchange value is inherently fetishistic and the result of a metaphysical abstraction. Yet fetishistic illusion is *practical* and not theoretical. As Žižek explains, in theory we are practical agents who use money for higher ends, but we *act as if* money possessed magic inherent properties (*Sublime*, 18)
15 See Regina Harrison, *Signs, Songs, and Memory in the Andes*; Tristan Platt, "Mirrors and Maize"; Gary Urton, *At the Crossroads of the Earth and the Sky*; Grimaldo Rengifo, "The Ayllu"; and Hillary S. Webb, *Yanantin and Masintin in the Andean World*.
16. See Regina Harrison, "Modes of Discourse."
17. Žižek calls this "biopolitical parallax," and he describes it in the following way: "How, precisely, in what mode of parallax, do these two aspects relate to each other? We should not succumb to the temptation of reducing capitalism to a mere form of appearance of the more fundamental ontological attitude of technological domination; the two levels, precisely insofar as they are two sides of the same coin, are ultimately incompatible: there is no meta-language that enables us to translate the logic of domination back into the capitalist reproduction-through-excess, or vice versa. The key question thus concerns the relationship between these two excesses: the 'economic' excess/surplus which is integrated into the capitalist machine as the force which drives it into permanent self-revolutionizing; the 'political' excess of power inherent in its exercise (the constitutive excess of representation over the represented)" (*Parallax View*, 298).

BIBLIOGRAPHY

Acosta, José de. *De procuranda Indorum salute*. In *Obras del Padre José de Acosta*, edited by Francisco Mateos. Madrid: Ediciones Atlas, 1954.

Acosta, José de. *Historia natural y moral de las Indias*. Edited by José Alcina Franch. Madrid: Dastin, 2002.

Acosta, José de. *Natural and Moral History of the Indies*. Edited by Jane E. Mangan. Translated by Frances López-Morillas. Durham: Duke University Press, 2002.

Adorno, Rolena. *The Polemics of Possession in Spanish American Narrative*. New Haven: Yale University Press, 2007.

Agia, Miguel. *Tratado Qve Contiene Tres Pareceres Graves En Derecho*. Lima: Ricardo, 1604.

Agricola, Georg, Herbert Hoover, and Lou H. Hoover. *De Re Metallica*. 1950.

Alves, André Azevedo, and J. M. Moreira. *The Salamanca School*. London: Continuum, 2010.

Aquinas, Thomas. *Commentary on Aristotle's* Metaphysics. Translated by John P. Rowan. Notre Dame, IN: Dumb Ox Books, 1995.

Aquinas, Thomas. *Commentary on Aristotle's* Nicomachean Ethics. Translated by C. I. Litzinger. Notre Dame, IN: Dumb Ox Books, 1993.

Aquinas, Thomas. *Commentary on Aristotle's* Physics. Notre Dame, IN: Dumb Ox Books, 1999.

Aquinas, Thomas. *Commentary on Aristotle's Politics*. Indianapolis: Hackett, 2007.

Aquinas, Thomas. *Political Writings*. Edited and translated by R. W. Dyson. Cambridge: Cambridge University Press, 2002.

Aristotle. *The Basic Works of Aristotle*. Edited by Richard McKeon. New York: Random House, 1941.

Aristotle. *Meteorologica*. Translated by Henry Desmond Pritchard Lee. Cambridge, MA: Harvard University Press, 2004.

Aspe Armella, Virginia, and Idoya Zorroza. *Francisco de Vitoria en la Escuela de Salamanca y su proyección en Nueva España*. Pamplona: Ediciones Universidad de Navarra, 2014.

Assadourian, Carlos Sempat. "Acerca del cambio de la naturaleza del dominio sobre las Indias: la mita minera del Virrey Toledo. Documentos de 1568–1571." *Anuario de Estudios Hispanoamericanos* 46 (1989): 3–70.

Assadourian, Carlos Sempat. "Fray Alonso de Maldonado: la política indiana, el estado de damnación del rey católico y la Inquisición." *Historia Mexicana* 38.4 (April 1989): 623–61.

Augustine. *The City of God against the Pagans*. Edited and translated by R. W. Dyson. Cambridge: Cambridge University Press, 1998.

Bakewell, Peter. J. *Miners of the Red Mountain: Indian Labor in Potosí, 1545–1650*. Albuquerque: University of New Mexico Press, 1984.

Bakewell, Peter. "Technological Change in Potosí: The Silver Boom of the 1570s." In *Mines of Silver and Gold in the Americas*, edited by Peter Bakewell, 91–92. Brookfield, VT: Variorum, 1997.

Barba, Álvaro Alonso. *Arte de los metales. Seguido de notas y suplementos al libro por un antiguo minero; juicios y comentarios*. 1640. Potosí: Editorial Potosí, 1967.

Bargalló, Modesto. *La amalgamación de los minerales de plata en hispanoamérica colonial*. Mexico, D.F.: Compañía Fundidora de Fierro y Acero de Monterrey, 1969.

Bargalló, Modesto. *La minería y la metalurgía en la América española durante la época colonial; con un apéndice sobre la industria del hierro en México desde la iniciación de la Independencia hasta el presente*. Mexico: Fondo de Cultura Económica, 1955.

Barnadas, Josep M. *Alvaro Alonso Barba, 1569–1662: investigaciones sobre su vida y obra*. La Paz: Biblioteca Minera Boliviana, 1986.

Barrera-Osorio, Antonio. *Experiencing Nature: The Spanish American Empire and the Early Scientific Revolution*. Austin: University of Texas Press, 2006.

Bauer, Ralph. *The Cultural Geography of Colonial American Literatures: Empire, Travel, Modernity*. Cambridge: Cambridge University Press, 2003.

Beasley-Murray, Jon. *Posthegemony: Political Theory and Latin America*. Minneapolis: University of Minnesota Press, 2010.

Bentancor, Orlando. "Francisco de Vitoria, Carl Schmitt, and Originary Technicity." *Política Común* 5 (2014). Available at http://quod.lib.umich.edu/p/pc/12322227.0005.002?view=text;rgn=main/.

Bentancor, Orlando. "Matter, Form, and the Generation of Metals in Álvaro Alonso Barba's *Arte de los metales*." *Journal of Spanish Cultural Studies* 8.2 (2007): 117–33.

Berthelot, Jean. *L'Exploitation des métaux précieux au temps des Incas*. Paris: A. Colin, 1978.

Berthelot, Jean. *Une Région minière des Andes péruviennes: 1480–1630*. Thèse 3ème cycle Histoire, Paris 10, EHESS, 1977.

Bobik, Joseph. *Aquinas on Matter and Form and the Elements: A Translation and Interpretation of the* De Principiis Naturae *and the* De Mixtione Elementorum *of St. Thomas Aquinas*. Notre Dame, IN: University of Notre Dame Press, 1998.

Brading, D. A. *The First America: The Spanish Monarchy, Creole Patriots, and the Liberal State, 1492–1867*. Cambridge: Cambridge University Press, 1991.

Bradley, Arthur. *Originary Technicity: The Theory of Technology from Marx to Derrida*. Houndmills: Palgrave Macmillan, 2011.

Brett, Annabel S. *Liberty, Right, and Nature: Individual Rights in Later Scholastic Thought*. Cambridge: Cambridge University Press, 1997.

Brett, Annabel. "Scholastic Political Thought and the Modern Concept of State." In *Rethinking the Foundations of Modern Political Thought*, edited by Annabel Brett and James Tully, 130–48. Cambridge: Cambridge University Press, 2007.

Brown, Kendall W. *A History of Mining in Latin America from the Colonial Era to the Present*. Albuquerque: University of New Mexico Press, 2012.

Brufau Prats, Jaime. *La Escuela de Salamanca ante el descubrimiento del Nuevo Mundo*. Salamanca: Editorial San Esteban, 1989.

Burchell, Graham, Colin Gordon, and Peter Miller, eds. *The Foucault Effect: Studies in Governmentality*. Chicago: University of Chicago Press, 1991.

Burgaleta, Claudio M. *José de Acosta, S.J., 1540–1600: His Life and Thought*. Chicago: Jesuit Way, 1999.

Calancha, Antonio. *Crónica Moralizada De Antonio De La Calancha*. Vol 3. Edited by Pastor I. Prado. Lima: Universidad Nacional Mayor de San Marcos, 1974.

Cañizares-Esguerra, Jorge. *Nature, Empire, and Nation: Explorations of the History of Science in the Iberian World*. Stanford, CA: Stanford University Press, 2006.
Capoche, Luis. *Relación general de la villa imperial de Potosí*. Madrid: Atlas, 1959.
Castro-Klarén, Sara. "Historiography on the Ground: The Toledo Circle and Guamán Poma." In *The Latin American Subaltern Studies Reader*, edited by Ana del Sarto, Alicia Ríos, and Abril Trigo, 143–71. Durham: Duke University Press, 2004.
Cavallar, Georg. *The Rights of Strangers: Theories of International Hospitality, the Global Community, and Political Justice since Vitoria*. Aldershot, England: Ashgate, 2002.
Châtellier, Louis. *The Religion of the Poor: Rural Missions in Europe and the Formation of Modern Catholicism, c. 1500–c. 1800*. Cambridge: Cambridge University Press, 1997.
Cicero, Marcus T., and H. Rackham. *De Natura Deorum: Academica*. London: W. Heinemann, 1951.
Cobo, Bernabé. *Obras del P. Bernabé Cobo*. Edited by Francisco Mateos. Madrid: Atlas, 1964.
Cole, Jeffrey A. *The Potosí Mita, 1573–1700: Compulsory Indian Labor in the Andes*. Stanford: Stanford University Press, 1985.
Colón, Cristobal. *Textos y documentos completos*. Madrid: Consuelo Varela, 1982.
Cosmas, Emmanuel Kanyama. *Participation as the Fundamental Basis of Analogy in St. Thomas Aquinas' Metaphysics*. Rome: Pontificia Universitas Urbaniana, 1996.
Crespo Rodas, Alberto. *La guerra entre vicuñas y vascongados (Potosí 1622–1625)*. La Paz: Collección Popular, 1969.
Cross, Harry, "South American Bullion Production and Export, 1550–1750." In *Precious Metals in the Later Medieval and Early Modern Worlds*, edited by John F. Richards, 397–423. Durham: Carolina Academic Press, 1983.
Dean, Carolyn. *A Culture of Stone: Inka Perspectives on Rock*. Durham: Duke University Press, 2010.
Deleuze, Gilles, and Félix Guattari. *A Thousand Plateaus: Capitalism and Schizophrenia*. Minneapolis: University of Minnesota Press, 1987.
Del Valle, Ivonne. "From José de Acosta to the Enlightenment: Barbarians, Climate Change, and (Colonial) Technology as the End of History." *The Eighteenth Century* 54.4 (2013): 435–59.
Del Valle, Ivonne. "Jesuit Baroque." *Journal of Spanish Cultural Studies* 3.2 (2002): 141–63.
Del Valle, Ivonne. "José de Acosta, Violence, and Rhetoric: The Emergence of

Colonial Baroque." *Calíope: Journal of the Society for Renaissance and Baroque Hispanic Society* 18.2 (2012): 46–72.

Des Chene, Dennis. *Life's Form: Late Aristotelian Conceptions of the Soul*. Ithaca: Cornell University Press, 2000.

Des Chene, Dennis. *Physiologia: Natural Philosophy in Late Aristotelian and Cartesian Thought*. Ithaca: Cornell University Press, 1996.

Des Chene, Dennis. *Spirits and Clocks: Machine and Organism in Descartes*. Ithaca: Cornell University Press, 2001.

Eichholz, David E. "Aristotle's Theory of the Formation of Metals and Minerals." *Classical Quarterly* 43.3–4 (1949): 141–46.

Elliott, J. H. *Empires of the Atlantic World: Britain and Spain in America, 1492–1830*. New Haven: Yale University Press, 2006.

Elliott, J. H. *Imperial Spain 1469–1716*. London: Penguin, 2002.

Elliot van Liere, Katherine. "Humanism and Scholasticism in Sixteenth-Century Academe: Five Student Orations from the University of Salamanca." *Renaissance Quarterly* 53.1 (Spring 2000): 57–107.

Ertman, Thomas. *Birth of the Leviathan: Building States and Regimes in Medieval and Early Modern Europe*. Cambridge: Cambridge University Press, 1997.

Fernández-Santamaría, J. A. *The State, War, and Peace: Spanish Political Thought in the Renaissance, 1516–1559*. Cambridge: Cambridge University Press, 1977.

Flórez Miguel, Cirilo, Maximiliano Hernández Marcos, and Roberto Albares Albares. *La primera escuela de Salamanca (1406–1516)*. Salamanca: Ediciones Universidad de Salamanca, 2012.

Foucault, Michel. *The Order of Things: An Archaeology of the Human Sciences*. New York: Vintage Books, 1994.

García de Llanos. *Diccionario y maneras de hablar que se usan en las minas y sus labores en los ingenios y beneficios de los metales (1609)*. La Paz: MUSEF Editores, 1983.

Gillespie, Michael Allen. *The Theological Origins of Modernity*. Chicago: University of Chicago Press, 2008.

González, Ángel Poncela, ed. *La Escuela de Salamanca: filosofía y humanismo ante el mundo moderno*. Madrid: Editorial Verbum, 2015.

González Casasnovas, Ignacio. *Las dudas de la corona: la política de repartimientos para la minería de Potosí (1680–1732)*. Madrid: Consejo Superior de Investigaciones Científicas, 2000.

González Holguín, Diego. *Vocabulario de la lengua general de todo el Perú llamada lengua quichua o del Inca*. Edited by Raúl Porras Barrenechea. Lima: Impr. Santa María, 1952.

Grellard, Christophe, and Aurélien Robert, eds. *Atomism in Late Medieval Philosophy and Theology*. Leiden: Brill, 2009.
Grice-Hutchinson, Marjorie. *The School of Salamanca: Readings in Spanish Monetary Theory, 1544–1605*. Oxford: Clarendon Press, 1952.
Hanke, Lewis. *Aristotle and the American Indians: A Study in Race Prejudice in the Modern World*. Bloomington: Indiana University Press, 1970.
Hanke, Lewis. *History of Latin American Civilization: Sources and Interpretations*. Boston: Little, Brown, 1967.
Hanke, Lewis. "Prólogo." In *Relación general de la villa imperial de Potosí*, 9–68. Madrid: Atlas, 1959.
Harris, Olivia. "The Sources and Meanings of Money: Beyond the Market Paradigm in an Ayllu of Northern Potosí." In *Ethnicity, Markets, and Migration in the Andes: At the Crossroads of History and Anthropology*, 297–328.
Harris, Olivia, and Enrique Tandeter, eds. *Ethnicity, Markets, and Migration in the Andes: At the Crossroads of History and Anthropology*. Durham: Duke University Press, 1995.
Harrison, Regina. "Modes of Discourse: The *Relación de Antigüedades Deste Reyno del Pirú* by Joan de Santacruz Pachacuti Yamqui Salcamaygua." In *From Oral to Written Expression: Native Andean Chronicles of the Early Colonial Period* (Latin American Series 4), edited by Rolena Adorno, 65–99. Syracuse: Maxwell School of Citizenship and Public Affairs, 1982.
Harrison, Regina. *Signs, Songs, and Memory in the Andes: Translating Quechua Language and Culture*. Austin: University of Texas Press, 1989.
Heidegger, Martin. *The Basic Problems of Phenomenology*. Translated by Albert Hofstadter. Bloomington: Indiana University Press, 1982.
Heidegger, Martin. *Nietzsche*. Translated by David Farrell Krell. San Francisco: Harper and Row, 1979.
Heidegger, Martin. *The Principle of Reason*. Translated by Reginald Lilly. Bloomington: Indiana University Press, 1991.
Heidegger, Martin. *The Question Concerning Technology, and Other Essays*. Translated by William Lovitt. New York: Harper and Row, 1977.
Hemming, John. *The Conquest of the Incas*. New York: Harcourt, Brace, Jovanovich, 1970.
Höpfl, H. M. "Scholasticism in Quentin Skinner's *Foundations*." In *Rethinking the Foundations of Modern Political Thought*, edited by Annabel Brett and James Tully, 113–29. Cambridge: Cambridge University Press, 2007.
Justinian. *The Digest of Roman Law: Theft, Rapine, Damage, and Insult*. Translated by C. F. Kolbert. Harmondsworth: Penguin, 1985.

Kamen, Henry. *Empire: How Spain Became a World Power, 1492–1763*. New York: HarperCollins, 2003.

Karatani, Kojin. *The Structure of World History: From Modes of Production to Modes of Exchange*. Translated by Michael K. Bourdaghs. Durham: Duke University Press, 2014.

Kaye, Joel. *Economy and Nature in the Fourteenth Century: Money, Market Exchange, and the Emergence of Scientific Thought*. New York: Cambridge University Press, 1998.

Koskenniemi, Martti. "Empire and International Law: The Real Spanish Contribution." *University of Toronto Law Journal* 61 (2011): 1–36.

Kusch, Rodolfo. *El pensamiento indígena y popular en América*. Buenos Aires: Hachette, 1977.

Lactantius. *De opificio Dei*. Turnhout: Brepols Publishers, 2010.

Langue, Frédérique, and C. Salazar-Soler. *Dictionnaire des termes miniers en usage en Amérique espagnole, XVIe–XIXe siècle = Diccionario de términos mineros para la América española, siglos XVI–XIX*. Paris: Editions Recherche sur les civilisations, 1993.

Larson, Magali Sarfatti. *Spanish Bureaucratic-Patrimonialism in America*. Berkeley: Institute of International Studies, University of California, 1966.

Las Casas, Bartolomé de. *Los tesoros del Perú*. Edited by Ángel Losada. Madrid: Consejo Superior de Investigaciones Científicas, Institutos Gonzalo F. de Oviedo y Francisco de Vitoria, 1958.

Lechtman, Heather. "Andean Value Systems and the Development of Prehistoric Metallurgy." *Technology and Culture* 25.1 (1984): 1–36.

Lechtman, Heather. "Technologies of Power: The Andean Case." In *Configurations of Power: Holistic Anthropology in Theory and Practice*, edited by John S. Henderson and Patricia J. Netherly, 244–80. Ithaca: Cornell University Press, 1993.

Levillier, Roberto. *Don Francisco de Toledo: Supremo Organizador Del Perú; Su Vida, Su Obra (1515–1582)*. Buenos Aires: Publisher not identified, 1935.

Levillier, Roberto. *Gobernantes del Perú: cartas y papeles, siglo XVI*. T 4. Madrid: Juan Pueyo, 1924.

Lovejoy, A. O. *The Great Chain of Being: A Study of the History of an Idea*. Cambridge, MA: Harvard University Press, 1936.

Lucretius Carus, Titus. *On the Nature of Things: De rerum natura*. Edited and translated by Anthony M. Esolen. Baltimore: Johns Hopkins University Press, 1995.

MacCormack, Sabine. *Religion in the Andes: Vision and Imagination in Early Colonial Peru*. Princeton, NJ: Princeton University Press, 1991.

Marx, Karl. *Capital: A Critique of Political Economy*. Vol. 1. Translated by Ben Fowkes. New York: Penguin Books, 1976.
Marx, Karl. *Capital: A Critique of Political Economy*. Vol. 3. Translated by David Fernbach. London: Penguin Books, 1981.
Matienzo, Juan de. *Gobierno del Perú*. 1567. Buenos Aires: Compañía Sudamericana de Billetes de Banco, 1910.
McInerny, Ralph. *Aquinas and Analogy*. Washington, DC: Catholic University of America Press, 1996.
Mignolo, Walter. "Commentary. José de Acosta's *Historia natural y moral de las Indias*: Occidentalism, the Modern/Colonial World, and the Colonial Difference." In *Natural and Moral History of the Indies*, edited and translated by Jane E. Mangan, 451–518. Durham: Duke University Press, 2002.
Mignolo, Walter. *The Darker Side of Western Modernity: Global Futures, Decolonial Options*. Durham: Duke University Press, 2011.
Molina, Cristóbal de, and Cristóbal de Albornoz. *Fábulas y mitos de los Incas*. Edited by Henrique Urbano and Pierre Duviols. Madrid: Historia 16, 1989.
Monardes, Nicolás, Modesto Bargalló, and Francisco Guerra. *Diálogo del hierro y de sus grandezas*. Mexico, D.F.: Compañía Fundidora de Fierro y Acero de Monterrey, S.A., 1961.
Money de Álvarez, Mary. *Oro y plata en los Andes: significado en los diccionarios de Aymara y Quechua, Siglo XVI–XVII*. La Paz: Maestría en Historias Andinas y Amazónicas, Post-grado de la Carrera de Historia, Universidad Mayor de San Andrés, 2004.
Montesinos, Fernando, and Francisco de Toledo. *Memorias antiguas historiales y políticas del Perú, por Fernando de Montesinos seguídas de las informaciones acerca del señorío los Incas, hechas por mandado de D. Francisco de Toledo, Virrey del Peru*. Madrid: Imprenta de Miguel Ginesta, 1882.
More, Anna Herron. *Baroque Sovereignty: Carlos de Sigüenza y Góngora and the Creole Archive of Colonial Mexico*. Philadelphia: University of Pennsylvania Press, 2013.
Mumford, Jeremy Ravi. *Vertical Empire: The General Resettlement of Indians in the Colonial Andes*. Durham: Duke University Press, 2012.
Mungello, David E. *Curious Land: Jesuit Accommodation and the Origins of Sinology*. Stuttgart: Steiner, 1985.
Murra, John V. *La organización económica del Estado inca*. Mexico: Siglo Veintiuno, 1978.
Norris, John A. "The Mineral Exhalation Theory of Metallogenesis in Premodern Mineral Science." *Ambix* 53.1 (2006): 43–65.

Newman, William R. *Promethean Ambitions: Alchemy and the Quest to Perfect Nature*. Chicago: University of Chicago Press, 2004.

Orrego Sánchez, Santiago. *La actualidad del ser en la "primera escuela" de Salamanca: con lecciones inéditas de Vitoria, Soto y Cano*. Pamplona: Ediciones Universidad de Navarra, 2004.

Pagden, Anthony. *The Fall of Natural Man: The American Indian and the Origins of Comparative Ethnology*. Cambridge: Cambridge University Press, 1982.

Pagden, Anthony. "The Forbidden Food: Francisco de Vitoria and José de Acosta on Cannibalism." *Terrae incognitae* 13.1 (1981): 17–29.

Pagden, Anthony. "Introduction." In Francisco de Vitoria, *Political Writings*, edited by Anthony Pagden and Jeremy Lawrance. Cambridge: Cambridge University Press, 1991.

Pagden, Anthony. *Lords of All the World: Ideologies of Empire in Spain, Britain, and France c. 1500–c. 1800*. New Haven: Yale University Press, 1995.

Pagden, Anthony. *Spanish Imperialism and the Political Imagination: Studies in European and Spanish-American Social and Political Theory, 1513–1830*. New Haven: Yale University Press, 1990.

Parker, Geoffrey. *The Grand Strategy of Philip II*. New Haven: Yale University Press, 1998.

Pena González, Miguel-Anxo. *De la primera a la segunda "Escuela de Salamanca": fuentes documentales y líneas de investigación*. Salamanca: Imprenta KADMOS, 2012.

Pena González, Miguel-Anxo. *La Escuela de Salamanca: de la monarquía hispánica al orbe católico*. Madrid: BAC, 2009.

Pérez Fernández, Isacio. *El anónimo de Yucay frente a Bartolomé de Las Casas: estudio y edición crítica del Parecer de Yucay, anónimo (Valle de Yucay, 16 de marzo de 1571)*. Cuzco: Centro de Estudios Regionales Andinos Bartolomé de Las Casas, 1995.

Pérez Fernández, Isacio. *Bartolomé de Las Casas en el Perú: el espíritu lascasiano en la primera evangelización del Imperio Incaico (1531–1573)*. Cuzco: Centro de Estudios Rurales Andinos Bartolomé de Las Casas, 1988.

Platt, Tristan. "Mirrors and Maize: The Concept of Yanantin among the Macha of Bolivia." In *Anthropological History of Andean Polities*, edited by John Murra, Nathan Wachtel, and Jacques Revel, 228–59. Cambridge: Cambridge University Press, 1986.

Rabasa, José. *Writing Violence on the Northern Frontier: The Historiography of Sixteenth Century New Mexico and Florida and the Legacy of Conquest*. Durham: Duke University Press, 2000.

Ramos, Demetrio. "La crisis indiana y la Junta Magna de 1568." *Jahrbuch für Geschichte von Staat, Wirtschaft und Gesellschaft Lateinamerikas* 23 (1986): 1–61.

Rengifo, Grimaldo. "The Ayllu." In *The Spirit of Regeneration: Andean Culture Confronting Western Notions of Development*, edited by Frederique Apffel-Marglin and PRATEC. London: Zed Books, 1998.

Robins, Nicholas A. *Mercury, Mining, and Empire: The Human and Ecological Cost of Colonial Silver Mining in the Andes*. Bloomington: Indiana University Press, 2011.

Salazar-Soler, Carmen. "Álvaro Alonso Barba: Teorías de la antigüedad, alquimia y creencias prehispánicas en las ciencias de la Tierra en el Nuevo Mundo." In *Entre dos mundos: Fronteras culturales y agentes mediadores*, edited by Berta Ares Queija and Serge Gruzinski, 269–99. Seville: Escuela de Estudios Hispano-Americanos, Consejo Superior de Investigaciones Científicas, 1997.

Salazar-Soler, Carmen. "Encuentro de dos mundos: las creencias acerca de la generación y explotación de los metales en las minas andinas del siglo XVI al XVIII." *Etnicidad y simbolismo en los Andes* (1992): 237–53.

Salazar-Soler, Carmen. "Las huacas y el conocimiento científico en el siglo XVI: a propósito del descubrimiento de las minas de Potosí." *Travaux de l'IFEA* (1997): 237–49.

Salazar-Soler, Carmen. "Magia y modernidad en las minas andinas: mitos sobre el origen de los metales y el trabajo minero." In *Tradición y modernidad en los Andes*, edited by Henrique Urbano, 197–219. Cuzco: Centro de Estudios y Debates Regionales Andinos "Bartolomé de Las Casas," 1993.

Scharff, Robert C., and Val Dusek, eds. *Philosophy of Technology: The Technological Condition, an Anthology*. Malden: Blackwell Publishers, 2003.

Schmitt, Carl. *The Nomos of the Earth in the International Law of the Jus Publicum Europaeum*. Translated by G. L. Ulmen. New York: Telos Press, 2003.

Schürmann, Reiner. *Heidegger on Being and Acting: From Principles to Anarchy*. Bloomington: Indiana University Press, 1987.

Seed, Patricia. *American Pentimento: The Invention of Indians and the Pursuit of Riches*. Minneapolis: University of Minnesota Press, 2001.

Seed, Patricia. *Ceremonies of Possession in Europe's Conquest of the New World, 1492–1640*. Cambridge: Cambridge University Press, 1995.

Sepúlveda, Juan Ginés de. *Demócrates Segundo; o, De las justas causas de la guerra contra los indios*. Edited by Ángel Losada. Madrid: Consejo Superior de Investigaciones Científicas, Instituto Francisco de Vitoria, 1951.

Sepúlveda, Juan Ginés de, and Bartolomé de Las Casas. *Apología de Juan Ginés de Sepúlveda contra Fray Bartolomé de Las Casas y de Fray Bartolomé de Las Casas contra Juan Ginés de Sepúlveda*. Edited and translated by Ángel Losada. Madrid: Editora Nacional, 1975.

Silverblatt, Irene. *Modern Inquisitions: Peru and the Colonial Origins of the Civilized World*. Durham: Duke University Press, 2004.

Skinner, Quentin. *The Foundations of Modern Political Thought*. 2 vols. Cambridge: Cambridge University Press, 1978.

Solórzano Pereira, Juan de. *Política indiana*. 5 vols. 1647. Madrid: Ediciones Atlas, 1972.

Sweezy, Paul M. *The Transition from Feudalism to Capitalism*. London: NLB, 1976.

Tandeter, Enrique. *Coercion and Market: Silver Mining in Colonial Potosí, 1692–1826*. Albuquerque: University of New Mexico Press, 1993.

Taussig, Michael T. *The Devil and Commodity Fetishism in South America*. Chapel Hill: University of North Carolina Press, 2010.

Taylor, Gerald. "Camay, camac et camasca dans le Manuscrit quechua de Huarochirí." *Journal de la Société des Américanistes* 63.1 (1974): 231–44.

Tierney, Brian. *The Idea of Natural Rights: Studies on Natural Rights, Natural Law, and Church Law, 1150–1625*. Atlanta: Scholars Press, 1997.

Truxillo, Charles A. *By the Sword and the Cross: The Historical Evolution of the Catholic World Monarchy in Spain and the New World, 1492–1825*. Westport: Greenwood Press, 2001.

Tuck, Richard. *Natural Rights Theories: Their Origin and Development*. Cambridge: Cambridge University Press, 1979.

Tung, Toy-Fung. "Vitoria's Ideas of Supernatural and Natural Sovereignty: Adam and Eve's Marriage, the Uncivil Amerindians, and the Global Christian Nation." *Journal of the History of Ideas* 75.1 (2014): 45–68.

Urton, Gary. *At the Crossroads of the Earth and the Sky: An Andean Cosmology*. Austin: University of Texas Press, 1981.

Van Kessel, Juan, and Dionisio Condori Cruz. *Criar la vida: trabajo y tecnología en el mundo andino*. Santiago, Chile: Vivarium, 1992.

Vieira, Monica Brito. "Mare Liberum vs. Mare Clausum: Grotius, Freitas, and Selden's Debate on Dominion over the Seas." *Journal of the History of Ideas* 64.3 (2003): 361–77.

Vilches, Elvira. *New World Gold: Cultural Anxiety and Monetary Disorder in Early Modern Spain*. Chicago: University of Chicago Press, 2010.

Vitoria, Francisco de. *Obras de Francisco de Vitoria: relecciones teológicas, edición crítica del texto latino, versión española, introducción general e introducciones con el estudio de su doctrina teológico-jurídica*. Edited by Teófilo Urdánoz. Madrid: Editorial Católica, 1960.

Vitoria, Francisco de. *Political Writings*. Edited by Anthony Pagden and Jeremy Lawrance. Cambridge: Cambridge University Press, 1991.

Vives, Juan L, Nero V. Del, Francisco Calero, María L. Arribas, and Pilar Usábel. *De Concordia Et Discordia in Humano Genere =: Sobre La Concordia Y La Discordia En El Género Humano; De Pacificatione = Sobre La Pacificación; Quam Misera Esset Vita Christianorum Sub Turca = Cuán Desgraciada Sería La Vida De Los Cristianos Bajo Los Turcos*. Valencia: Ajuntament de València, 1997.

Wallerstein, Immanuel. *The Modern World-System I: Capitalist Agriculture and the Origins of the European World-Economy in the Sixteenth Century*. Berkeley: University of California Press, 2011.

Watson, Alan. *The Digest of Justinian*. Philadelphia: University of Pennsylvania Press, 1998.

Webb, Hillary S. *Yanantin and Masintin in the Andean World: Complementary Dualism in Modern Peru*. Albuquerque: University of New Mexico Press, 2012.

Wey Gómez, Nicolás. *The Tropics of Empire: Why Columbus Sailed South to the Indies*. Cambridge, MA: MIT Press, 2008.

Wilson, Catherine. *Epicureanism at the Origins of Modernity*. Oxford: Clarendon Press, 2008.

Zimmerman, Michael E. *Heidegger's Confrontation with Modernity: Technology, Politics, and Art*. Bloomington: Indiana University Press, 1990.

Žižek, Slavoj. *The Parallax View*. Cambridge, MA: MIT Press, 2006.

Žižek, Slavoj. *The Sublime Object of Ideology*. London: Verso, 1989.

INDEX

abstraction, 8, 10, 23, 181, 201, 215, 341, 376n14; violence of, 39

Acosta, José de, 31, 35, 152–53, 186–87, 233, 235, 251, 264–65, 289; on amalgamation, 210–12; on Amerindian labor, 205–10, 311; on "barbarians," 160, 163–68, 170, 181; on circulation and evangelization, 32, 151, 153, 158, 162, 164–65, 175–78, 180–81, 204, 212; commodity fetishism of, 32, 182, 200–1; critique of Sepúlveda, 162–63; defense of Spanish Empire, 156–58, 164–66, 204, 215, 287, 297, 347; *De procuranda Indorum salute*, 31–32, 151, 153, 155–58, 162, 165, 176, 182, 191, 204, 206, 212, 235; on discipline and the Indians, 155, 159–61, 163, 167, 175, 212–13; on the *encomienda*, 167–70, 172–73, 204; on division of labor in the New World, 173, 178, 180; *Historia natural y moral de las Indias* (*Natural and Moral History of the Indies*), 31–32, 151, 163, 155–56, 175, 177, 182–83, 187, 189, 194, 204, 206, 210–13, 235, 252; hylomorphism of, 180, 182, 191–92, 194, 213, 254; influence on Solórzano, 291, 301; on metals, 37, 188–91, 193–95, 200, 202–3, 207, 215, 282, 327, 331, 348, 364n7; metaphysical instrumentalism of, 32–33, 150, 152, 155, 162–63, 168, 174–75, 200–1, 204, 206, 209–10, 216–17, 284, 291, 347; on mining, 32, 151–52, 154, 180, 202–3, 205–10, 216, 303; on the *mita*, 170–73, 204, 207, 302; on money, 196–97, 199; on natural slavery, 157, 160, 163, 166; onto-theology and, 154–57, 173, 175–77, 204–5; pragmatism of, 33,

150, 154–55, 169, 175, 215, 217, 291, 348; Thomism of, 152, 161, 164, 168, 368n15; utilitarianism of, 150, 175, 215, 347
Adorno, Rolena, 100–3, 182, 191, 214, 223–25, 363n2 (Ch. 2), 366nn16–17; on Las Casas, 111, 143; on the School of Salamanca, 43; on Sepúlveda, 117, 125, 364n5; on Vitoria, 363n34, 365n9
Adorno, Theodor, 26, 279, 281, 349
Agia, Miguel, 288
Albornoz, Cristóbal de, 190, 331, 337
amalgamation (*beneficio*), 2, 33–34, 37, 201, 217, 261–63, 283, 302, 304, 308, 322, 327–29, 333, 368n16, 372n23, 375n10; Acosta on, 210–13; introduction of, 258, 262–63, 269, 275, 280, 286, 289, 348; mining and, 34; *mita* and, 34, 36, 153, 227, 254, 267–70, 275, 279, 281, 283, 301–2, 304, 314, 316, 319; Toledan reforms and, 33, 153, 218–19, 226–27, 231–32, 237, 254–55, 266–71, 275, 349. *See also* mercury; Toledan reforms
Amerindians, 31, 43, 47, 49, 52, 143, 154, 206; Acosta on, 158–61, 165–75, 178–81; conversion of, 55, 287; enslavement of, 97–98, 102, 183; as imperfect matter, 27, 252; inhumanity of, 102 125–28; Las Casas on, 99, 139–40; liberal arts and, 34, 138–40; mechanical arts and, 96, 139, 242, 244, 298; *mita* and, 98; nature of, 97, 100–1, 131, 160, 165, 179, 355n8, 363n33; property of, 162; right to self-government of, 96; salvation of, 4, 27, 99, 128, 150, 157–58, 160, 180, 203, 207–8, 212–13, 235, 264–65, 304, 348–49; Sepúlveda on, 102, 125, 129–31, 191, 363n33, 365n12; as slaves by nature, 30, 41–42, 95, 97, 119, 131; sovereignty of, 143, 162; subjection of, 46, 89–90, 100; Vitoria on, 88–91, 360n19; wage labor and, 246

Andean beliefs, 39, 330, 333, 340; ambiguity of, 351–52; on the life of metals, 37–38, 285, 329, 333, 374n8; as pure difference, 286; in return of the Inca, 39, 315–16, 342, 351; as retroactive effect of metaphysical instrumentalism, 341, 345, 351; Spanish beliefs about metal and, 329–30. *See also* vitalism

Anónimo de Yucay frente a Bartolomé de Las Casas, 33, 218, 226–236, 249, 265, 275, 348, 371n19; Matienzo and, 244, 249

Aquinas, Thomas, 3–15, 60–62, 66–67, 71, 103, 113, 115, 120–22, 124, 136, 152, 178, 183, 192; Acosta on, 167, 169; Aristotelianism of, 125; *De regimine principum*, 14, 61, 64–65, 124, 299; commentaries on Aristotle, 8, 176, 195, 269, 361n27, 365–66n15; on discipline, 174, 361n23; instrumentalism in, 4, 14, 25, 114, 236, 275, 345, 357n12; Las Casas and, 99; metaphor in, 12, 14; metaphysics of, 3, 340; on money, 196, 241, 349; philosophy of, 3–4, 196; on principle, 4–7, 11; Prime Mover in, 11, 71, 356n10; Scholasticism of, 25; *Secunda secundae*, 61, 182; Sepúlveda and, 96, 104–7, 109–10,

112, 114, 120–21, 347, 364n6;
Solórzano and, 291, 298–99, 306,
328; *Summa Theologica*, 3, 8, 13, 41,
52, 60, 83–84, 147–48, 174, 203–4,
293–94, 361n23, 370–71n17; Toledan
ideology and, 218, 224, 226; Vitoria on, 46, 53, 60, 91, 101. *See also*,
metaphysics: Aristotelian-Thomist;
Thomism

Aristotle, 3–4, 14, 16, 86, 102–3, 178,
183, 298, 328; Acosta on, 171–72,
180; Aquinas on, 72, 148, 360n21;
doctrine of (four) causes, 61; dualism of, 34; Heidegger on, 185,
362n29; Las Casas on, 99, 134,
136–38, 142; Marx on, 368n17; on
mastery, 112; on matter, 9; *Metaphysics*, 360n27; metaphysics of
handiwork and, 19; on money,
196; on natural slavery, 41, 97, 125,
133; on natural subordination,
119; *Nicomachean Ethics*, 65, 176,
195, 206, 356–57n11; *Physics*, 8–10,
16–18, 61, 67, 71, 176, 182; *Politics*,
148; on prime mover, 10–11, 71;
principle, 4–6; Sepúlveda on, 104,
166; on technique, 67; Vitoria on,
60–62, 65, 86, 112

Aristotelianism, 4, 323, 355n6, 374n8
Assadourian, Carlos Sempat, 222
autctoritas (authority), 45, 77
Augustine, 4, 57, 105, 157, 362n30;
Acosta on, 159, 162, 164, 202; *City
of God*, 3, 102, 124; Sepúlveda on,
105–6, 121
autonomy, 22, 30–31, 64–66, 85, 149,
192, 325; of civil power, 81, 83; of the
commonwealth, 81; indigenous, 31,
54, 101, 145; political, 37, 110;
relative, 58
azogueros (mine owners), 288–89, 306

Barba, Álvaro Alonso, 37, 322–25, 331,
374n8, 375n10; *Arte de los metales*,
37, 322, 374n8
Barrera-Osorio, Antonio, 154, 176
Berthelot, Jean, 331–32
biopolitics, 26, 279, 281
Brading, D.A., 219, 225, 250, 291–92,
354n2
Brett, Annabel, 44–45
Brown, Kendall, 270, 332
Burgaleta, Claudio M., 153, 187, 192

caciques, 154, 240–41, 245–47, 253
Cañizares-Esguerra, Jorge, 154, 175–76, 215
Cano, Melchor, 3, 43
capital, 22–24, 197, 270, 282, 344–45;
financial, 369n20; mobile, 40, 219;
movement of, 26, 281; valorization
of, 38
capitalism, 23, 25, 200–1, 280–82, 336,
344, 358n22, 376n17; mercantile, 57,
201, 245, 254, 261–62, 264, 270, 279,
282; non-extractivist, 343;
technological, 39, 333, 351–52
Capoche, José Luis, 34, 226, 256–75,
279–81, 301, 348–49; defense of
rescate (recovery and sale) of stolen
metals, 257, 273–74, 280; *Relación
general de la villa imperial de Potosí*,
34, 207, 218, 226, 250, 256–74,
368n16
Cardano, Girolamo, 322, 331
Casa de Contratación, 41, 129

Catholic Church, 42, 47, 75, 80, 83, 164; expansionism of, 4, 202
Cerro Rico (Rich Hill), 1, 248, 256, 258, 261–62, 280
Charles V, 40–42, 44, 93, 37, 100, 292, 363n34; Cortés's letter to, 123; reign of, 228; Vives and, 3, 362n30
Cicero, 4, 63, 70, 361n21
civil law, 3, 111, 290, 292–93, 295
civil power, 6, 89, 155, 219, 236, 240, 306, 345; Acosta on, 174; Las Casas on, 145, 347; Vitoria on, 29–30, 46, 48, 50, 57, 59–61, 63–64, 69, 71–73, 75, 77, 79–83, 85–86, 91, 96, 101, 118, 145, 149, 174, 204, 254
coca, 249–50
coercion, 102, 159, 161, 173, 212, 287, 300
colonial administration, 26, 32, 57, 83, 263–64, 268; Acosta on, 154–60, 167–70, 172–75, 178, 180–82, 188, 192, 206–8, 210–11, 213, 215–16; in the *Anónimo de Yucay*, 235–36; Toledan reforms and, 217–18, 226; Solórzano on, 297, 313, 374n4; Vitoria on, 75, 89, 91, 291
colonialism, 101, 128, 153–54, 183–85, 352, 355n8; anti-colonialist emancipation, 39, 351
Columbus, Christopher, 40, 88, 197, 355n8
commodities, 20–25, 34, 45, 264, 309, 358n18, 369n17, 369n20; accumulation of, 251; circulation of, 23–24, 50, 52, 55, 93, 254, 262, 268, 279–80, 346, 349; money and equality of, 196, 198–200; nature as source of, 101; production of, 213; souls as, 181

commodity fetishism, 20, 22, 26, 182, 193, 198, 376n14; Marxian critique of, 282
common good (*utilidad común*), 4, 35, 93, 118, 167, 227, 277, 279; Acosta on, 156–58, 168–70, 177–81, 196, 204; in *El anónimo de Yucay*, 231; Aquinas on, 51, 113, 121–22; Capoche on, 250, 266, 276, 281; of the Indians, 146–47, 227; Matienzo on, 241, 243, 245; Solórzano on, 300, 312, 317–20; Vitoria on, 51–52, 55, 58, 68–69, 72, 74–75, 85, 345
commonwealth, 231, 277, 345; Indian, 347; Vitoria on, 30, 52, 55–59, 68, 71–83, 85, 88, 91–92, 96, 118, 145, 148–49, 177, 347
Comunero revolt, 43–44
conquest of the Americas, 2–4, 32, 35, 91, 104, 119, 141, 228, 292; Acosta on, 151, 157, 164, 179, 215, 297; Charles V's suspension of, 100, 221, 223; final cause of, 86; justifications of, 2, 29, 41–44, 123, 128, 128, 143, 216, 293, 297, 345; Las Casas on, 99; Matienzo on, 238, 240–41; metaphysical principles of, 48–49; Solórzano on, 374n4; violence of, 232; Vitoria on, 53–55
corporatism, 35, 69, 121, 244, 298
Council of Castile, 221, 223
Council of Indies, 98–100, 143, 221, 290
Crown (*see* Spanish Crown)
credit, 25, 41, 199, 201, 204, 218, 344; capitalism and, 358n19; entrepreneurial, 369n20; manure and, 335; network of, 2, 41, 262, 264, 319, 336

Dean, Carolyn, 338–41
debt, 199, 344; fertility and, 335; Spanish imperial, 286
decolonial theory, 37–38, 330, 352
De la Gasca, Pedro, 98, 100
Del Valle, Ivone, 153–54, 156
Demócrates Segundo, o de las Justas Guerras contra los Indios (Sepúlveda), 31, 95, 99, 102, 104–31, 132, 161
Descartes, René, 192, 354n4; dualism of, 192–93
Des Chene, Dennis, 192, 354n4
divine law, 47, 75, 106, 362n30; Acosta on, 157; Aquinas on, 8, 60, 147; Las Casas on, 140; Matienzo on, 240–41; Sepúlveda on, 107; Toledan circle and, 278; Vitoria on, 49–50, 68, 71–73, 78–79, 84, 149
Divine Providence, 106, 135, 152, 208; mining and, 107, 200, 202–5, 208, 303
Dominican order, 42, 96–97, 99–100, 152, 223
dominion, 29–31, 47, 59, 87–88, 112–14, 116–18, 142, 144, 226, 251; indigenous, 54–55, 88, 131, 147, 250; as mastery, 61–62, 91; metaphysics of, 78; ontology of, 89; principle of natural subordination and, 46, 54; reason as ground of, 44, 48; Spanish, 29, 33, 42, 104–5, 112, 124, 130, 150, 152, 162, 169, 191, 216, 221–24, 228, 239, 241, 253–54, 290–97, 320, 345–47, 360n19, 363n34; technical, 83; time as ground of, 295–96
dominium, 44–45, 57, 87, 130, 295, 363n4; imperial, 147

efficient cause, 12–13, 15, 18, 85, 102–3, 144, 201, 267, 325, 334; Acosta on, 189, 192; Aquinas on, 67; Las Casas on, 31, 96, 144–45, 147, 347, 366n17; Scholasticism and, 183, 186; Solórzano and, 329; Thomism and, 193, 200, 277, 366–67n5; Vitoria on, 70–73, 79, 145, 345
empiricism, 3, 155, 186, 213, 331, 353n2
encomienda system, 41–42, 91, 97–98, 167–69, 170–73, 218, 222; Acosta's defense of, 204, 215; Las Casa's attack on, 143; mita and, 162, 169, 171, 178, 196; Solórzano's defense of, 374n4
encomenderos, 97–99, 143, 154, 265, 286; abuses of, 222–23, 310; Acosta on, 168; Matienzo on, 247; mining policies of, 42
enframing, 18–19, 25, 349
Enlightenment, 27, 45, 182; dialectic of, 279
Erasmus, 3, 42, 99
Espinosa, Diego de, 223, 228, 370n12
evangelism, 158–59
evangelization, 4, 30, 32, 45, 179, 202, 227, 348, 354n2; Acosta on, 151, 153–54, 164–65, 167, 169, 174, 179, 207, 211–12, 287, 347; *Anónimo de Yucay* on, 231, 234–37, 275; Las Casas on, 99, 139, 143; mining and, 34, 56, 213, 216, 235, 265, 348; *mita* and, 35–36; right to, 49, 55, 57, 59, 91, 229; Solórzano on, 312; Valladolid debates and, 100; Vitoria on, 101, 124, 254
exchange, 21–24, 50, 55, 86, 146, 177, 196, 262, 268; Acosta on, 195;

Andean economy and, 334–36, 338; Aristotle on, 195; *encomienda* system and, 167–68; direct, 369n20; gift, 336; logic of, 38, 336; metals and, 205–6, 216; *mita* and, 372n26; money and, 152, 196–200, 216, 241, 243, 333, 358n18; Vitoria on, 360n19; wage labor and, 174, 219, 233, 241, 245–46, 251, 273, 301, 348
exchange value, 128, 200–1, 204, 308, 333–36, 344, 348; fetishism and, 376n14
extractivism, 200, 343

Ferdinand the Catholic, 42, 293
fertility, 340; debt and, 335; mining and, 337; money and, 337, 339
fetishism, 20, 22, 183, 191, 200, 209, 214; in Acosta, 152; in Andean beliefs, 333–34; in Capoche, 266; in decolonial theory, 285–86; in Matienzo, 252; in metaphysical instrumentalism, 197, 340, 346; in Sepúlveda, 128; in Vitoria, 80. See *also* commodity fetishism
final cause, 35, 67–70, 83, 85–86, 116, 134, 190, 345; Aquinas on, 10, 13; Las Casas on, 145–47; Solórzano on, 305; Vitoria on, 61–65, 73, 75–76, 79, 108
formal causality, 13, 70, 77, 118, 324, 345
formal subsumption, 270
Foucault, Michel, 26, 189, 277, 281, 349

General Resettlement, 226, 246, 265, 276–78, 372n24
gold, 138–39, 145–46, 190, 193, 202, 230–31, 238, 314–15; Andean beliefs and, 331–32, 338–39, 342; Caribbean, 40–41; centrality to colonial society of, 207; Divine Providence and, 107, 208, 234, 298; as end of nature, 322; evangelization and, 234; exchange value of, 129–30; Marx on, 21; Monardes on, 88; as money, 183, 197; persecution of Indians and, 122; promotion of silver over, 336; Spanish extraction of, 50–51, 180, 233, 318–19, 322, 327; tax on, 56; value of, 200, 307
González Casanovas, Ignacio, 42, 222, 238
González Holguín, Diego, 342, 371n22
governmentality, 277–79, 373n29
Guaman Poma, 220, 255

Harris, Olivia, 333–38, 340–44
Heidegger, Martin, 15–19, 23–26, 185–86; on Aristotle, 61, 67; on technology, 60
hierarchy, 69, 73, 116, 340, 367n11; of the arts, 7; of being, 158, 170, 183, 191, 193–94, 197; of cognitive functions, 154; of forms, 169; natural, 7, 173, 181, 204, 302; of natural physics, 103; rational, 302; of subordination, 192
Hispaniola, 40–42, 96, 100, 207
Historia natural y moral de las Indias (*Natural and Moral History of the Indies*) (Acosta), 31–32, 151, 163, 155–56, 175, 182–83, 235; Amerindian idolatry in, 189; description of amalgamation in, 211–12; description of mines in, 206, 210; metals

in, 194–95; ontotheological ground of, 177, 187–88; principle of natural subordination in, 204, 252; violence in, 213

Huancavélica, 232, 255, 291, 308, 328, 374n6

huayras/guaira smelting, 209, 258–62, 371n22; transition to amalgamation and, 34, 258, 269–70, 275, 280

humanism, 111, 152–54, 225, 251, 365n10

humanist rhetoric, 117, 119

human law, 52, 84, 147, 182, 316; divine law and, 70–71, 84, 163; natural law and, 64, 84, 108

hylomorphism, 76, 132, 174, 178, 183, 189; Acosta on, 162–63, 165, 192, 215; doctrine of, 8, 117, 356n9; Las Casas on, 96, 254, 347; metaphysical instrumentalism and, 127; Sepúlveda's use of, 111, 115, 141, 254, 346–47; Vitoria and, 72, 182

ideology, 20, 23, 27, 93, 197–98, 213, 258, 320, 357–58n16, 369n19, 372n25; capitalist, 336, 343, 345, 376n14 (*see also* commodity fetishism); of the circle of Toledo, 218, 226, 276, 348; of expansion, 86; imperial, 2, 25, 29–30, 90–91, 93, 95, 101–3, 119, 127–28, 131–32, 135, 143, 146–48, 186, 201, 209, 226, 232, 243, 252, 285, 289, 311, 346, 354n2, 367n11; Inca, 339; metaphysical, 3, 216; in Las Casas, 223; in Sepúlveda, 116, 127, 146, 225; Spanish, 2, 4, 32, 146, 223, 270, 362n30; in Vitoria, 148

imperial reason, 27, 36, 48, 83, 86, 131, 134, 150, 157, 175, 213, 216, 362n29; Acosta and, 157, 159, 162–63, 168–70, 197; contradictions of, 183, 227, 313; critique of, 44, 149; ground of, 188, 226; principle of subordination and, 113, 116, 205; Roman, 185–86; Vitoria and, 297

Inca empire, 225, 253, 256, 276, 372n26

Incas, 88, 97, 144, 171–74, 230–31, 256, 276, 339–40; despotism/tyranny of, 34, 156, 218, 224–25, 228, 238–42, 245–47, 250–54, 265, 278, 282, 305, 372n24; metals and, 338; restitution of lands/wealth, 223–24, 316; vitalism and, 39

indigenous: beliefs, 3, 39, 286, 329–31, 341, 351–52 (*see also* Andean beliefs; vitalism); culture, 336; episteme, 37; labor, 27, 32, 35, 38, 41, 143, 150–51, 205, 215, 231, 233, 249, 252, 256, 282, 286–88, 303–10, 313–14, 354n2 (*see also* labor: Amerindian; labor: compulsory; *mita*; *repartimientos de labor*); miners, 270, 333; oppression, 370n16; past, 225; peoples, 29, 34, 42, 48, 54, 56, 87, 145, 147, 167, 218–19, 229, 233, 259, 281, 299, 302, 374n4; restitution, 150, 228; ritual use of metals, 285; technological agency, 261, 269

indios varas (contract labor Indians), 259, 261

ingenios (amalgamating mills), 34, 195, 197, 255, 268–69, 275; dangerous conditions of, 272–73; polysemy of, 195, 368n16

instrumentalism, 37, 82, 150, 154, 183, 218, 226, 236, 282, 284, 313, 350–51;

in Acosta, 162, 172; Andean vitalism and, 38; in Aquinas, 148, 328; contradictions of, 82; disavowed, 36, 83, 89, 204; hylomorphic, 31; impasses of, 26, 34, 39, 150, 226, 254, 275, 282, 347, 350; imperial, 26–27, 31, 36, 183; inconsistencies of, 319; martial, 110; mining and, 275; nature and, 325; pragmatic, 33; in Solórzano, 314; teleology and, 21, 132, 294; utilitarian, 38–39; in Vitoria, 285, 296–97. *See also* metaphysical instrumentalism
iron, 88, 129–30, 307, 326, 374n4

Junta Magna, 221–23, 225
just war, 29, 35, 46, 60, 80, 254, 290; Acosta on, 156–57, 160, 162, 166; causes of, 95; Las Casas on, 139; right to declare, 86, 91, 122, 124; Sepúlveda on, 93, 122, 124, 155, 174, 191; Vitoria on, 30, 45, 47–49, 53, 55, 57–58, 92

Kamen, Henry, 218–19, 221–22, 226, 290, 302, 370n2; on mining, 40; on networks of commerce and credit, 199, 213, 277, 336; on Toledan reforms, 292
kurakas (native lords of Peru), 143, 228, 238, 253, 265

labor, 22, 65, 72, 198, 200, 205, 213, 278, 364n4, 368–69n17; Amerindian, 27, 35, 41–42, 98–101, 139, 143, 150–51, 159, 162, 166–70, 173, 175, 180, 182, 193, 205–6, 208–10, 233, 244, 248, 250, 256, 259, 270, 282, 286–89, 301, 313–14, 321, 346, 354n2; as commodity, 24; compulsory, 2, 28, 166, 171–72, 217, 223, 237–38, 245, 251–52, 258, 267–68, 282, 297, 300–2, 304, 306, 310–11, 314, 318 (*see also* mita; *repartimientos de labor*); deskilled, 252; division of, 171, 173–74, 177–78, 180, 259, 298, 302, 305; intellectual, 172, 181, 183, 194, 298; living, 270, 273, 344; manual, 85, 172, 178, 181, 183, 205, 259, 298; slave, 368n17; technique and, 92; tributary, 34, 166, 231, 238, 297; wage, 34, 246, 261, 282, 309–10, 336
labor levies, 98–99, 153, 170–72, 206 (*see also* mita)
Lactantius, 63–64, 361n21
Las Casas, Bartolomé de, 31–33, 100, 149, 221, 366n17; Acosta and, 154, 157, 160, 179; advocacy for Amerindian rights, 96–97, 99–101, 121; *Anónimo de Yucay* and, 228–30, 233, 235; anti-imperial politics of, 146–47, 150; on barbarism, 95–96, 132–38; on Christian love, 137; critique of hylomorphism, 96, 141–47, 163, 254, 347; canon law and, 111; debate with Sepúlveda, 44, 93, 95, 101, 103, 151, 160, 275, 297, 313, 347, 363n2, 366nn16–17 (*see also* Valladolid debate); *De thesauris* (*On the Tomb Treasures of Peru*), 142–44, 147; on the *encomienda* system, 42, 223; followers of, 97; *Historia de Indias*, 100; on law, 106, 366n16; on principle of natural subordination, 31, 93–94, 139–41, 152, 297; reduction ad absurdum and, 165; on restitu-

tion, 31, 123, 146, 150, 163, 223–25, 228; Thomism of, 287

law: canon, 111, 143; reason and, 294; Roman, 4, 290, 295, 306, 366n17; will of the prince and, 6, 291, 293–94, 296. *See also* divine law; human law; natural law

Laws of Burgos, 42, 96

laws of nations (*ius gentium*), 29, 52, 91, 133, 138

Lechtman, Heather, 338–41

León Pinelo, Antonio de, 289, 293

Lucretius, 63–64, 68

MacCormack, Sabine, 187, 192, 225

Machiavelli, Niccolò, 3, 42

Machiavellianism, 3, 153–54, 219–20, 355n6

machina mundi, 185, 192, 196, 204

machine, 255, 262; introduction to the production of silver and, 265, 267, 270; Heidegger on, 19, 185, 362n29; Scholastic conception of, 30, 92, 184–86, 192–93, 204–5, 213, 216, 275, 277

Mair, John, 41, 97

Manco Inca, 97, 219

Marx, Karl, 20–24, 28, 196, 270, 279, 281–82, 358n18, 368n17; *Capital*, 20–24, 196–99, 282, 344; critique of commodity fetishism, 26; on ideology, 92, 197–98, 358n16, 369n19

mastery, 18, 30, 36, 48, 54–55, 57, 59, 61–62, 115, 118; Aristotle on, 112, 367n5; artificial, 2, 14–15, 79, 103, 183, 216, 268, 345; natural, 14, 79, 86, 117, 183, 216, 227, 268; political, 28, 30, 91–92, 152, 345; technical, 14, 30, 59, 85, 92, 101, 149, 227, 357n14

material cause, 13, 31, 324; commonwealth as, 118, 345, 347; Vitoria on, 72–73, 75–76, 78–79, 96, 145, 149

materialism, 28, 68, 76, 236, 266, 364n6; dialectical, 201, 213; Vitoria's attack on, 79–80

Matienzo, Juan de, 34, 166, 225, 233, 237–54, 259, 275; denigration of Amerindians, 244–45; *Gobierno del Perú*, 34, 166, 218, 225–26, 237–50, 252, 282; influence on Solórzano, 301; influence on Toledan reforms, 252–53, 348

matter: force of, 62–65, 68–70, 73, 75, 78; passive, 193, 195, 269, 305, 334; privation of, 174–75, 352

Mendoza, Antonio de, 100, 218

mercury, 153, 210–11, 231–32, 237, 255, 266, 324, 331; circulation of, 261; mining of, 280, 291, 308, 326. *See also* amalgamation

metals: composition of, 151; instrumental role of, 32; life of, 37, 189–91, 197, 201, 203, 285, 322, 325, 327, 329–31, 333, 337, 341, 343; self-reproductive/self-generative power of, 37–38, 285, 321–27, 329, 331, 333, 335, 341, 343–44, 351; as standing reserve, 195

metaphysical instrumentalism, 15, 21, 90, 92–93, 96, 250, 265, 278–79, 281, 366n5, 372n25; Acosta and, 32, 152, 155, 162, 168, 184, 200, 204, 206, 209, 216, 348; Andean beliefs and, 38, 284, 325, 329–30, 351; *Anónimo de Yucay* and, 231; Aquinas and, 46, 121, 132, 204, 215, 226, 229, 275,

282, 294, 296, 300, 350; Aristotelian physics and, 102; capitalism and, 282, 350; contradictions of, 26, 28, 34–35, 95, 102, 143, 150, 152, 183, 201, 209, 218, 281, 284, 350; disavowed, 268; exhaustion of, 350–51, 374n8; hylomorphism and, 127; impasses of, 26–27, 29–30, 32, 38–39, 95, 149, 216, 218, 296, 351, 370n13; imperial, 3–4, 27, 150, 186, 197, 226, 232, 245, 270, 276, 345–46, 352; logical efficacy of, 266; Matienzo and, 244, 251–52; Mignolo and, 358n22; mining and, 275; modernity and, 19; nature and, 3, 92, 270, 320; paradoxes of, 2, 36, 38, 236, 269, 352; radicalization of, 216, 297; principle of natural subordination and, 15, 84, 127, 155, 206; Sepúlveda and, 119, 131, 225, 237, 252; teleology and, 127; Solórzano and, 296–97, 321, 329; Thomist, 69; Toledan reforms and, 309; Vitoria and, 30, 46, 57, 63, 80, 89, 94, 101, 111, 146–47, 150, 291

metaphysics: Aristotelian-Thomist, 10, 91, 155; of handiwork, 17–19, 25, 346; imperial, 3, 134, 182–83, 187, 191–94, 213–14, 279, 304, 308, 331; Western, 3, 16–17, 19, 61, 182, 330, 340, 346. *See also* metaphysical instrumentalism

Mexico, 126, 212, 221, 286 (*see also* New Spain)

Mignolo, Walter, 37, 330, 332, 341, 358n22

mita, 2, 33, 98–99, 217, 219, 222, 226, 237, 239, 256, 282, 285–89, 294, 348–50, 372n26; Acosta on, 162, 170, 172–73, 178, 196, 201, 204, 206, 215; amalgamation and, 34, 36, 153, 227, 254, 267–70, 275, 279, 281, 283, 301–2, 304, 314, 316, 319; arguments against, 35–36, 252–253, 297, 305–14, 316, 318, 320; arguments in favor of, 35, 297–305, 314, 316, 320; Capoche on, 258, 274–75, 279; *encomienda* and, 169, 171–73, 178, 196, 204; evasion of, 272, 315; justification of, 32, 151, 170, 172–73, 196, 205; scandal and, 255; Solórzano's solution to, 315, 318–20; as tributary labor, 33–34, 124, 217, 231. *See also* Toledan reforms

modernity, 19, 23, 25–27, 76, 192, 276–77, 350, 352; Acosta on, 216; capitalist, 1, 27, 46; colonial origins of, 218, 281; indigenous beliefs and, 38–39, 286, 351; nominalism, 355n6; Scholastic roots of, 2, 354n4; Sepúlveda on, 150; technological, 19, 27, 39, 46, 362n29

Molina, Luis de, 43, 152

Monardes, Nicolás de, 88, 129, 322, 365n13

Montesinos, Fernando, 41, 229

More, Anna, 288, 292

More, Thomas, 42, 307, 318

Mumford, Jeremy, 221, 251, 281, 373n29; on Matienzo, 246; on Toledan reforms, 218, 222–23, 225, 246, 265, 276–79, 372n24

multitude, 70, 73, 75–76, 78, 114, 121

natural causality, 15, 80, 83, 90, 92, 191

natural history, 1, 29, 150, 153, 182, 188, 193

natural law, 1, 36, 55, 83, 91, 183, 214–15, 296, 350, 365n11; Acosta and, 154, 171–72, 175, 178–79, 202; Aquinas on, 365n15, 370n17; contradiction of, 86; ground of, 186, 191, 316; hierarchies and, 204; human law and, 64, 84; ideology and, 93; just titles and, 48; Las Casas and, 99, 135, 140, 142, 144, 146, 149, 223–24; mining and, 26–28; ontological origin of, 101–3; principle of natural subordination and, 278–79; Sepúlveda and, 93, 95, 106–7, 110–11, 113, 116–18, 120, 124, 130–31; violent origin of, 85; Vitoria and, 35, 44, 48–52, 57, 70–72, 74, 77, 79–80, 119, 148–49, 209, 284, 292, 320, 345, 360n19; will of the monarch and, 294
natural rights, 59, 133, 207, 292, 295, 359n6; theory of, 44, 102, 209
natural servitude, 76, 87, 97, 119, 125, 169, 254; Acosta on, 204; doctrine of, 174; Matienzo on, 238–44
natural slavery, 101–2, 111–12, 139, 141, 158, 365n9; doctrine of, 100, 125; origin of, 87; principle of, 157, 163, 166, 363n34
necessity, 62–63, 65, 266; final causality and, 69; as ideological illusion, 266; of the law, 71, 85, 215; mastery and, 112; technical, 218, 268, 270, 282, 349
New Laws, 98–99, 218, 230
New Spain, 43, 100, 123, 153, 237, 255, 280; inhabitants of, 126, 305

Ockham, William of, 3, 355n6
Ondegardo, Juan Polo de, 98, 225, 227–29, 231
Ovando, Juan de, 221, 225
Ovando, Nicolás de, 41–42

pacifism, 3, 99, 104, 163, 362n30
Pagden, Anthony, 77, 87–88, 90, 101, 355n8, 360n17, 363–64n4; on Acosta, 158, 368n15; on Sepúlveda, 111, 364n6, 365n12; on Solórzano, 292–95; on the Valladolid debates, 44; on Vitoria, 43, 45, 292, 359n4, 360n19, 361n26, 368n15
Palacios Rubios, Juan de, 42, 96
papal bull of donation (concession), 43, 100, 131, 145, 229, 284, 292
Pérez Fernández, Isacio, 226–30, 233 (*see also* Anónimo de Yucay)
perfect community, 147, 149, 179, 291, 312, 314; Amerindian labor and, 180; Aquinas on, 113, 298; Las Casas on, 142; mine illness and, 308, 312; multitude and, 73; state as, 277, 301; Vitoria on, 58–59, 83, 254
perfection, 66, 102, 126, 189, 355n7, 364n7; Aquinas on, 356n10, 361n23; form and, 114, 116, 126, 177, 245; Indians and, 170, 172, 233–34, 237; Las Casas on, 134, 136; metals and, 201, 203; metaphysics of, 110; New World as deprived of, 237; prime matter and, 76; Vitoria on, 58, 82
Philip II, 143, 218–19, 221–23, 232, 286, 292; Acosta and, 153; Sepúlveda and, 99; Toledo and, 252–54
piety, 105, 220, 314, 316, 320

Pius V, 223, 232
Pizarro, Francisco, 42–43, 240
Pizarro, Gonzalo, 98
Platonism, 4, 355n6
Pliny, 129, 301, 307–8, 326
Política indiana, 35, 156, 173, 284, 294–95, 321–22, 327–28, 373–74n4; as exhaustion of metaphysical instrumentalism, 345, 350
political subordination, 8, 82, 85, 114, 149
political theology, 104, 156, 215, 294
potestas (capability), 45, 77, 155, 181
Potosí, 1, 150, 216, 245, 262, 285–86; Acosta on, 32–33, 183, 204–13, 216, 348; amalgamation and, 195, 210, 212–13, 254–55, 319, 322; Amerindian labor and, 98, 151, 206–8, 242, 256, 287, 317; Capoche on, 257, 260, 264–67, 269, 271, 273, 280, 368n16; discovery of mines at, 43, 215; economic crisis of, 237, 286–87, 348–49; *huayras* at, 371n22; Matienzo on, 247, 249–50; *mita* and, 153, 288, 319; money and, 333; principle of natural subordination and, 285, 311; Sandoval y Guzmán on, 288–89; self-generation of metals and, 325, 375n10; silver from, 231, 238, 280, 282, 286, 336; recovery (*rescate*) of metals at, 272; as symptom of imperial instrumentalism, 36; *tacana* at, 372n23
prime matter, 9–10, 12–13, 92, 141, 186, 251, 341; Aquinas on, 8–9, 356n10; form and, 76, 192, 245, 302, 323, 366n5; perception of the New World and, 39

principle of natural subordination, 13, 30, 84–85, 92, 125–26, 149, 152, 227, 237, 245, 270, 357n12; Acosta and, 32–33, 186, 201, 210, 213, 215–16; Aquinas on, 60; contradictory interpretations of, 101; exhaustion of, 35; instrumentalism and, 284; Las Casas on, 297; Matienzo and, 255; mining and, 285, 311; natural law and, 52; Sepúlveda's use of, 102, 131, 148, 364n7; Solórzano and, 316; Vitoria on, 59, 216
principle of subordination, 4
private property, 238, 240–41, 245–46, 275
providentialism, 4, 150, 232, 349, 361n21
prudence, 109, 116, 118, 120–21, 123–24, 126, 133, 138, 142, 181, 292; Jesuit, 173; political, 118, 121, 139, 141

reason, 57, 65–67, 71–72, 74, 96, 109–10, 117, 359n6; Acosta on, 183, 187, 194–95; Amerindian use of, 99, 126–27, 132–33, 191, 243, 253; Aquinas on, 293–94, 361n23, 364n15; authority of, 16; Christian, 167; cynical, 369n22; dominion and, 44; instrumental, 84, 186; limits of, 44, 48, 150; moving by, 8, 60; natural, 49, 106, 135, 168; natural law and, 48, 80, 106, 108, 113, 148, 172, 293, 345; power and, 75; practical, 108, 178, 211; principle of, 61, 132, 179; right, 84, 106–8, 133, 139, 316; sin and, 122; of state, 3, 153, 220, 316, 369n1; supreme, 84, 103, 203–4; violence and, 53. *See also* imperial reason

reducciones, 33, 217, 278, 287, 291, 348
reductio ad absurdum, 140, 165
repartimientos de labor (compulsory Indian labor system), 98, 289
requerimiento, 96, 219, 226, 363n1
right to commerce, 49
right to evangelize, 49, 143, 229
right to repel force with force, 66, 96, 108, 117, 220; Acosta on, 162, 179; Las Casas on, 228, 347; Sepúlveda on, 106, 108, 119, 149; Vitoria on, 29, 53, 55–56, 59, 73–74, 76, 85–86, 91–92, 109, 174
right to travel, 48–49, 51, 53, 55
Robins, Nicholas A., 218, 280–81
Roman Empire, 44, 105, 124, 164, 309
Roman jurisprudence, 3, 125, 290–91, 293, 295–97
royal fifth (*quinto*), 41, 57, 86, 235, 263–64, 303, 343; life of metals and, 325–27; *mita* and, 222

Saint Paul, 72, 106, 137, 160, 298
Salazar-Soler, Carmen, 38, 331–32, 368n16
Santo Tomás, Domingo de, 97, 99
Sarmiento de Gamboa, Pedro, 156, 225, 228, 238, 246, 252; *Historia Índica*, 225, 238
Schmitt, Carl, 45, 359n15
Scholasticism, 3–4, 19, 182, 185–86, 354n4, 355n8; Aquinas and, 25, 353n2; colonialism and, 3, 184; corporatism and, 244; humanism and, 119, 153, 365n10; liberalism and, 46; natural right and, 360n19; revival of, 43; technique and, 15; Vitoria and, 61–62

Scholastics, 44, 152, 369n20; Spanish, 3, 200
School of Salamanca, 3, 43, 111, 155, 192, 229, 290, 354n4, 359n2; analogical thinking in, 364n4; First, 152–53; Second, 187
Schürmann, Reiner, 16–18, 357n14
science, 16, 32, 182, 215, 299, 352; empirical, 154, 176; history of, 2, 27, 154, 330; imperial, 2, 27, 353n2; physical, 61, 176, 178
self-determination, 73, 76, 97, 141, 239–40, 259
Sepúlveda, Ginés de, 30–34, 96, 99–102, 133, 138–43, 150, 250; Acosta on, 158, 160, 165–66; on Amerindians, 363n33, 365n12; *Anónimo de Yucay* and, 230, 233; *Apología*, 101; debate with Las Casas, 44, 93, 95, 101, 103, 151, 160, 275, 297, 313, 347, 363n2, 366nn16–17 (*see also* Valladolid debate); *Demócrates Segundo*, 31, 95, 99, 102, 104–31, 132, 161; humanist ideology of, 225; literary aspect of, 364n6; Matienzo and, 239–40, 243, 245–46; principle of natural subordination and, 148, 191, 347; Vitoria and, 346, 364n5
silver, 21, 50, 88, 122, 190, 193, 197, 204, 222, 289; Acosta on, 207–8, 210–12; Andean beliefs and, 202, 331–32, 338–39; *Anónimo de Yucay* on, 230–31, 233; circulation of, 41, 280, 286; extraction of, 1–2, 36, 107, 180, 205, 247, 270, 273, 286, 288, 348; *huayra* and, 258, 261–62; lack of, 262–64; Las Casas on, 138–39, 145–46; in

Peru, 177; Potosí and, 1, 43, 249–50, 282, 336; production of before 1570, 248–49; self-generation of, 325; Sepúlveda on, 128, 130; Solórzano on, 314, 316, 318–19, 326–28; tax on, 56. *See also* amalgamation

Skinner, Quentin, 44–45, 359n14

slave by nature, 116, 118, 125, 137, 160 (*see also* Amerindian: as slaves by nature; natural servitude; natural slavery)

slavery, 86, 101–2, 111–12, 137; abolition of Amerindian, 98; civil, 111, 125, 142, 365n9; as punishment, 180, 310. *See also* Amerindians: as slaves by nature; natural servitude; natural slavery; slave by nature

social body, 35, 74, 122, 171, 244, 301–2, 305–6, 308

social relations, 22, 198, 214, 369n17

Society of Jesus (Jesuits), 152–53, 288

Solórzano Pereira, Juan de, 35–37, 284–85, 289–310, 312–23, 326–29, 331, 374n4; Acosta and, 156; on alchemy, 327–29; demonization of indigenous, 315–16; impasse of imperial reason and, 313–21; instrumentalism and, 314, 345; on the life of metals, 321–22, 326–29; on the *mita*, 255; *Política indiana*, 35, 156, 173, 284, 294–95, 321–22, 327–28, 350, 373–74n4; on the royal fifth, 343

Soto, Domingo de, 3, 43, 100, 152

sovereignty, 6, 42, 57, 91, 144–45, 182, 355n8; Amerindian, 29, 60–61, 101, 131, 140, 143, 146, 155–56, 161–63, 224, 231, 315–16; crisis of, 264;

ground of, 80; political, 68; secular, 45; Spanish, 1, 4, 28–29, 31, 35, 88, 95–96, 224–25

Spanish Empire, 1–3, 29, 36, 40, 59, 92, 152, 176, 178, 217–18, 287, 359n6, 363n35; apologists of, 27, 289–92, 312, 345, 350; capitalism and, 46; Christian ideology of, 223; conservation of, 172, 265, 268, 292, 301–5, 311; decline of, 214, 288; ecclesiastical power in, 3, 86, 232, 291, 373–74nn3–4; economic crises of, 29, 33, 143, 237, 254–55, 263, 266–68, 273, 284–86, 289, 348; justification of, 35, 226, 360n19; legitimacy of, 132; metaphysical instrumentalism and, 4, 146, 270; providential design of, 167; violent origins of, 32, 150–51, 156–57, 165, 175, 181, 183, 193, 287, 296–97, 320

Suárez, Francisco, 3, 43–44, 152, 187, 192–93, 354n4

surplus value, 24–26, 32, 35, 181, 218, 334, 344–45; generation of, 343; logic of, 281; ownership of subsoil and, 375n11

Taussig, Michael, 336, 344

Tawantinsuyu, 97, 99

technical subordination, 8, 12, 82–84, 92, 203

technical manipulation, 2, 7–8, 10, 80, 156, 226–27, 325; Aristotle on, 367n8; limits of, 328

technique, 14, 30, 67–69, 88, 106, 110, 118, 158, 175, 183, 196, 213, 215–16, 261, 267–70, 277, 279, 364n6; Amerindian, 103, 139; artisanal, 68–70,

199, 214, 216; *beneficio* as, 301; development of, 129; disavowal of, 83, 85, 103, 186; imperial reason and, 188; metaphysical instrumentalism and, 275; metaphysics and, 346; money as, 197; natural law and, 85, 103; of penitence, 212; philosophy and, 367n5; technological essence and, 339; transposition to nature of, 10, 14–15, 28, 36, 59, 69–70, 83–85, 89–90, 92, 102–3, 149, 172, 186, 191, 268, 299, 301, 320, 325, 335, 345, 351, 366n5

teleology, 68, 127, 132, 172, 182, 202–4, 234, 356; Aristotelian, 16, 18; natural, 2, 93

Third Council of Lima, 153, 225

Thomism, 3–4, 41, 152, 192, 287; Aristotelian, 60, 67, 103, 107, 155, 164, 182, 189, 288, 355n6, 356n9 (*see also* metaphysics: Aristotelian-Thomist); inverted, 341

Tierney, Brian, 44–45

titles for ruling the New World, 41, 46, 91, 95, 147, 345, 360n19; Acosta on, 156–57, 162, 181; just, 48, 56–57, 59, 80, 95; Matienzo on, 239; Solórzano on, 374n4; unjust, 48, 179; Vitoria on, 275, 292, 296, 320

Toledan reforms, 33, 36, 153, 217–19, 221–23, 226–27, 252, 270, 275, 348; Andean culture and, 226, 276–78; biopolitics and, 279; Capoche on, 263, 268; contradictions of, 286; ideological framework of, 33, 216, 226; Matienzo and, 237–38; metaphysical instrumentalism and, 284, 349; natural subordination and, 254; Solórzano on, 310; Spanish imperial politics and, 292; state power and, 278. *See also* General Resettlement; *mita*; amalgamation

Toledo, Francisco de, 33, 153, 156, 166, 219–27, 233, 238, 287, 290, 372n24; Andean culture and, 276, 278; Capoche on, 265–67, 271; on Inca removal, 252–54; intellectual circle of, 153, 157, 218, 224–26, 250, 278, 303, 348; at Potosí, 255–56, 262–63; Solórzano and, 290

Toledo, García de, 224, 227, 229, 233

Tupac Amaru, 219–20

University of Paris, 3, 41

University of Salamanca, 41, 43, 97, 99–100, 290, 359nn4–5

use value, 21, 24, 201

Vaca de Castro, Cristóbal, 97–98

value, 18, 21, 25–26, 32, 129, 125, 197–201; valorization of, 24–25, 199. *See also* exchange value; surplus value; use value

Valladolid debate, 30–31, 94–95, 99–101, 217, 296–97, 313, 346; Acosta on, 161; inconsistencies of metaphysics and, 151, 217; Toledo's continuation of, 225

Velasco, Pedro de, 212, 237, 255

Vilches, Elvira, 176, 196, 199–200, 354n4, 369n20

violence, 53, 63, 69–71, 123, 128; of abstraction, 39; Acosta on, 153–54, 156–57, 161–63, 165, 170, 173, 181, 210, 212–13; Aquinas on, 174; artificial, 206, 320; colonial, 176,

244; of the conquest, 55, 232, 241; illegitimate, 109, 150, 174; mining and, 33, 303, 313; origins of empires and, 32, 156–57, 165, 175, 183 (*see also* Spanish Empire: violent origins of); Sepúlveda on, 106, 108–9, 117; Spanish, 97, 153–54, 156, 183, 273, 312; symbolic, 174–75; Vitoria on, 119

vitalism, 39, 331, 338, 341; Andean, 37–39, 285–86, 330, 339, 351–52

Vitoria, Francisco de, 3, 29–34, 43–66, 68–93, 95–97, 228, 230, 301; Acosta and, 151–52, 157, 165, 168, 170, 174–79, 181; on Amerindian sovereignty, 101; *Anónimo de Yucay* and, 229; commentaries on Aquinas, 46, 294; Aristotelian-Thomism of, 182; *On the American Indians* (*De indis*), 29, 31, 45–46, 80, 90, 109, 112, 147, 359n5; on Aristotle, 112; on civil power, 118, 145, 218, 254; imperial ideology and, 36, 101, 128, 143, 147–50, 346; impasses in, 345; on just war, 108–9, 119–20, 360n19; on law, 204; metaphysical frame of, 151, 217; metaphysical instrumentalism of, 226, 236, 291, 347; on natural law, 209, 224, 292; on natural slavery, 363n34, 365n9; *On Civil Power* (*De potestate civili*), 30, 46, 52, 57, 60, 79, 91; *On the Laws of War* (*De iure belli*), 29, 46, 53, 57, 91, 109; on political power, 58; principle of natural subordination and, 144, 158, 215; *On the Power of the Church*, 30, 46, 80, 91; *relecciones* of, 29, 43, 46, 57, 72, 80, 88, 91, 359n4; Schmitt and, 359n15; Sepúlveda and, 346, 364n6, 366n16; Solórzano and, 292, 296, 316, 319–20

Vives, Juan Luis, 3, 99, 362n30

Wallerstein, Immanuel, 23, 199
Wey Gómez, Nicolás, 184–86, 196, 355n8

yanaconas, 98, 237, 245, 247–49, 258, 271

www.ingramcontent.com/pod-product-compliance
Lightning Source LLC
Chambersburg PA
CBHW032023290426
44110CB00012B/649